Children, Young People and Dark Tourism

This book is the first of its kind to offer an innovative examination of the intersecting influences, contexts, and challenges within the field of children's dark tourism. It also outlines novel conceptualizations and methods for scholarship in this overlooked field.

Presently, tourism research, and in dark tourism specifically, relies primarily on adult-centered theories and data collection methods. However, these approaches are inadequate for understanding and developing children's experiences and perspectives. This book seeks to inform and inspire research on children's experiences of dark tourism. Designed to appeal to students and scholars, it brings together insights from leading experts. The book focuses on five themes, to explore the conceptual and historic origins of children's dark tourism, developmental contexts, child perspectives, specific contexts relevant to children's encounters, and methodological approaches.

This book is aimed at an international array of scholars and students with inherent research interests in the contemporary commodification of death and 'difficult heritage' within the visitor economy. Thus, the book will provide a multidisciplinary scope within the fields of history, heritage studies, childhood studies, psychology, education, sociology, human geography, and tourism studies. The volume is primarily intended for undergraduate and postgraduate study, as well as scholars and tourism professionals.

Mary Margaret Kerr is Professor of Health and Human Development at the University of Pittsburgh, where she founded the Children and Dark Tourism research project. Dr. Kerr's internationally recognized team, which includes youth as researchers, studies the experiences of young tourists at dark tourism sites. In addition to contributing her expertise on youth coping with mass trauma, Dr. Kerr has pioneered research in child-centered research methods for the tourism field, which historically has overlooked children and youth.

Philip R. Stone is Executive Director of the Institute for Dark Tourism Research at the University of Central Lancashire (UK). He is an internationally recognized scholar in the field of 'dark tourism' and 'difficult heritage' and has published extensively about the subject. Philip is also a media consultant on dark tourism, with clients including the BBC, CNN, The New York Times, *The Guardian*, and The *Washington Post*. His latest book, the first-ever tourist guidebook dedicated to dark tourism – *111 Dark Tourism Places in England You Shouldn't Miss* (2021) – brings dark tourism scholarship to the public market.

Rebecca H. Price writes about the novice researcher experience. She frequently collaborates across disciplines and settings to explore how individuals seek answers to their questions. Her work can be found in library, education, and tourism outlets.

Routledge Cultural Heritage and Tourism Series
Series editor: Dallen J. Timothy
Arizona State University, USA

The Routledge Cultural Heritage and Tourism Series offers an interdisciplinary social science forum for original, innovative, and cutting-edge research about all aspects of cultural heritage-based tourism. This series encourages new and theoretical perspectives and showcases groundbreaking work that reflects the dynamism and vibrancy of heritage, tourism, and cultural studies. It aims to foster discussions about both tangible and intangible heritages, and all of their management, conservation, interpretation, political, conflict, consumption and identity challenges, opportunities, and implications. This series interprets heritage broadly and caters to the needs of upper-level students, academic researchers, and policy-makers.

The Economics and Finance of Cultural Heritage
How to Make Tourist Attractions a Regional Economic Resource
Vincenzo Pacelli and Edgardo Sica

Resilience, Authenticity and Digital Heritage Tourism
Deepak Chhabra

Cultural Heritage and Tourism in Japan
Takamitsu Jimura

Medieval Imaginaries in Tourism, Heritage and the Media
Jennifer Frost and Warwick Frost

Children, Young People and Dark Tourism
Edited by Mary Margaret Kerr, Philip R. Stone, and Rebecca H. Price

Tourism and Development in the Himalaya
Social, Environmental and Economic Forces
Edited by Gyan Nyaupane and Dallen J. Timothy

Cultural Tourism and Cantonese Opera
Jian Ming Luo

For more information about this series, please visit: https://www.routledge.com/Routledge-Cultural-Heritage-and-Tourism-Series/book-series/RCHT

Children, Young People and Dark Tourism

**Edited by Mary Margaret Kerr,
Philip R. Stone, and Rebecca H. Price**

Routledge
Taylor & Francis Group

LONDON AND NEW YORK

First published 2023
by Routledge
4 Park Square, Milton Park, Abingdon, Oxon OX14 4RN

and by Routledge
605 Third Avenue, New York, NY 10158

Routledge is an imprint of the Taylor & Francis Group, an informa business

British Library Cataloguing-in-Publication Data
A catalogue record for this book is available from the British Library

Library of Congress Cataloging-in-Publication Data
A catalog record has been requested for this book

ISBN: 978-0-367-46942-9 (hbk)
ISBN: 978-1-032-29169-7 (pbk)
ISBN: 978-1-003-03219-9 (ebk)

DOI: 10.4324/9781003032199

Typeset in Times New Roman
by SPi Technologies India Pvt Ltd (Straive)

Contents

Figures

Tables

Contributors

Roy Ballantyne is an emeritus professor at the University of Queensland, Australia. His research interests include environmental learning in free-choice learning environments, as well as visitor learning, behavior, and interpretation more broadly. He has worked with ecotourism and wildlife tourism sites in Australia, the USA, Canada, China, and South Africa, and his work in dark tourism has included sites in Asia, South Africa, and Australia.

Tobias Broughton has taught in primary and secondary schools and served as Head of Department and Head of School across the Arts, STEM, and Social Sciences. He holds a Bachelor of Education, Bachelor of Arts (Hons) in Art History, is an exhibiting visual artist, and is currently undertaking his PhD in Art History.

Laura Burns is a member of the University of Pittsburgh's Children and Dark Tourism research project, where she studies overnight school trips. A former history teacher and now high school principal, she has traveled extensively with adolescents, including overnight excursions to post-disaster destinations.

Antonia Canosa is a social anthropologist and research fellow at the Centre of Children and Young People, Southern Cross University (Australia) with an interest in children's rights, participation, and wellbeing. Her work is grounded in social justice and critical pedagogy with an emphasis on participatory, collaborative, and creative methodologies to empower children and young people to contribute to research, policy, and practice. Antonia has worked across a number of areas, including children's rights in the tourism industry; identity and belonging in childhood; youth activism for sustainable tourism; and ethical research involving children.

Andrea Croom is a licensed psychologist at Spiegel Freedman Psychological Associates. Dr. Croom has taught at the graduate level, lectured to medical professionals, and conducted research on shared decision-making and how families and couples cope with chronic/life-limiting illnesses. In addition to her clinical career, Dr. Croom has published research on parent–child conversations at death-related exhibits and children's comments at a mass trauma site.

Sue Dockett is emeritus professor, Charles Sturt University, Australia. Dr. Dockett has a particular interest in rights-based participatory research involving young children. Her work has focused on refining and critiquing research approaches as well as conceptual and theoretical analysis of what is meant by engaging children and young people's perspectives within research agendas.

Jennifer Frost is an associate professor in the Department of Management, Sport and Tourism at La Trobe University, Australia. Her research interests are in well-being, events, and cultural heritage. With Warwick Frost, she is the Co-Editor-in-Chief of the *Journal of Heritage Tourism*, Foundation Editor of the *Routledge Advances in Events Research* series, and the co-author of *Commemorative Events: Identity, Memory, Conflict* (Routledge, 2013).

Warwick Frost is Professor of Tourism, Heritage and the Media in the Department of Management, Sport and Tourism at La Trobe University, Australia. Originally a historian, his research interests are now in heritage tourism, events, zoos and national parks, and media-induced tourism. With Jennifer Frost, he is the Co-Editor-in-Chief of the *Journal of Heritage Tourism*, the Foundation Editor of the *Routledge Advances in Events Research* series, and the co-author of *Commemorative Events: Identity, Memory, Conflict* (Routledge, 2013).

Cecilia Greene is Director of Special Education in Mechanicsburg Exempted Village School District (US). For 31 years, she has organized and directed multiday and overnight school excursions with large groups of adolescents, including students with disabilities and serious medical conditions. She has also co-designed and contributed to multiple studies of youth and tourism, including ethnographic and archival studies of children's experiences with difficult heritage sites. Her most recent study involves youth as researchers of their classmates' experiences while visiting a terrorist attack memorial.

Karen Hughes is an associate professor in tourism at the University of Queensland Business School, Australia. She has taught and published in the areas of interpretation, visitor management, sustainable tourism, tourist behavior, and wildlife tourism. Karen has a particular interest in the design of interpretive signs and experiences.

Daniel E. Keller has been a member of the Children and Dark Tourism research project at the University of Pittsburgh for the past six years. He has worked extensively with youth researchers, most recently in his role as Research Team Coordinator. His projects also include pedagogical strategies for teaching oral history research methods to children and to undergraduate students.

Margee Kerr is a sociologist and author. She earned her PhD in 2009 from the University of Pittsburgh and currently teaches and conducts research on fear, specifically how and why people engage in 'scary' experiences like haunted attractions, horror movies, and paranormal investigations. She enjoys working as a consultant for attractions and museums and is the author of *SCREAM: Chilling Adventures in the Science of Fear* named as a must-read by the *Washington Post*. Her latest book with co-author Linda Rodriguez McRobbie *OUCH: Why Pain Hurts and Why It doesn't Have to* was published in 2021. Her work has been featured in the *New York Times*, *Parade*, *Atlantic Monthly*, and NPR's *Science Friday*, among other outlets.

Mary Margaret Kerr is Professor of Health and Human Development at the University of Pittsburgh, where she founded the Children and Dark Tourism research project. Her internationally recognized team, which includes youth as researchers, studies the experiences of young tourists at dark tourism sites. In addition to contributing her expertise on youth coping with mass trauma, she has pioneered research in child-centered research methods for the tourism field, which historically has overlooked children and youth.

Michael Lovorn is Dean of the School of Education at Saint Mary's University of Minnesota. He has visited over 70 countries and taught in over 30. His areas of expertise include trends and issues in contemporary education, critical thinking, education theory and philosophy, qualitative research methods, and historiographical analysis. He has conducted local, regional, and international historiography field trips and arranged service-learning opportunities for teacher candidates that center on historical preservation.

R. Scott Marsh is a middle school social studies teacher at the Mechanicsburg Exempted Village School District (US). For 20 years, he has co-directed multiday and overnight school excursions with large groups of children and adolescents. He has also co-designed and contributed to multiple studies of youth and tourism, including ethnographic and archival studies of children's experiences with difficult heritage sites. His most recent work involved teaching tourism and heritage research methods to youth, who then designed and conducted original research at two dark tourism sites.

Kristen Marsico is a school psychologist at Howard County Public Schools. While at the University of Pittsburgh, she examined artifacts left by children at dark tourism sites, including the Flight 93 and Pentagon Memorials. Her research demonstrates that children experience important learning and growth at dark tourism sites, and her aim is making trips to these sites more accessible for all children, including youth with disabilities.

Jan Packer is a research fellow at the University of Queensland Business School, Australia. Her research focuses on applying principles from educational, environmental, and positive psychology to understand and improve visitor experiences at natural and cultural tourism attractions such as museums, zoos and aquariums, botanic gardens, national parks, ecotourism, and wildlife tourism attractions.

Rebecca H. Price writes about the novice researcher experience. She frequently collaborates across disciplines and settings to explore how individuals seek answers to their questions. Her work can be found in library, education, and tourism outlets.

Gopika Rajanikanth has been a member of the Children and Dark Tourism research project at the University of Pittsburgh for the past several years. Her specialization is teaching research methods to adolescent researchers. She has also worked collaboratively on projects around children's understanding of death and mass trauma.

Cristina Restrepo-Harner is a BCBA Coordinator for Washington, DC, Public Schools, specializing in classroom and family supports for children with disabilities. Prior to this position, she was a member of the Children and Dark Tourism project, where she examined young children's reactions to the events of September 11th through qualitative coding of drawings, messages, and letters. She facilitated the creation of a special travel guide for families whose children have emotional and behavioral challenges.

Philip R. Stone is Executive Director of the Institute for Dark Tourism Research at the University of Central Lancashire (UK). He is an internationally recognized scholar in the field of 'dark tourism' and 'difficult heritage' and has published extensively about the subject. Philip is also a media consultant on dark tourism, with clients including the BBC, CNN, *The New York Times*, *The Guardian*, and The *Washington Post*. His latest book, the first-ever tourist guidebook dedicated to dark tourism – *111 Dark Tourism Places in England You Shouldn't Miss* (2021) – brings dark tourism scholarship to the public market.

Timothy M. Wagner serves as a high school principal for the Upper St. Clair School District in Pittsburgh, Pennsylvania, where he has spent time as an elementary and middle school teacher, a secondary gifted education coordinator, and the middle school English Language Arts Curriculum Leader. Additionally, he holds an appointment as an adjunct faculty member in the Education Department at Washington & Jefferson College. Trained also as a developmental psychologist, he has traveled extensively on educational excursions with adolescents.

Gregory Wittig is a middle school language art and US History teacher of 30 years at the Falk Laboratory School, University of Pittsburgh. He has focused on postmodern readings of history and literature, focusing on the unraveling of media texts as a way to name the world in which students live. Mr. Wittig has organized and conducted multiple school excursions to sites of mass trauma.

Daniel W. M. Wright, lecturer in tourism management at the University of Central Lancashire, has a PhD in post-disaster tourism management and development. He has published widely about tourism futures in the academic literature, using futurology as a transdisciplinary field of study to forecast, anticipate, and provoke the future of tourism in global visitor economies. Daniel is a member of the Institute for Dark Tourism Research (iDTR), as well as an Editorial Board member for the *Journal of Tourism Futures*. He teaches tourism-related subjects at undergraduate and post-graduate levels.

Acknowledgments

We first want to thank Dallen Timothy, who encouraged us to write this book. Thank you for believing in this project.

Writing about children, young people, and dark tourism requires a multidisciplinary and multigenerational team willing to venture beyond current conceptions of tourism. We are deeply grateful to those who joined us in this journey:

- Each contributing author forged new pathways, skillfully blending their unique expertise across disciplines to inform and guide our readers. Together, they have envisioned a new future in tourism that welcomes children and youth into their rightful places.
- Families and educators entrusted us with their children so that we might learn from them as we traveled together.
- Our families supported us so that we could bring this volume forth. Mary Margaret thanks Bruce, Cristina, and Rob Perrone and Tommy Kerr. Rebecca thanks Jeff, Lily, and Lena Laing.
- Colleagues, students, and reviewers not only read drafts but gave us feedback that sent us in better directions.
- Interpreters, destination managers, and museum educators welcomed our questions and visits.
- Aaron Porter, graphic illustrator, painstakingly adjusted our figures for production.
- Faye Leerink, our commissioning editor, wisely advised and undoubtedly improved this work. Editorial Assistant Medha Malaviya skillfully ushered the manuscript through production. I would also like to thank Jashnie Jabson, our production editor, who has been looking through the production process of the book.
- Mechanicsburg Exempted Village School District – whose gracious leaders, staff, families, and students welcomed us into their schools and their excursions – inspired and facilitated our research.

Most importantly, we appreciate every child and young person whose tourism experience is represented here. We wrote this book for you.

Mary Margaret Kerr, Philip R. Stone, and Rebecca H. Price

Preface

Why do we need a tourism book specifically about young tourists? It is because the tourism field has largely overlooked their experiences. Even the term 'tourist' connotes adults only. Despite calls from leading tourism researchers, children have been marginalized in the literature. As a result, generations of tourism students complete their training without learning about this important, large market. Yet, children are our adult tourists of tomorrow.

In dark tourism and difficult heritage, the problem is more pronounced. While adults fully understand death concepts, children's understanding of death develops over years. As a result, children touring medical museums, cemeteries, memorials, concentration camps, or battlefields may not fully comprehend but nevertheless react to what they see. Meanwhile, interpreters struggle to explain death and suffering to their young guests.

Moreover, children explore sites through their senses. Unlike adults, they run, climb, hide, and touch. They may even laugh or shout to their friends as they reenact war or torture. Children's emotional immaturity creates yet another dilemma for tourism sites. How can a dark tourism site accommodate families with young children? How do curators protect and engage the young while not diminishing the experience for others? For example, grim artifacts on lower shelves are at eye level for a frightened young child. And adults reflecting on tragedy may resent immature young people speeding through an exhibit. Destination managers wonder how to balance children's interests with those of adults.

Because of its focus on adults, tourism research has left us with many unanswered questions: How can we ensure that young tourists have a meaningful experience? What do children and young people think, feel, and believe when visiting a dark tourism site? How can we document these overlooked encounters? Which research methods can we use ethically? How do we unite traditional tourism researchers with childhood scholars and practitioners? Should children and young people themselves join us as evaluators and researchers? If so, how?

This book offers answers. Here the reader will find a series of interdisciplinary essays. These offer developmental context and, in so doing, provide conceptual underpinnings for scholarship related to children and dark tourism.

Parts I and II address two fundamental questions that form the basis for child dark tourism research. First, Part I answers the question, 'How should we think about children and young people as tourists?' We begin by generally examining children and young people in tourism, Then, we offer a new conceptual framework for understanding their experiences at dark sites.

Part II answers the question, 'Who are young tourists?' Drawing upon developmental and cognitive psychology, education, sociology, tourism, and thanatology studies, experts offer succinct background illustrated with engaging examples from tourism.

Part III addresses fundamental interrelationships between dark tourism and [re]presentation, when considered from a child's perspective. Authors draw on empirical cases, demonstrating how dark tourism is interpreted for young tourists and the implications and consequences thereof.

Part IV draws upon multifaceted dimensions of consuming dark tourism, including young people's experiences with entertainment, edutainment, educational excursions, and historiography. This part also presents a new conceptual model for understanding the experiences of children living in dark tourism locations.

Part V articulates research methods specifically related to children's travel. This part outlines how to establish and maintain research partnerships with schools. Experienced child tourism researchers explain age-appropriate, ethical data collection and outline multiple innovative strategies suitable for children in tourism studies. Original case illustrations from researchers traveling with children appear in each chapter. New to the field are accounts from young people conducting their own research at dark tourism sites.

In summary, this volume seeks to inform and encourage research about children's dark tourism experiences. We hope you will find inspiration and answers in its pages.

Mary Margaret Kerr
Philip R. Stone
Rebecca H. Price

Part I

Dark Tourism and Childhood

1 'Seen but Not Heard'

Children in (Dark) Tourism Research Agendas

Philip R. Stone

Introduction

Recorded in a 15th-century collection of homilies titled *Mirk's Festial* by Augustinian clergyman John Mirk, the Old English proverb '*a mayde schuld be seen, but not herd*' was specifically aimed at young unmarried women who were expected to remain quiet during adult conversations. With its origins in medieval religious culture, the etymology of the proverb evolved to '*children should be seen and not heard*' whereby a youth's naivety or ignorance of adult matters has largely remained to this day. With this in mind, the dearth of attention paid to children in tourism scholarship (Poria and Timothy, 2014) largely reflects children who are very much seen within the visitor economy, but their touristic experiences have not been fully interrogated and, consequently, their voices remain unheard. Tourism is generally perceived – and researched – as an adult affair full of hedonic experiences. As a social, cultural, and economic phenomenon, the study of tourism is often framed within the context of mobility that connects broader patterns of tourist flows with individual (adult) life trajectories (Hall, 2004). Meanwhile, dark tourism is where our significant dead occupy touristic landscapes and, as such, 'ghosts' of the noteworthy deceased haunt our collective conscience through memorials, museums, and visitor attractions (Stone, 2020a). In reality, adults and children often roam dark tourism traumascapes together and mutually consume tragic memories. Yet, this seemingly invisibility of children and their tourist experiences, both within mainstream tourism generally and dark tourism in particular, draw researchers to scrutinise the practice as mainly adult-orientated. For instance, the recent *Routledge Handbook of the Tourist Experience* (Sharpley, 2022) invites leading scholars to create a comprehensive tome of contemporary research on the tourist experience. Yet, despite its extensive multidisciplinary synthesis of tourist behaviours and motivations, the dynamic parameters of the volume omit children – both as touristic consumers and external mediators – in creating tourist experiences bespoke to youth.

The purpose of this chapter, therefore, is to highlight this critical gap in the tourism literature generally and, more specifically, identify the challenges and

DOI: 10.4324/9781003032199-2

strictures for studying children within dark tourism experiences. In so doing, I offer an overview of relevant scholarship before outlining potential reasons for the scarcity of studies on children within tourism. Consequently, I suggest how children's touristic experiences should be revealed and why children's voices are paramount not only to tourism research agendas, but also to dark tourism encounters. Ultimately, I argue that children should be liberated and their voices unmuted within tourism research if the aim of scholarly inquiry is to conceptualise 'tourist experiences' more comprehensively and responsibly. As a special note, despite socio-cultural, biological, or legal definitions, children (or minors) may be generally defined as young human beings below the age of puberty or below the legal age of majority. Hence, children may be specifically defined here as those 14 years old or under, and youths as aged between 15 and 18 years old.

'Missing Children': A Review of Children in Tourism Scholarship

In an editorial for the *Annals of Tourism Research* almost four decades ago, Nelson Graburn called for the expansion of social anthropology within tourism studies and, in particular, for the inclusion of children within tourism research. Graburn (1983: 3) observed the lack of children in tourism studies and states:

> Most lacking, perhaps, are studies of the effects of tourism on the historical, natural, and geographical awareness of children: the trajectory and interrelationship of their touristic and recreational experiences, and the relations of these to their adult life styles and to their subsequent recreational and vocational behaviours.

While there have been efforts during the intervening years to examine children within tourism and leisure, particularly within family tourism research (Li, Lehto, and Li, 2020) and to include youth identity and belonging in tourism destinations (Canosa, 2014, 2016), children and childhood experiences in tourism remain neglected and absence (Small, 2008; Poria and Timothy, 2014). Tourism is recognised as a powerful force for change for host destinations, with both positive and negative consequences, yet the focus is often upon adults. The silence of children in tourism research is further highlighted by Canosa, Brent, and Wray (2016), who advocate new 'voice-generative' methods to discover children's agency in tourism. However, despite numerous calls to action, tourism scholarship over recent years has focussed on other 'new' types of tourist experience (Sharpley and Stone, 2011, 2012). For instance, travellers with disabilities have been explored within the context of embodiment, whereby social power structures and commitments to emancipation are at the core of the tourist experience (Small and Darcy, 2010). Other types of tourists have also been explored, including gay tourism and experiences of the LGBTQ+ community. Studies are often focussed on relationships between gay tourism, (Western) homosexual culture and leisure space, the erotic, sexual politics and hegemony, or sexual diversity (Pritchard, Morgan, and Sedgley, 2002; Waitt and Markwell, 2006; Bailey, 2021). The elderly are also often the focus of tourism scholarship, with studies including, but not limited to, quality of life and wellbeing (Kim, Woo, and Uysal, 2015), senior tourism and the impacts of aging on destinations (Alén, Losada, and Domínguez, 2016), and aging populations and tourism economics and marketing (Isa, Ismail, and Fuza, 2020; Pak, 2020). Moreover, with regard to the elderly tourist, Bauer (2021) examines the role of travel medicine and psychological travel health care in 'dark tourism' (also see Bauer, 2015; and Pratt, Tolkach and Kirillova, 2019). Tourism scholarship has also recently become gendered focussed, with specific studies examining the researcher and masculinities (Porter, Schänzel, and Cheer, 2021), as well highlighting feminist voices and women's contributions in tourism research (Correia and Dolnicar, 2021).

Undoubtedly, scholarly attention has been paid to other tourist groups and even social-scientific tourism researchers, yet children and their experiences are a crucial consumer group within tourism, but are conspicuous by their absence in research. Indeed, tourism as a form of consumption for families both in the developed and developing world, parents are spending increasing amounts of money (as well as time) on travel with their children (Shuxia, 2018; Song, Park, and Kim, 2020). Moreover, children have fundamental influence on destination and activity decision-making within family units, whereby Therkelsen (2010) suggests that mothers and children hold equal weight in vacation decision-making. Parents are also sensitive to their children's 'best and worst' choices, with parents willing to pay extra money on touristic activities that fulfil children's preferences (Curtale, 2018). Additionally, the satisfaction of children in their tourist experience affects the satisfaction and behavioural intentions of parents, with younger children having a greater influence (Thornton, Shaw, and Williams, 1997).

Despite limited studies available based on children's voices, early studies focussed on children's influences on parental decision-making within travel and tourism, as well as children's influence on adult tourism experiences (Jenkins, 1978; Filiatrault and Ritchie, 1980; Howard and Madrigal, 1990; Fodness, 1992; Tagg and Seaton, 1994; Thompson, Pinney, and Schibrwosky, 1996; Thornton et al., 1997). Other early studies stress the importance of children as a distinct market segment which, in turn, focused on the benefits for the tourism industry (Ryan, 1992; Swarbrooke and Horner, 1999). This focus on children as consumers – or 'toddler tourism' – and implications for marketing strategy and buyer motivations has continued with later studies examining children as co-decision-makers in holidays and leisure (Nickerson and Jurowski, 2001; Turley, 2001; Connell, 2005; Gram, 2007; Wu and Wall, 2017). Meanwhile, Cullingford (1995) began the task of examining children's actual 'holiday experiences', where familiarities and differences of the young tourist experience were considered important. Indeed, Cullingford (1995: 127) suggests children's awareness of differentiation 'is the side of their experience that has remained unexploited'.

Similarly, Gamradt (1995) attempted to capture 'little narratives' (after Lyotard, 1984) of young tourist experiences, particularly within broader notions of nationhood and neo-colonialism. Likewise, relationships between children's geographical knowledge and travel experiences are examined by Poria, Atzaba-Poria, and Barrett (2005), who argue that industry operators should better develop (young) tourist experiences that enhance children's knowledge (also see Rhoden, Hunter-Jones, and Miller, 2016). Influences of children on vacation travel patterns are also examined by Nickerson and Jurowski (2001), who demonstrate that children not only have higher survey response rates than adults, but they provide unique perspectives about destination planning and development that can increase child satisfaction (and, therefore, parent satisfaction). Additionally, the role of children and tourism is explored by Buzinde and Manuel-Navarrete (2013), who examine the social production of space within tourism enclave resorts and, in particular, children's perceptions of tourism boundaries (also see Yang, Yang, and Khoo-Lattimore, 2020). In doing so, Buzinde and Manuel-Navarrete (2013) argue that the tourism sector creates liminal inclusions and exclusions for the geographies of childhood and, consequently, a better understanding of lived socio-spatial experiences for children is required (also see Canosa and Wray, 2013).

Much of family tourism research, albeit related to children, adopts adult perspectives whereby children are classified as passive actors (Obrador, 2012). However, recent research has begun to take more child-centred approaches, which has dignity and the voice of children at its core (Canosa, Graham, and Wilson, 2018a). For example, issues of empowering children, particularly with regard to responsible and sustainable tourism development, have been explored. Indeed, Sèraphin, Yallop, Seyfi and Hall (2020) highlight the significance of nurturing children to have the essential skills, education, and experience to be sustainability thinkers, actioners, and transformers (also see Koščak et al., 2021). Meanwhile, Wu, Wall, Zu, and Ying (2019) examine family tourism in China – the world's most populous country – and explore memorable tourist experiences, with picture drawing and story sharing as methods to elicit children's voices (also see Zhong and Peng, 2021). Similarly, Qeidari, Shayan, Solimani and Ghorooneh (2021) conducted a phenomenological study of learning experiences of Iranian children in rural tourism destinations (also see Barbieri, Stevenson, and Knollenberg, 2019). They use paintings as a research method to bring out children's voices and experiences of rurality and environmental awareness, which Poris (2006) calls 'empowering fun' (also see Ertaş, Ghasemi, and Kuhzady, 2021).

Recently, Elmi et al. (2020) examined children and travel memories and compared what children remember and what children desire in their tourism experiences. They heed the call by Poria and Timothy (2014) to investigate children's representations of tourism and work 'with them' and not 'about them'. Hence, through interactive research methods, Elmi et al. (2020) appraise two different, but complementary aspects of memory – firstly, past experiences or *retrospective memory* and, secondly, ideal expectations or *prospective memory*. In order to discover children's voices, the researchers scrutinise children and their 'ideal' tourism trips, which reflect children's wishes and desires about ideal journeys that they wish to undertake. In turn, Elmi et al. (2020) indicate children's prospective memories have profound implications for tourism marketing strategies.

Though academia may be paying limited attention to children and despite implications for broader place-making, the industry recognises the importance and value of children within tourism. In particular, at the time of writing, COVID-19 enveloped the world as a global pandemic and, as a result, ruptured visitor economies. Some scholars were quick to realise that tourism and, specifically, children in tourism could serve as key stakeholders in a post-COVID world (Stone, 2020b; Sèraphin, 2021). Furthermore, Sèraphin (2021) argues that despite all the negative impacts of COVID-19, the pandemic has provided an opportunity to review old practices and adopt new ones. By outlining an engaging research method toolkit for the study of children in tourism, Sèraphin (2021) suggests that children be empowered to become tourists of tomorrow.

Similarly, Radic (2019) examines children within the cruise industry and suggests that children while on board cruise ships have certain levels of autonomy. Consequently, children can co-create their own memorable cruise experience with further implications for cruise company management (Radic, 2019). To that end, children and youths are progressively being recognised within the visitor economy and have specific marketing and management requirements (Sèraphin and Gowreesunkar, 2020). Sèraphin and Green (2019) go on to note that children as young customers of today are our impending consumers. Consequently, children are significant in their contribution to conceptualising and branding tourism destinations of the future (Sèraphin and Green, 2019). In terms of modern tourism development, Koščak, Fabjan and O'Rourke (2018) even suggest that children should be an integral part of participatory destination planning. The relationship between children and tourism development is also examined by Leonard (2019) whereby young teenagers are involved in co-constructed tours of Belfast, Northern Ireland. As a place of contestation, Leonard (2019) demonstrates how children's local spatial knowledge challenges adult discourses and how children can contribute to touristic understandings of divided cities. In particular, Leonard (2019) argues that existing tourism frameworks are extended to incorporate children's own ways of seeing and experiencing their everyday spatial lives.

Hence, children are becoming increasingly important in heritage tourism research and how children make sense of heritage (Frost and Laing, 2017; Wu and Wall, 2017). Much of this is related to understanding children's engagement with interpretation at cultural heritage attractions (Sutcliffe and Kim, 2014). Indeed, it is the circumstances under which interpretative techniques are utilised by children, and how children interact with their peers which affect the level of understanding of different (hi)stories (Sutcliffe and Kim, 2014). For instance, in terms of heritage and diaspora tourism, post-colonial or post-disaster destinations rely heavily on their diaspora for survival of their visitor economies. Therefore, heritage links with members of dispersed populations are challenged by keeping connections with younger generations (either second or third diaspora generations). As Sèraphin (2020) notes, childhood tourist experiences of the country of origin are a transformative tool which can lead to either 'dediasporisation' (if negative), or 'transnational attachment' (if positive). Children's engagement and experiences at cultural heritage attractions remain a priority for tourism scholarship, but so too are children's experiences at difficult heritage or dark tourism sites. It is here within sites *of* death or sites *associated with* disaster that children wander with adults and consume stories of pain and shame. It is to 'heritage that hurts' and children experiences thereof that I now turn.

'Lost Children': Dark Tourism and Child Encounters

Dark tourism as a multidisciplinary field of study attempts to capture contemporary (re)presentations of our notable dead. In turn, dark tourism permits the dead to become contemporary commodities, and for tragic memories to be retailed in socially sanctioned tourist environments (Stone, 2020a). The dominion of dark tourism offers a selective voice and records tragedy across time, space, and context and, subsequently, can provide reflectivity of both place and people. Yet, different cultural, political, and linguistic representations of dark tourism and varying interpretive experiences are complex and multifarious and cannot be taken at face value. Instead, dark tourism in its many guises offers visual signifiers and multiplicity of meanings within touristic landscapes as global visitor sites function as retrospective witnesses to acts of atrocity or tragedy (Stone, 2018a). Contemporary memorialisation is played out at the interface of dark tourism, where consumer experiences can catalyse sympathy for the victims or revulsion at the context. We gaze at dark tourism in the knowledge that the victims are already dead, though the precise context and history of the victims can never be truly understood (Hartmann et al., 2018). Ultimately, dark tourism and its difficult heritage is about death and the dead, but through its touristic consumption, perhaps tells us more about life and the living (Stone and Sharpley, 2008). It is here that dark tourism attracts families and children as together they experience 'mortality moments' and fatality of the Other (Stone, 2012a,b).

However, a cautionary note to examining 'dark tourist' experiences is offered by Seaton (2022) in that 'experience' raises a perennial problem in social science inquiry. In short, the insider/outsider dilemma raises questions of whether judgements of human behaviour can be made by authorised or empowered external observers. Or, should judgements be reached by the self-defining claims and assertions (and, on occasion, denials) by people about themselves (Seaton, 2022)? Nevertheless, Stone (2018b: 510) suggests that 'there can never be a so-called dark tourist as a defined taxonomy because to consume (dark) tourism is to consume experiences'. Despite methodological issues of the insider/outsider dilemma, Stone (2018b: 510) goes on to argue that dark tourism as a social phenomenon involving the 'mass movement of people across global traumascapes', including children, 'cannot be fully understood without a critical examination of the meaning or significance of dark tourism to tourists themselves' (also see Iliev, 2020).

Of course, whilst the general call for research into young tourist experiences remains, there are now emerging and potentially fruitful studies which examine children and young people within certain tourist environments – namely dark tourism. For instance, Darlington (2014) as a young undergraduate, offers an autobiographical account of her childhood visits to a dark tourism site and her subsequent reconciliation with bereavement. During a school history visit to Flanders Fields – a major theatre of battle during World War I – Darlington recalls her grief of losing a friend in an accident some months earlier, who was unconnected with the Flanders site. It is here that Darlington recognises *dark tourism doesn't need dark tourists – it just needs people who are interested in learning about this life and this world* (Darlington, 2014 – quoting Philip Stone in Coldwell, 2013: online). In particular, Darlington suggests her youthful visit to Flanders and her ensuing exposure to wartime suffering and death, where a million soldiers from more than 50 countries were wounded or killed in action, became a mediated journey of secular catharsis for her deceased friend (also see Fallon and Robinson, 2017). Moreover, her dark tourism visit gave way to an emotional release of grief. As a result, it provided an intersection of childhood bereavement and mediation of mortality for a (private) death unrelated to a (public) dark tourism site (Stone and Sharpley, 2008; Stone, 2012a, 2020a). As Darlington (2014: 45) notes:

> My catharsis went much further than just being allowed the chance to cry over the deceased… my senses were heightened [at the site], and I could truly understand the pain of those left behind, the sorrow, the social unity, and the cultural identity that the war had shaped for those connected. I guess this is empathy at its most obvious. I could feel the aches and pains really in me… As a result I was able to understand my own feelings [of grief], and found great comfort in not being alone… I could use my imagination to conjure up people who had felt the same. The battlefields made me empathise [and] empathy, perspective, catharsis and emotional release were my personal outcomes. Dark tourism and experiences like this connect the living with the dead through emotion and understanding. I became at peace with my own feelings and with the reality of death and my friend being gone.

While Darlington (2014) is articulate in recalling her teenage grief, other studies have begun to capture mortality moments of children in other disaster or dark tourism environments. Most notably, Kerr et al. (2017a,b) offer empirical studies of post-disaster situations and children's reactions to terrorist attacks and, in doing so, examine children's letter writing. They go on to note that in the case of the Flight 93 airline crash during the 9/11 terrorist atrocity, children's letters proved a valuable research instrument to ascertain children's emotional state. These letters, written at school in the immediate aftermath of the 9/11 attack and addressed to victims' families, recovery volunteers, or first responders, are archived at the Flight 93 National Memorial in Pennsylvania, USA. In them, children reveal specific coping strategies, recollections of the tragedy, and a sense of community and American pride (Kerr et al., 2017a). These unique perspectives go on to form part of 'interpreting terrorism' at a national memorial site, whereby children's letters and subsequent comment cards become a bedrock of the 'child's voice' (also see Stone, 2012b; Croom et al., 2018). As Kerr et al. (2017b: 86) state in relation to the Flight 93 National Memorial:

> …we allowed children who were not the beneficiaries of formal interpretation to 'talk with' us through their comment cards. We listened to their voices in order to understand their personal meaning-making of the tragic events commemorated, but not formally interpreted, for them at this memorial site.

Similarly, Price and Kerr (2018) probe children's exploratory play at war memorials and the meaning-making processes behind such behaviour. While young children's play behaviour at particular dark tourism sites is often a focus for social media moral commentary, the cultural context of children's behaviour at memorials and other difficult heritage sites is less understood (Kerr and Price, 2016, 2018; Price and Kerr, 2018; Kerr, Stone, and Price, 2021). Indeed, the notion of respect and moral codes are inherent at memorial tourist sites, and children's playful behaviour at such sites exposes not only design and interpretation issues, but also issues of values, deep-seated feelings, and beliefs. In particular, Dresler and Fuchs (2021) explore moral geographies of using Auschwitz as a teaching space for German students. As a moral space imbued with moral judgements and where secular morality can be constructed (Stone and Sharpley, 2013), the Auschwitz death camp was experienced by the youths as affirming collective identity, emotional engagement, and moral reflection (Dresler and Fuchs, 2021). It is here that child's and youth experiences of dark tourism raise complex issues and manifold challenges.

Much of this revolves around dark tourism and sense-making, whereby dark tourism's potential capacity to inform and reflect contemporary interpretative and pedagogic theory is unique (Roberts, 2018). Indeed, as Roberts (2018) points out, the emotional, sensory, and relational aspects of dark tourism and their relevance to interpretative processes and experiential learning is substantial (also see Sun and Lv, 2021; Martini and Buda, 2020; Oren, Shani, and Poria, 2021). Yet, while children have different cognitive conceptions of mortality and trauma, dark tourism site interpretation is very much adult-oriented while children audiences consume the same adult messaging and narratives (Kerr, Stone and Price, 2020).

The narratives of a contested heritage site – the Battle of the Boyne Visitor Centre in Ireland – and children's experiences of such difficult (adult) heritage are explored by Roche and Quinn (2017). In particular, they suggest children's preconceptions of dark tourism are informed by the site's location, media and social media images, as well as social interaction with their peers and tour guides (Roche and Quinn, 2017; also see Leonard, 2019). While present-day dissonance has been studied elsewhere within dark tourism (see, for example, Hartmann et al., 2018), Halevi (2020) offers a unique historical examination of the travel experiences of a small group of adolescent girls as they toured 'dark sites' of the early United States between 1782 and 1834. Halevi (2020) goes on to suggest that such sites were viewed as places which exemplified civic sacrifice or benevolence. Consequently, the young girls' visits to such early dark tourism sites were part of a collective desire to foster female civic virtue in the new Republic (of the USA) (Halevi, 2020).

There is little doubt that dark tourism is full of moral ambiguities and ethical challenges for the dark tourism researcher. Notwithstanding the subject lacking the voice of children and their experiences, scholars have made tentative but vital steps into dark tourism and youthful encounters. That said, however, incorporating children's voices into future (dark) tourism research agendas remains a priority, and it is this that I now turn.

'Children's Voices': Incorporating Children into (Dark) Tourism Scholarship

Despite the rather scant literature base for childhood experiences within (dark) tourism, a number of methodological and research design issues are evident. These issues often revolve around the child's voice within the research process and, as such, applying appropriate methods to extract, listen, and interpret young tourists' experiences (Khoo-Lattimore, 2015). With the complex realities of family life and domesticity (Canosa, 2018), the invisibility of children's experiences in tourism research means that specific research methods and approaches are required (Obrador, 2012; Frost and Laing, 2017). Indeed, the popularity of the post-modernist paradigm often calls for individual voices to be captured and heard. As noted earlier, other 'new' groups of tourist experiences are beginning to be unmuted, yet children's experiences largely remain silent. Therefore, this deficiency of child-centred research limits researchers' ability to fully conceptualise social, economic, and cultural constructs of tourism (Carr, 2011).

Consequently, Khoo-Lattimore (2015) outlines a number of methodological challenges and concerns when conducting studies of young tourist experiences. With regard to dark tourism and children, many innovative research designs are inherent within this book and the focus of numerous chapters (for example, see Chapter 2). Without repeating them here, these issues often focus on the ethical nuances of conducting participatory research with children. As such, Canosa, Graham, and Wilson (2018b) suggest an increased ethical responsibility from researchers which requires a reflexive and relational approach. Researchers also require specific data collection techniques used for interacting with children at different developmental stages. Moreover, many tourism scholars do not have academic grounding in childhood studies or cognitive-behavioural backgrounds and, subsequently, lack requisite skillsets to specifically research children. Crucially, intrinsic safeguarding issues around the welfare and/or exploitation of children (WTO, 2001) may complicate or even prevent research into young tourist experiences (Carr, 2011; Poria and Timothy, 2014). Likewise, university research ethics committees, the search for parents' consent, or statutory governmental requirements may be too challenging for many tourism scholars – thus negating any ventures into the field. Additionally, theoretical frameworks used to conceptualise tourism may not be relevant to children's travel experiences. These paradigmatic assumptions are often based on tourism and notions of escape, exposure to culture and social realities of the Other, existential authenticity, performativity, and mobilities – all of which may be significantly less applicable to children.

The idea is that children are the best informants about themselves (Measelle et al., 1998; Atzaba-Poria, Pike, and Deater-Deckard, 2004). Importantly, a lack of tourism scholars adopting 'post-disciplinary' research approaches may mean some researchers lack requisite abilities, disciplinary knowledge, or collaborative research partners to deal with children (Coles, Hall, and Duval, 2009; Stone, 2011, 2013). Thus, tourism scholars need to go beyond their disciplinary strictures and engage in local, national, and international research partnerships and scholarly collaborations with those with expert training and grounding in childhood studies. For example, the work of Kerr, Stone, and Price (2021) is a manifestation of such a research collaboration, whereby child psychology experts and social science scholars teamed up to investigate children's experiences within (dark) tourism. Consequently, whilst still limited, voices of childhood experiences are increasingly emerging in different contexts and cultures, including within tourism (Canosa and Graham, 2020). Echoes of Nelson Graburn's (1983) original call for children to be included in the social anthropology of tourism scholarship can still be heard, but progressive research into a complex and ethically fraught area is slowly being made.

Conclusion

Children are not young human beings to be moulded but are people to be unfolded. In other words, and as I have argued in this chapter, children and their voices should now be heard within the touristic experience. Without such voices, extracted through appropriate research methodologies and designs, tourism as a social and cultural phenomenon is only partially constructed. In many ways, by outlining key scholarship within tourism and children studies, I have added to the call for research action. Yet, by outlining inherent issues that many researchers may face, particularly within the tourism academy, researching children's experiences within tourism remains highly problematic.

With this in mind, and as a potential solution to reconnoitre future research avenues into children and tourism, tourism scholars must fully collaborate in new research partnerships. In short, driven by a post-disciplinary research agenda, subject gatekeepers need to be challenged by new times and new methods. In so doing, established and emerging tourism researchers should further connect and affiliate with scholars who are expert in children and youth studies. With post-disciplinary research alliances, traditional subject borders and interests will be transgressed and, consequently, research approaches will be characterised by increased reasonableness, expertise, flexibility, and inclusivity. While this is important for tourism in general, it is crucial for dark tourism in particular. With global traumascapes dominating dark tourism, children are increasingly exposed to explicit or abstract fatality narratives. Our significant dead are made spectacular by dark tourism and spectacles of the deceased surround us. We need to know how children make sense of this dominion of the commemorated/commercialised dead. Ultimately, we need to know whether children's experiences of dark tourism and its difficult heritage helps 'monsters' to be tamed or conjured.

References

Alén, E., Losada, N., & Domínguez, T. (2016) The impact of ageing on the tourism industry: An approach to the senior tourist profile. *Social Indicators Research*, 127: 303–322.

Atzaba-Poria, N., Pike, A., & Deater-Deckard, K. (2004) Do risk factors for problem behaviour act in a cumulative manner? An examination of ethnic minority and majority children through an ecological perspective. *Journal of Child Psychology Psychiatry*, 45(4): 707–718.

Bailey, E.G. (2021) Is gay tourism more than tourism? A case study of Puerto Vallarta, Mexico. *Humanity & Society*, May. DOI: 10.1177/01605976211014013.

Barbieri, C., Stevenson, K.T., & Knollenberg, W. (2019) Broadening the utilitarian epistemology of agritourism research through children and families. *Current Issues in Tourism*, 22(19): 2333–2336.

Bauer, I.L. (2015) Looking over the fence – How travel medicine can benefit from tourism research. *Journal of Travel Medicine*, 22(3): 206–207.

Bauer, I.L. (2021) Death as attraction: The role of travel medicine and psychological travel health care in 'dark tourism'. *Tropical Diseases, Travel Medicine and Vaccines*, 7(24). DOI: 10.1186/s40794-021-00149-z.

Buzinde, C.N., & Manuel-Navarrete, D. (2013) The social production of space in tourism enclaves: Mayan children's perceptions of tourism boundaries. *Annals of Tourism Research*, 43: 482–505.

Canosa, A. (2014) *Silent Voices: A Critical Analysis of Tourism Social Impact Studies. CAUTHE Conference 2014: Tourism and Hospitality in the Contemporary Word – Trends, Changes & Complexity, Conference Proceedings*, 10–13 February.

Canosa, A. (2016) *Voices from the Margin: Youth, Identity and Belonging in a Tourist Destination*. Unpublished PhD Thesis, Centre for Children and Young People, School of Education, Southern Cross University, Lismore, Australia.

Canosa, A. (2018) 'Mummy, when are we getting to the fields?' Doing fieldwork with three children. In Porter, B.A., & Schänzel, H.A. (Eds) *Femininities in the Field: Tourism and Transdisciplinary Research*. Bristol: Channel View Publications, pp. 84–95.

Canosa, A., Brent, M.D., & Wray, M. (2016) Can anybody hear me? A critical analysis of young residents' voices in tourism studies. *Tourism Analysis*, 2(2–3): 325–337.

Canosa, A., & Graham, A. (2020) Tracing the contribution of childhood studies: Maintaining momentum while navigating tensions. *Childhood*, 27(1): 25–47.

Canosa, A., Graham, A., & Wilson, E. (2018a) Reflexivity and ethical mindfulness in participatory research with children: What does it really look like? *Childhood*, 25(3): 400–415.

Canosa, A., Graham, A., & Wilson, E. (2018b) Child-centred approaches in tourism and hospitality research: Methodological opportunities and ethical challenges. In Nunkoo, R. (Ed) *Handbook of Research Methods for Tourism and Hospitality Management*. Cheltenham: Edward Elgar Publishing Ltd, pp. 519–529.

Canosa, A., & Wray, M. (2013) *The 'Voiceless' Population: Understanding the Social Impacts of Tourism and their Influence on the Development of Young People in Tourist Destinations*. International Conference: Sustainability Issues and Challenges in Tourism, Istanbul, Turkey, 3–5 October, Bogaziçi University Printhouse, pp. 53–57. DOI: 10.1080/13032917.2013.861733.

Carr, N. (2011) *Children's and Families' Holiday Experience*. London: Routledge.

Coldwell, W. (2013) Dark tourism: Why murder sites and disaster zones are proving popular. *The Guardian*, 31 October, Retrieved from https://www.theguardian.com/travel/2013/oct/31/dark-tourism-murder-sites-disaster-zones [Accessed: 30 May 2018].

Coles, T.E., Hall, C.M., & Duval, D.T. (2009) Post-disciplinary tourism. In Tribe, J. (Ed) *Philosophical Issues in Tourism*. Bristol: Channel View Publications, pp. 80–100.

Connell, J. (2005) Toddlers, tourism and Tobermory: Destination marketing issues and television-induced tourism. *Tourism Management*, 26: 763–776.

Correia, A., & Dolnicar, S. (Eds) (2021) *Women's Voices in Tourism Research: Contributions to Knowledge and Letters to Future Generations*. Brisbane: University of Queensland Press. DOI: 10.14264/817f87d.

Croom, A.R., Squitiero, C., & Kerr, M.M. (2018) Something so sad can be so beautiful: A qualitative study of adolescent experiences at a 9/11 memorial. *Visitor Studies*, 21(2): 157–174.

Cullingford, C. (1995) Children's attitudes to holidays overseas. *Tourism Management*, 16: 121–127.

Curtale, R. (2018) Analyzing children's impact on parents' tourist choices. *Young Consumers*, 19: 172–184.

Darlington, C. (2014) Dark tourism: A school visit to Flanders. *Bereavement Care*, 33(2): 44–47.

Dresler, E., & Fuchs, J. (2021) Constructing the moral geographies of educational dark tourism. *Journal of Marketing Management*, 37(5–6): 548–568.

Elmi, B., Bartoli, E., Fioretti, C., Pascuzzi, D., Ciucci, E., Tassi, F., & Smorti, A. (2020) Children's representation about travel. A comparison between what children remember and what children desire. *Tourism Management Perspectives*, 33. DOI: 10.1016/j.tmp.2019.100580.

Ertaş, C., Ghasemi, V., & Kuhzady, S. (2021) Exploring tourism perceptions of children through drawing. *Anatolia*. DOI: 10.1080/13032917.2021.1883079.

Fallon, P., & Robinson, P. (2017) 'Lest we forget': A veteran and son share a 'warfare tourism' experience. *Journal of Heritage Tourism*, 12(1): 21–35.

Filiatrault, P., & Ritchie, J. (1980) Joint purchasing decisions: A comparison of influence structure in family and couple decision-making units. *Journal of Consumer Research*, 7(September): 131–140.

Fodness, D. (1992) The impact of family life cycle on the vacation decision-making process. *Journal of Travel Research*, 31: 8–13.

Frost, W., & Laing, J.H. (2017) Children, families and heritage. *Journal of Heritage Tourism*, 12(1): 1–6.

Gamradt, J. (1995) Jamaican children's representations of tourism. *Annals of Tourism Research*, 22(4): 735–762.

Graburn, N.H. (1983) Editor's page. *Annals of Tourism Research*, 10(1): 1–3.

Gram, M. (2007) Children as co-decision makers in the family? The case of family holidays. *Young Consumers*, 8(1): 19–28.

Halevi, S. (2020) In sunshine and in shadow: Adolescent girls and thanatourism in the early American Republic. *Journal of Tourism History*, 12(1): 71–85.

Hall, C.M. (2004) Tourism and Mobility. *CAUTHE 2004: Creating Tourism Knowledge*. Australia: University of Brisbane. Available online at http://hdl.handle.net/10523/738 [Accessed: 21 September 2021].

Hartmann, R., Lennon, J., Reynolds, D.P., Rice, A., Rosenbaum, A.T., & Stone, P.R. (2018) The history of dark tourism: A round table discussion. *Journal of Tourism History*, 10(3): 269–295.

Howard, D.R., & Madrigal, R. (1990) Who makes the decision: The parent or the child? The perceived influence of parents and children on the purchase of recreation services. *Journal of Leisure Research*, 22(3): 244–258.

Iliev, D. (2020) Consumption, motivation and experience in dark tourism: A conceptual and critical analysis. *Tourism Geographies*. DOI: 10.1080/14616688.2020.1722215.

Isa, S.M., Ismail, H.N., & Fuza, Z.I.M. (2020) *Elderly and Heritage Tourism: A Review. IOP Conference Series: Earth & Environmental Science*, Vol 447, 6–7 November.

Jenkins, R.L. (1978) Family vacation decision-making. *Journal of Travel Research*, 16(4): 2–7.

Kerr, M.M., Fried, S.E., Price, R.H., Cornick, C., & Dugan, S.E. (2017a) Rural children's responses to the Flight 93 crash on September 11, 2001. *Journal of Rural Mental Health*, 41(3): 176–188.

Kerr, M.M., & Price, R.H. (2016) Overlooked encounters: Young tourists' experiences at dark sites. *Journal of Heritage Tourism*, 11(2): 177–185.

Kerr, M.M., & Price, R.H. (2018) 'I know the plane crashed': Children's perspectives in dark tourism. In Stone, P.R., Hartmann, R., Seaton, T., Sharpley, R., & White, L. (Eds) *The Palgrave Handbook of Dark Tourism Studies*. London: Palgrave Macmillan, pp. 553–583.

Kerr, M.M., Price, R.H., Savine, C.D., Ifft, K., & McMullen, M.A. (2017b) Interpreting terrorism: Learning from children's visitor comments. *Journal of Interpretation Research*, 22(1), Retrieved from https://www.interpnet.com/NAI/nai/_publications/JIR_v21n1_Kerr.aspx [Accessed: 30 May 2018].

Kerr, M.M., Stone, P.R., & Price R. (2021) Young tourists' experiences at dark tourism sites: Towards a conceptual framework. *Tourist Studies*, 21(2): 198–212.

Khoo-Lattimore, C. (2015). Kids on board: Methodological challenges, concerns, and clarifications when including young children's voices in tourism research. *Current Issues in Tourism*, 18(9): 845–858.

Kim, H., Woo, E., & Uysal, M. (2015) Tourism experience and quality of life among elderly tourists. *Tourism Management*, 46: 465–476.

Koščak, M., Fabjan, D., & O'Rourke, T. (2018) No one asks the children, right? *TOURISM: An International Interdisciplinary Journal*, 66(4): 396–410.

Koščak, M., Knežević, M., Binder, D., Pelaez-Verdet, A., Işik, C., Mićić, V., Borisavljević, K., & Šegota, T. (2021) Exploring the neglected voices of children in sustainable tourism development: A comparative study in six European tourist destinations. *Journal of Sustainable Tourism*. DOI: 10.1080/09669582.2021.1898623.

Leonard, M. (2019) The teenage gaze: Teens and tourism in Belfast. *Childhood*, 26(4): 448–461.

Li, M., Lehto, H., & Li, H. (2020) 40 years of family tourism research: Bibliometric analysis and remaining issues. *Journal of China Tourism Research*, 16(1): 1–22.

Lyotard, J.F. (1984) *The Postmodern Condition: A Report on Knowledge*. Minneapolis: University of Minneapolis Press.

Martini, A., & Buda, D.M. (2020) Dark tourism and affect: Framing places of death and disaster. *Current Issues in Tourism*, 23(6): 679–692.

Measelle, J.R., Ablow, J.C., Cowan, P.A., & Cowan, C.P. (1998) Assessing young children's views of their academic, social and emotional lives: An evaluation of the self-perception scales on the Berkeley puppet interview. *Child Development*, 69: 1556–1576.

Nickerson, N.P., & Jurowski, C. (2001) The influence of children on vacation travel patterns. *Journal of Vacation Marketing*, 7(1): 19–30.

Obrador, P. (2012) The place of the family in tourism research: Domesticity and thick sociality by the pool. *Annals of Tourism Research*, 39(1): 401–420.

Oren, G., Shani, A., & Poria, Y. (2021) Dialectical emotions in a dark heritage site: A study at the Auschwitz Death Camp. *Tourism Management*, 82. DOI: 10.1016/j.tourman.2020.104194.

Pak, T.Y. (2020) Old-age income security and tourism demand: A quasi experimental study. *Journal of Travel Research*, 59(7): 1298–1315.

Poria, Y., Atzaba-Poria, N., & Barrett, M. (2005) The relationship between children's geographical knowledge and travel experience: An exploratory study. *Tourism Geographies*, 7(4): 389–397.

Poria, Y., & Timothy, D.J. (2014) Where are the children in tourism research? *Annals of Tourism*, 47: 93–95.

Poris, M. (2006) Understanding what fun means to today's kids. *Young Consumers*, 7(1): 14–22.

Porter, B.A., Schänzel, H.A., & Cheer, J.M. (2021) *Masculinities in the Field: Tourism and Transdisciplinary Research*. Bristol: Channel View Publications.

Pratt, S., Tolkach, D., & Kirillova, K. (2019) Tourism & death. *Annals of Tourism Research*, 78. DOI: 10.1016/j.annals.2019.102758.

Price, R.H., & Kerr, M.M. (2018) Child's play at war memorials: Insights from a social media debate. *Journal of Heritage Tourism*, 13(2): 167–180.

Pritchard, A., Morgan, N., & Sedgley, D. (2002) In search of lesbian space? The experience of Manchester's gay village. *Leisure Studies*, 21(2): 105–123.

Radic, A. (2019) Towards an understanding of a child's cruise experience. *Current Issues in Tourism*, 22(2): 237–252.

Rhoden, S., Hunter-Jones, P., & Miller, A. (2016) Tourism experiences through the eyes of a child. *Annals of Leisure Research*, 19(4): 424–443.

Roberts, C. (2018) Educating the (Dark) masses: Dark tourism and sensemaking. In Stone, P.R., Hartmann, R., Seaton, T., Sharpley, R., & White, L. (Eds) *The Palgrave Handbook of Dark Tourism Studies*. London: Palgrave Macmillian, pp. 603–638.

Roche, D., & Quinn, B. (2017) Heritage sites and schoolchildren: Insights from the Battle of the Boyne. *Journal of Heritage Tourism*, 12(1): 7–20.

Ryan, C. (1992) The child as visitor. *World Travel and Tourism Review*, 2: 135–139.

Seaton, T. (2022) Chapter 24: Remembracing, remembrance gangs and co-opted encounters: Loading and reloading dark tourism experiences. In Sharpley, R. (Ed) *Routledge Handbook of the Tourist Experience*. Abington, Oxon: Routledge, pp. 269–281.

Sèraphin, H. (2020) Childhood experience and (de)diasporisation: Potential impacts on the tourism industry. *Journal of Tourism, Heritage & Services Marketing*, 6(3): 14–24.

Sèraphin, H. (2021) COVID-19 and the acknowledgement of children as stakeholders of the tourism industry. *Anatolia*, 32(1): 152–156.

Sèraphin, H., & Gowreesunkar, V. (Eds) (2020) *Children in Hospitality and Tourism: Marketing and Managing Experiences*. Berlin/Boston: De Gruyter.

Sèraphin, H., & Green, S. (2019) The significance of the contribution of children to conceptualising and branding the smart destination of the future. *International Journal of Tourism Cities*, 5(4): 544–559.

Sèraphin, H., Yallop, A.C., Seyfi, S., & Hall, C.M. (2020) Responsible tourism: The 'why' and 'how' of empowering children. *Tourism Recreation Research*. DOI: 10.1080/02508281.2020.1819109.

Sharpley, R. (2022) Routledge Handbook of the Tourist Experience. Routledge: London.

Sharpley, R., & Stone, P.R. (Eds) (2011) *Tourist Experience: Contemporary Perspectives*. Abington, Oxon: Routledge.

Sharpley, R., & Stone, P.R. (Eds) (2012) *Contemporary Tourist Experience: Concepts and Consequences*. Abington, Oxon: Routledge.

Shuxia, W. (2018) Children tourism investigation and analysis in Zhejiang Province, China. *Asia Pacific Journal Multidisciplinary Research*, 6: 74–81.

Small, J. (2008). The absence of childhood in tourism studies. *Annals of Tourism Research*, 35(3): 772–789.

Small, J., & Darcy, S. (2010) Understanding tourist experience through embodiment: The contribution of critical tourism and disability studies. In Buhalis, D., & Darcy, S. (Eds) *Accessible Tourism: Concepts and Issues*. Bristol: Channel View Publications, pp. 73–97.

Sojasi Qeidari H, Shayan H, Solimani Z, Ghorooneh D. A phenomenological study of the learning experience of children in rural tourism destinations. *Tourist Studies*. 2021;21(2):235–259. DOI:10.1177/1468797620985781.

Song, H., Park, C., & Kim, M. (2020) Tourism destination management strategy for young children: Willingness to pay for child-friendly tourism facilities and services at a heritage site. *International Journal of Environmental Research and Public Health*, 17(19): 7100. DOI: 10.3390%2Fijerph17197100.

Stone, P.R. (2011) Dark tourism: Towards a new post-disciplinary research agenda. *International Journal of Tourism Anthropology*, 1(3/4): 318–332.

Stone, P.R. (2012a) Dark tourism and significant other death: Towards a model of mortality mediation. *Annals of Tourism Research*, 39(3): 1565–1587.

Stone, P.R. (2012b) Dark tourism as 'mortality capital': The case of Ground Zero and the significant other dead. In: Sharpley, R., & Stone, P.R. (Eds) *Contemporary Tourist Experience. Concepts and Consequences*. London: Routledge, pp. 71–94.

Stone, P.R. (2013) Dark tourism scholarship: A critical review. *International Journal of Culture, Tourism and Hospitality Research*, 7(3): 307–318.

Stone, P.R. (2018a) Dark tourism in an age of 'Spectacular Death'. In Stone, P.R., Hartmann, R., Seaton, T., Sharpley, R., & White, L. (Eds) *The Palgrave Handbook of Dark Tourism Studies*. London: Palgrave Macmillan, pp. 189–210.

Stone, P.R. (2018b) The 'Dark Tourist' experience. In Stone, P.R., Hartmann, R., Seaton, T., Sharpley, R., & White, L. (Eds) *The Palgrave Handbook of Dark Tourism Studies*. London: Palgrave Macmillian, pp. 509–513.

Stone, P.R. (2020a) Dark tourism and 'spectacular death': Towards a conceptual framework. *Annals of Tourism Research*, 83. DOI: 10.1016/j.annals.2019.102826.

Stone, P.R. (2020b) Dark tourism memorial sites will help us heal from the trauma of coronavirus. *The Conversation*. Available online at https://theconversation.com/dark-tourism-memorial-sites-will-help-us-heal-from-the-trauma-of-coronavirus-139164 [Accessed: 24 September, 2021].

Stone, P.R., & Sharpley, R. (2008) Consuming dark tourism: A thanatological perspective. *Annals of Tourism Research*, 35(2): 574–595.

Stone, P.R., & Sharpley R. (2013) Deviance, dark tourism and 'dark leisure': Towards a (re)configuration of morality and the taboo in secular society. In Elkington, S., & Gammon, S. (Eds) *Contemporary Perspectives in Leisure: Meanings, Motives and Lifelong Learning*. Abington, Oxon: Routledge, pp. 54–64.

Sun, J., & Lv, X. (2021) Feeling dark, seeing dark: Mind-Body in dark tourism. *Annals of Tourism Research*, 86. DOI: 10.1016/j.annals.2020.103087.

Sutcliffe, K., & Kim, S. (2014) Understanding children's engagement with interpretation at a cultural heritage museum. *Journal of Heritage Tourism*, 9(4): 332–348.

Swarbrooke, J., & Horner, S. (1999) *Consumer Behaviour in Tourism*. Oxford: Butterworth-Heinemann.

Tagg, S., & Seaton, T. (1994) How different are Scottish Family Holidays from English? In Seaton, T. (Ed) *Tourism: The State of the Art*. Chichester: John Wiley & Sons, pp. 540–548.

Therkelsen, A. (2010) Deciding on family holidays—Role distribution and strategies in use. *Journal of Travel & Tourism Marketing*, 27(8): 765–779.

Thompson, W., Pinney, K., & Schibrwosky, J. (1996) The family that gambles together: Business and social concerns. *Journal of Travel Research*, 34: 70–74.

Thornton, P.R., Shaw, G., & Williams, A.M. (1997) Tourist group holiday decision-making and behaviour: The influence of children. *Tourism Management*, 18(5): 287–297.

Turley, S.K. (2001) Children and the demand for recreational experiences: The case of zoos. *Leisure Studies*, 20(1): 1–18.

Waitt, G., & Markwell, K. (2006) *Gay Tourism: Culture and Context*. New York: Routledge.

WTO. (2001) *The Incidence of Sexual Exploitation of Children in Tourism*. United Nations World Tourism Organisation, UNWTOeLibrary. Retrieved from www.e-unwto.org/doi/abs/10.18111/9789284405008 [Accessed: 30 May 2018].

Wu, M.Y., & Wall, G. (2017) Visiting heritage museums with children: Chinese parents' motivations. *Journal of Heritage Tourism*, 12(1): 36–51.

Wu, M.Y., Wall, G., Zu, Y., & Ying, T. (2019) Chinese children's family tourism experiences. *Tourism Management Perspectives*, 29: 166–175.

Yang, M.J.H., Yang, E.C.L., & Khoo-Lattimore, C. (2020) Host-children of tourism destinations: Systematic quantitative literature review. *Tourism Recreation Research*, 45(2): 231–246.

Zhong, S., & Peng, H. (2021) Children's tourist world: Two scenarios. *Tourism Management Perspectives*, 38. DOI: 10.1016/j.tmp.2021.100824.

2 Young Tourists' Experiences at Dark Tourism Sites[1]

Toward a Conceptual Framework

Mary Margaret Kerr, Philip R. Stone, and Rebecca H. Price

While dark tourism aimed at adults reminds them of past tragic fights, faults, and follies, thousands of children and youth also consume inherent memorial messages at dark tourism sites. This paper addresses these unnoticed childhood encounters, about which scholarly discourse remains conspicuously silent. At present, dark tourism research focuses almost exclusively on adults and does not adequately explain young tourists' experiences. How children experience dark tourism sites has much to do with their understanding of death. Because younger children may not possess an adult-like knowledge of death, they are unable to experience a site as *dark*. Other theoretical disparities include children's limited agency in choosing their destinations and their unique and often playful exploration of dark places.

To address the inadequacy of current dark tourism conceptualizations, we propose a new framework to encourage scholarly interrogation of children's experiences at dark tourism sites. Drawing from multiple sources including archival studies and original research with youth, we offer a rationale for considering four major, intersecting influences on a young tourist's experience: Understanding of death, visit preparation (at home or in school), site and interpretation features, and dynamics of the specific visit (e.g., group membership, norms, and itinerary). Ultimately, this paper uncovers potential research avenues to bring children's perspectives and experiences to the core of dark tourism research.

Introduction

Dark tourism has become an internationally recognized taxonomy denoting touristic travel to sites *of* or *associated with* death and "difficult heritage" (Hartmann, 2014; Lennon and Foley, 2000; Stone, 2006; Stone et al., 2018). While dark tourism aimed at adults reminds them of past tragic fights, faults, and follies, thousands of young tourists also consume inherent memorial

DOI: 10.4324/9781003032199-3

messages at dark tourism sites. This paper addresses these unnoticed child-hood encounters, about which scholarly discourse remains conspicuously silent. Poria and Timothy (2014: 93) argue that "children's voices ought to be heard if the aim of scholarly inquiry is to conceptualize the tourist experi-ence more comprehensively and responsibly." Accordingly, we introduce the topic in a child's own words.

One morning in 2002, Aaron, the young son of one of the authors, slid an envelope under the author's door. Addressed to "Daddy and Mummy," the envelope contained a *memento mori* [reminder that one shall die] from a young boy coming to terms with mortality:

> *To Daddy and Mummy*
> *I like you very much and I wanted to live with you all the time but when you die I won't forget about you.*
> *Love*
> *Aaron*
> *xxxxxxxxxxxxx*

In fact, many young children think about death, although adults may not realize it.

> The reality is this… Children encounter death often. Their awareness with death and dying may come from real-life and/or vicarious experi-ences. Whether their familiarity comes from video games, children's lit-erature, cartoons, feature films, while watching the evening news, or from personal experience in their very own homes and/or communities, one thing remains constant: Death is tangible.
>
> (Perkins and Mackey, 2008: 13)

Dark tourism, with its death narratives and fatality encounters, also makes death "tangible" – certainly within popular culture (Penfold-Mounce, 2018; Walter, 2009). The remembered dead exert agency while "edutaining" tour-ists, occupying touristic traumascapes roamed by children as well as adults (Roberts, 2018; Stone, 2012; Stone and Sharpley, 2008; Stone et al., 2018; Walby and Piché, 2015). Despite taxonomical debates between the term "her-itage tourism" and "dark tourism" (Light, 2017), remembrance through dark tourism, offering a selective interpretive voice, records tragedy across time, space, and context, and subsequently causes one to reflect on both places and people. Hence, different cultural, political, and linguistic representations of dark tourism and varying revelatory experiences are complex and multifarious for younger tourists (Kerr and Price, 2018).

Young tourists' experiences within specific dark tourism encounters remain largely unstudied. That is especially pertinent considering their developing notions of death, their vulnerability, and their increasing visitation to places of the remembered dead (Kerr and Price, 2016). Thus, we propose an original conceptual framework that provides a "scholarly route map" for the academic interrogation of young tourists' experiences at dark tourism sites, with a focus on sites commemorating death (i.e., heritage sites). We recognize that young tourists also visit dark sites known for their entertainment, but that experience requires its own conceptualization. Although a position paper cannot cover every conceptual or practical eventuality, we offer a judicious framework that shines scholarly spotlights on pertinent areas worthy of examination.

Ultimately, our paper uncovers potential research pathways that situate young tourists and their visit dynamics at the core of dark tourism research. Firstly, however, we review "young tourists in tourism research" as a foundational basis for our subsequent "young tourists at dark tourism sites" conceptual framework. For brevity, we use "young tourists" to refer to those aged 5–14, while denoting developmental differences that further define these subgroups.

Young Tourists in Tourism Research

Over 35 years ago, in *Annals of Tourism Research*, then-editor Nelson Graburn called for the expansion of social anthropology within tourism studies and, specifically, for the inclusion of young tourists within tourism research. Graburn noted:

> Most lacking, perhaps, are studies of the effects of tourism on the historical, natural, and geographical awareness of children: the trajectory and interrelationship of their touristic and recreational experiences, and the relations of these to their adult life styles and to their subsequent recreational and vocational behaviors.
>
> (1983: 3)

More recently, noting Graburn's charge to the field, Small (2008) observes that studies of children and childhood in tourism remain largely absent. This conclusion comes despite some early studies focusing on children's influences on parental decision-making within travel and tourism, and children's influence on adult tourism experiences (e.g., Tagg and Seaton, 1994). Other research stresses young tourists as a distinct tourism industry market segment (Ryan, 1992; Swarbrooke and Horner, 1999). This focus on children as consumers – or "toddler tourism" – and its implications for marketing strategy and buyer motivations, continues with studies examining children as co-decision-makers in holidays and leisure (e.g., Cicero and Teichert, 2018; Ji et al., 2014; Schänzel and Carr, 2016; Wu and Wall, 2017).

Meanwhile, some researchers imagine the multiple ways in which children might make meaning of a tourism experience. For instance, Cullingford (1995) begins the task of examining young tourists' actual holiday experiences by suggesting that they might vary as widely as those of adult tourists. Likewise, Poria et al. (2005) argue that industry operators should improve tourist experiences that enhance knowledge and examine relationships between young tourists' geographical knowledge and travel experiences. Similarly, Elmi et al. (2020) explore young tourists' representations of travel and offer a comparison between what they remember and what they desire with regard to tourism.

Young tourists are also a key feature of the general tourism product and experience. This phenomenon is particularly so in marketing and consumer behavior, family tourism, or social tourism (Minnaert, 2014; Schänzel et al., 2012). Marketing to young tourists is also noticeable at large and well-traveled dark tourism destinations, which become increasingly "kitschified" (Potts, 2012). Importantly, however, there are potential issues of meaning-making within dark tourism. Specifically, the tourist experience depends on the availability and communication of narratives and stories. Much of this has depended on host sites as the main designers and promoters of touristified narratives, while the role of the tourist is often ignored in the process. Therefore, an issue of co-constructing meaning depends on tourists' involvement, willingness, and ability to actively participate in the storytelling (Chronis, 2012).

Yet, much of the co-construction of meaning is focused on adult experiences, and children's meaning-making of heritage remains overlooked. Despite increasing interest in tourist experiences and meaning-making (Bosangit et al., 2015), the tourism research agenda largely ignores young tourists (Frost and Laing, 2017; Poria and Timothy, 2014; Small, 2008). As Mooney et al. (2017: 106) protest, "children are relegated to the side-lines, as if unworthy of being in full view because their purchasing power relies on influencing their parents."

Why have researchers been reluctant to study young tourists? To begin, tourism training programs typically do not adopt a "post-disciplinary" and child-focused approach (Stone, 2011). As a result, researchers may lack requisite skillsets, disciplinary knowledge, or research partners to enter the complex world of young tourists (Coles et al., 2009). Moreover, laws governing ethical research with children require researchers to submit clearances and document suitable data collection methods in order to minimize risk. Obtaining informed consent can be arduous and time-consuming. Faced with these additional constraints, scholars hesitate to engage children in their research protocols (Poria and Timothy, 2014).

Despite the barriers, emerging studies examine young tourists' experiences within dark tourism (Kerr and Price, 2018). Indeed, studies have begun to capture mortality moments of young tourists in some dark tourism environments, such as 9/11 sites, Holocaust exhibits, and war memorials (Kerr et al., 2017b; Price, 2018; Price and Kerr, 2018). Israfilova and Khoo-Lattimore (2018) examine the educational importance of dark tourism, focusing on young tourists' experiences after their exposure to dark sites. They go on to suggest that visits to a thanatological attraction enlighten students and motivate them to study history as well as develop a deeper sense of patriotism. However, the cultural context of young tourists' behavior at memorials and other difficult heritage sites is still not well understood. Emotional aspects are also explored by Martini and Buda (2020), who within a framework of "emotion and affect" suggest that while most young tourists felt sorrow at dark tourism sites, others enjoyed the experience for its fascination and the chance to see places with "their own eyes."

Indeed, "with their own eyes" might characterize a recent development in tourism research. Canosa et al. (2019) propose an infusion of principles and practices from childhood studies into tourism research. Methods from childhood studies offer direct access to children's thoughts, experiences, and motivations. Recent developments in childhood studies embrace children's voices and empower children to research their own experiences. Such research prioritizes children's observations and experiences (Canosa et al., 2016, 2019; Kellett, 2005, 2010; Prout and James, 1997; Woodhead, 2008), and in so doing could provide valuable first-hand information about how children view tourist experiences as consumers. Taking into account these principles, we continue our review of the literature by examining the complexities of childhood and adolescence in dark tourism.

Dark Tourism and the Complexities of Young Tourists

To fully examine child dark tourism one must consider three complexities that characterize young tourists: *What* they comprehend about death, *why* they visit dark sites, and *how* they explore destinations. The first complexity emerges when one considers that dark tourism frameworks presume a mature adult understanding of death. However, "children who have not reached a mature understanding of these death concepts may find it difficult to engage with and understand death-related material presented in a museum exhibit" (Patterson, 2007: 58). Even when children understand death intellectually, they may not be emotionally ready to cope with it, as Israfilova and Khoo-Lattimore noted (2018).

Children must cognitively master five subconcepts to attain a mature understanding of death. *Universality* refers to the concept that all living creatures die (Speece and Brent, 1984). Children's initial understanding of this concept may express itself in terms of plants or insects but not humans. *Finality*, or *irreversibility*, conveys the notion that after death, a living thing cannot come back to life (Rosengren et al., 2014). For example, children may believe that humans come back to life because make-believe characters do.

Non-functionality describes the inability of the dead to engage in functions such as speaking, hearing, and thinking. For example, young children attempt conversations with mummies or taxidermy displays. *Inevitability* is an understanding that death is unavoidable. Adults witness this confusion when children propose escape plans for victims commemorated at memorials. *Causality* connotes different causes of death, including illness, accidents, and intentional acts. Scholars often refer also to a sixth concept, *personal mortality*, or the acknowledgment that one's own death is inevitable and unavoidable (Bianucci et al., 2015). Adolescents' awareness of personal mortality can evidence itself in *death anxiety* (Croom et al., 2018).

Each child's developmental timeline is influenced by their cognitive development and ability to understand abstract concepts. Children's books, films, and television clarify or confuse their thinking (Gutiérrez et al., 2014), as do "thanatechnology" (Sofka, 1997, 2009), social media, and internet gaming. Personal experiences with the death of a pet or extended family member can accelerate children's comprehension of death (Hunter and Smith, 2008). Although researchers cannot pinpoint exactly when children learn each concept, Table 2.1 offers a general timeline, with implications for child dark tourism.

A second childhood complexity arises when considering *why* young tourists visit dark tourist sites. Contemporary dark tourism theories focus on what motivates adult visits, but "the general field of tourism research and the specific field of tourism motivation seem to lack a systematic interest in children's motives" (Larsen and Jenssen, 2004: 45). For example, teachers – not students – typically plan itineraries for school trips (Cooper and Latham, 1988). As Walter (2009: 53) observes, "children find themselves being taken to battlefields by teachers and parents, just as children in this and other cultures find themselves being taken to church or temple."

As Shavanddasht and Schänzel (2019: 308) suggest, "there is an urgent need to understand adolescents' needs, motivations, and role in the tourism industry, particularly the factors that may affect their tourism satisfaction." Arguably, this charge takes on enhanced importance within a dark tourism experience, where a young tourist's motivations may shape how they respond to a particular traumascape in childhood and, in turn, influence their future travel decisions.

A third complexity reveals itself when one observes *how* young tourists explore sites in uniquely youthful ways. Young children's play at dark sites may reflect their immature understanding of the death and destruction remembered there. Stevens and Franck (2015) observed, for example, a young child jumping on and off rocks at the National AIDS Memorial.

Reenacting death through more elaborate play may reflect an older youth's comprehension of the events or their desire to "assert ownership" (e.g., Knudsen, 2011). To illustrate, Bowman and Pezzullo (2010) and Carr (2011) observed young tourists playing war at battlefields and abandoned bunkers. Similarly, Roche and Quinn (2017) watched young tourists play on the structures and rubble of former battlefields. Yet, youthful play may provoke censure by adult critics. Conflicting views on children's behaviors emerged in Cui et al.'s 2019 analysis of visitor comments at the Memorial of the Victims of the Nanjing Massacre. Some visitors viewed young tourists' actions as annoying, disruptive, and disrespectful. Indeed, notions of respect and morality are inherent at memorial tourist sites, and play at such sites exposes not only design and interpretation issues, but also conflicts of values, beliefs, and deep-seated feelings (Price and Kerr, 2018). As Sutcliffe and Kim conclude:

> Research tends to examine how accompanying adults interact with children, teach children about appropriate behaviour in heritage/museum venues, or types of exhibits favoured by children. These approaches assume

Table 2.1 Overview of Children's Understanding of Death

Age group	Typical Understanding of Death[a]	Manifestations at Dark Tourist Sites
5-year-olds	Recognize that living things die (e.g., plant, flower, insect, bird): • Differs between types of plants- trees and flowers dying seems more definite than weeds View human death as sleep: • "Dad, look she's sleeping" – child at Bog People exhibit (Patterson, 2007: 55) View death as reversible: • "When will they come back to life?" Understanding of finality of death: • Occurs around age five • Can differ from adult understanding because children can compare to death being like sleep • Can also compare death to religious ideas	Play among tombstones or on memorials or during solemn ceremonies Run in open spaces Puzzled by adult expressions of sadness or rules to be quiet Ask "What?" questions that may not be related to death Explore memorials through touch, even when prohibited
6–8-year-olds	More likely to grasp concepts of universality, finality, and non-functionality than younger children: • Age doesn't play a significant role in whether children perceive the subconcept of inevitability of death Engage in "fuzzy thinking" about inevitability and universality Confused about time sequence due to inability to seriate: • "I hope you [Holocaust victim] come to play at my house." Still emotionally immature and vulnerable: • "Scared and never want to feel it again." • "I felt really sad when I saw the Nazis burning the synagogue. I didn't want to come here cause I thought that it was going to make me cry and be sad." "I want to go home." Non-functionality and causality are the last concepts to be understood: • Recognize non-functionality more readily in animals than in humans	Play with audio tour equipment without listening to recordings If bored by exhibits and interpretive texts if designed only for adults, may play to entertain themselves and attract censure from others Focus on easy-to-read words in large print May want to leave if sad or scared Ask "What?" questions related to representations of death May show an interest in "the helpers" (first responders)

Table 2.1 (Continued)

Age group	Typical Understanding of Death[a]	Manifestations at Dark Tourist Sites
9–11-year-olds	Show a better understanding of universality, causality, and inevitability: • "As I look over the fields and try to [under]stand all the letters, writings and so on I realize that this is life and a hard part of it," – 11 year-old at 9/11 crash site Begin to learn about personal mortality but lack a complete understanding of the concept Express more curiosity surrounding the causes of death: • "I will never forget the experience I had here in each and every exhibit. The most depressing part was hearing about cremation, gassing and death marches."	Ask "Why?" questions relating to representations of death Express empathy for victims and their families Distressed by graphic photographs and films (Israfilova and Khoo-Lattimore, 2018) Begin to ascribe blame to perpetrators Spontaneously leave memorial tributes including personal belongings, notes, and drawings Begin to express opinions about the memorial design itself
12–14-year-olds	Comprehend universality, causality, inevitability, and personal mortality Try to connect new death understandings to concrete examples in the world around them New realization of personal mortality may increase death anxiety as they enter adolescence: • Could result in topic avoidance. • "This museum has taught me to appreciate every second of my life, as it may be the last."	May deflect anxiety through humor related to death Seek peers to study and talk about aspects of exhibits; may prefer interactions with peers to interactions with tour guides Exhibit interest in taking action to prevent wrongful death and suffering Use reenactments to understand the dark event: • "Let's build a dam!" –students approaching a creek near site of mass trauma caused by a dam failure Show stronger identification with socio-cultural aspects of sites and events commemorated (Israfilova and Khoo-Lattimore, 2018) Express curiosity about religious symbols and other abstract representations of death (e.g., markings on headstones, architectural features of memorials, lettering, personal artifacts left at gravesites) Make connections between what they view and things they have learned through past experiences (i.e., school, visiting other sites, something they watched on television) Relate experiences and to religious or secular views (Israfilova and Khoo-Lattimore, 2018)

a Sources that inform this column include: Hunter and Smith, 2008; Jackson and Colwell, 2001; Koocher et al., 1976; Lazar and Torney-Purta, 1991; Mak, 2011; Nguyen and Gelman, 2002; Nguyen and Rosengren, 2004; Noppe and Noppe, 1997; Rosengren et al., 2014.

children to be adults in the making, and rarely approach the child's visit in the way that a child likely would, and that is from the aspect of play.

(2014: 335)

Due to the aforementioned complexities, current dark tourism theories do not illuminate young tourists' experiences. Therefore, we set out to build a framework that encourages more relevant paths for studying young tourists.

Young Tourist Experiences at Dark Tourism Sites: An Initial Conceptual Framework

As McCabe (2005: 103) observes, "to develop a meaningful engagement within the sociology of tourism with tourists, we will have to recognize the cultural and interactional contexts in which we engage with our subjects." Mindful of the complexity of these contexts in childhood, we first examined hundreds of archived tributes and post-visit comments authored by children and youth at two 9/11 sites, two Holocaust museum exhibits, and a medical history museum (Croom et al., 2018; Kerr and Price, 2018; Kerr et al., 2016, 2017a, 2017b). The texts and artwork shed light on how children of different ages interpreted and reacted to death and dying. We also observed the impact of adult influences on children's developing understanding of death in dark tourism contexts (Kerr and Price, 2018).

Mindful that children and youth should increasingly take a more active research role, as "experts of their own lives" (Pinter and Zandian, 2014: 64), we next sought school partnerships to engage youth in writing, photographing, and discussing their own dark tourism experiences. These partnerships allowed us to travel on a day or overnight excursion while observing and interacting with youth ages 11–15 years (302 total). We deemed such collaborations necessary to "enter the field, that is, to mingle with their research participants, breathe and feel the same air, taste the same food, and hear the same sounds" (Matteucci and Gnoth, 2017: 20). The youth participated in motorcoach interviews, wrote post-visit comments, and took photographs (Burns, 2018; Price, 2018).

What emerged from our "mobile ethnography" are three additional influences that frame our conceptualization of young tourists' experiences. Altogether, the framework shown in Figure 2.1 encompasses (a) children's

Influences on a Young Tourist's Experience at a Dark Site

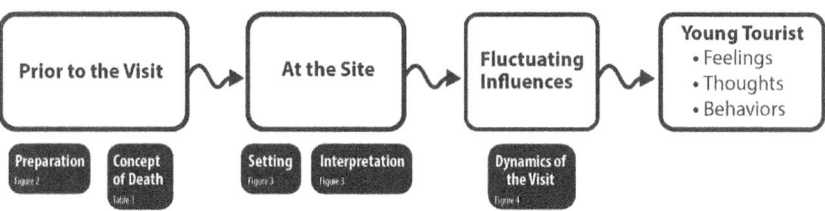

Figure 2.1 Overview.

prior understanding of death, (b) preparation before a visit, (c) site attributes, including interpretation, and d) dynamics of the visit.

Preparation for a Visit

While traveling, we interviewed site managers, tour guides, museum interpreters, parents, and educators. These conversations resulted in the next part of the framework: The preparation a young tourist receives before visiting a dark tourism site, as shown in Figure 2.2.

Informal preparation includes how adults explain tragedy to children. Intentionally or not, family conversations may shape children's experiences at dark sites (Frost and Laing, 2017; Kerr and Price, 2016, 2018). Our archival studies confirmed the influence of such family dialogues. Many comments at a Holocaust museum referred to family conversations, such as this one: "I have grown up, hearing about the Holocaust my whole life. I lost many of my family members to the Holocaust."

Moreover, family religious discussions affect children's understanding and beliefs about deadly events (Rosengren et al., 2014). Gutiérrez et al. (2014: 60) note that in both European and American contexts, "parents responded not just with information but also with reassurance when their children asked questions about death, with religious reassurance figuring especially prominently." As expressed by a 9th-grade girl who visited a genocide memorial, "objects strengthened my feelings of sadness toward people who were victims of this event. This trip also enhanced my perceptions about those who do not have faith and lack good moral habits" (Israfilova and Khoo-Lattimore, 2018: 8).

Social media posts or television also influenced children's meaning-making (Falk and Dierking, 2018). In this context, dark tourism sites offer online activities, videos, and readings (e.g., the U.S. Smithsonian Latino Center's

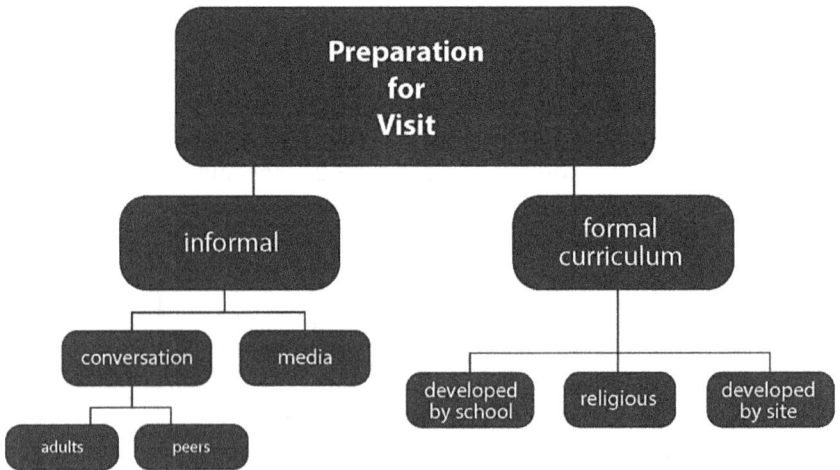

Figure 2.2 Preparation for Visit.

Theater of the Dead and the Greenwich, UK Royal Museum's Death in the Museum Trail) – an influence documented by Roche and Quinn (2017), who note that a site's media and social media images inform pre-teens' preconceptions of dark tourism.

While *formal instruction* specifically about death and dying is uncommon in some countries (King-McKenzie, 2011), teachers may require students to read about the deaths commemorated (Kucan et al., 2017; Lovorn, 2012). When done well, such instruction increases engagement. For example, students who had studied an 1889 flood disaster eagerly asked many questions, while poring over artifacts and exploring gravestones with intense interest. In contrast, poor preparation reveals itself in children's visitor comments such as these visitor log entries collected at the U.S. Holocaust Memorial Museum:

- "What they teach us at school are just numbers, but this showed me the stories."
- "When I am at school my teachers have never told me about this. Once I go back I'm going to teach my class about all of this."

Formally or informally, how a nation prepares its children – how it defines the "nationhood" that binds its people together – is reflected in what, how, when, and whom it chooses to memorialize. Therefore, young tourists visiting dark tourism destinations may experience a reinforcement of values that they are expected to uphold (Park, 2010).

In summary, dark tourism researchers have yet to explore how, when, and where young tourists learn about dark sites or events, even though these prior messages may be more influential than the site itself (Roche and Quinn, 2017). As a starting point, researchers might consider comments young tourists record on social media and visitor book pages (Kerr et al., 2017b). Collaborations with schools offer a more comprehensive view of preparation. Our own collaborations, observations, and visitor comment analyses led us to a third, complex component of our framework – the site, including its exhibits and interpretation, as shown in Figure 2.3.

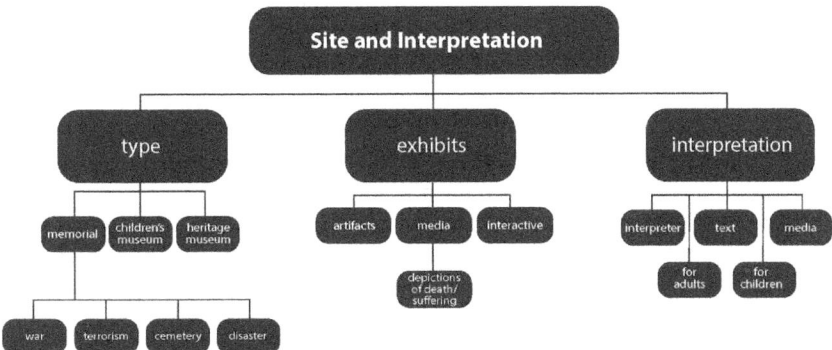

Figure 2.3 Site and Interpretation.

Site and Interpretation

Site attributes influence the tourist experience (Virgili et al., 2018). Our observations captured young tourists exploring memorials, Holocaust and war museums, cemeteries, and exhibits focused on death and disaster (Burns, 2018; Price, 2018). These "physical contexts" (Falk and Dierking, 2018) are incorporated as site type, exhibits, and interpretation.

In dark tourism, the *types* of sites may include museums, visitor attractions, memorials, cemeteries, battlefields, penitentiaries, and concentration camps – each with many spaces for children to explore. Moreover, children and youth who visit a site during a memorial event honoring a loved one would likely experience this as a dark event (Laing and Frost, 2013). In addition, temporary "dark *exhibits*" might create a dark tourism experience in an unexpected venue. "The Bog People" exhibition (of mummified human bodies) hosted by a natural history museum provides one example (Patterson, 2007). The Indianapolis Children's Museum offers a permanent exhibit, "The Power of Children," where live theatre communicates the stories of children, including Ryan White, who died of HIV/AIDS. To enhance the experience, the exhibit includes White's completely recreated bedroom showing his bed, toys, and clothes (Simon, 2009).

Such visual exhibits merit special consideration. Because younger children may not comprehend abstract ideas or text in interpretative panels (Moscardo et al., 2007), they often focus on visuals, such as films or photographs. Yet, cognitive psychology research reveals that disturbing images can have an emotional impact on children of any age. Young children, in particular, may view geographically or temporally remote images in videos as present-day threats (Hirsch and Holmes, 2007). This research is pertinent to understanding the young tourist experience because news videos figure prominently in some museums. Similarly, photographs could upset young visitors, as an adult visitor recalled in a Holocaust museum: "Today I saw a little boy sobbing over a picture of other little boys, his age, emaciated and being experimented on" (U.S. Holocaust Memorial Museum visitor log, November 1, 2018).

At present, we have only such anecdotal reports to tell us how young tourists feel when confronted with death and suffering at sites where the victims include children. Examples include the Oklahoma City Bombing Memorial and the Pentagon 9/11 Memorial, where the designs highlight each victim's age. Also, consider the display of child remains at the genocide memorial site at Ntarama in Rwanda (Caplan, 2007) and the Roman Dead exhibit of child skeletons at the Museum of London Docklands.

While exhibits may frighten some young tourists, others may *prefer* intense images. Walter (2004: 619) records in his study at the Korperwelten/Body Worlds exhibition one youth's comment that the exhibit "is very educational but there should be more blood and gutsy stuff it is not very scary or evolting [sic]." Walter (2004: 619) further notes that "though several guestbook entries by those coming without children make predictions of children having nightmares, comments about actual pre-adolescent children refer to their positive engagement in a factual, rather than a gothic, gaze."

Further complicating the very young "tourist's gaze" (Urry and Larsen, 2011) are accommodations for individuals with disabilities. At dark sites, short captions intended for those with visual impairments may project large-print, grim words that young children can read (e.g., BABY DEAD at a kidnapping exhibit). Moreover, placing some objects at lower heights to accommodate wheelchair users inadvertently situates gruesome artifacts directly at a young child's eye-level. These immediately accessible artifacts become particularly salient for young tourists, because "children naturally seek opportunities to explore and learn through their senses" (Shaffer, 2016: 43).

The *interpretation* adopted by the site also influences young tourists' experiences. For example, they may find no interpretation (e.g., London's 7 July bombing memorial), or listen to tours written solely for adults (see Walter, 2004). Conversely, when young tourists visit the Anne Frank House Museum and Memorial in Amsterdam, they read and hear a child's words. A similar child-centered interpretation awaits young tourists at "Daniel's Story," the children's exhibit of the U.S. National Holocaust Memorial. Notwithstanding well-rehearsed arguments of authenticity within dark tourism (Sharpley and Stone, 2009), site interpretation (or lack thereof) in this context warrants more research. For example, Roche and Quinn (2017) found that young tourists most easily recalled items that a tour guide specifically pointed out or described with hand gestures. Observations during such school trips informed the framework's final section, *dynamics of the visit*, as illustrated in Figure 2.4.

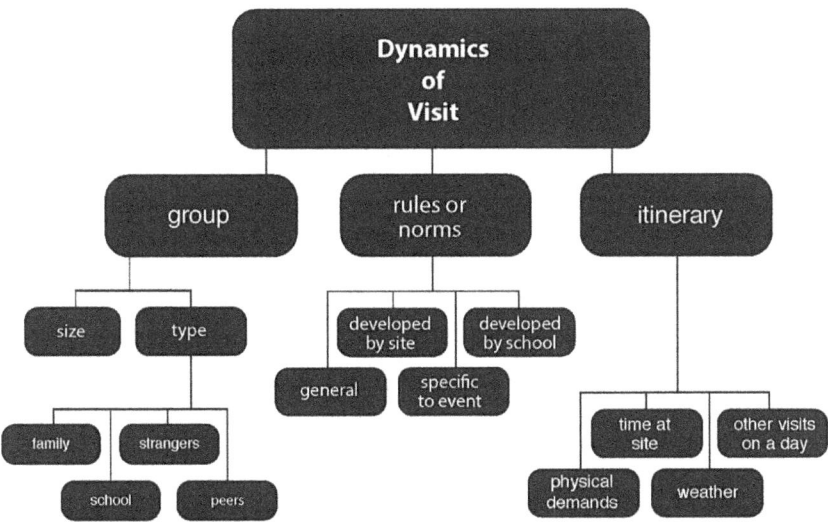

Figure 2.4 Dynamics of Visit.

Dynamics of the Visit

Fluctuating influences that may be present during a visit include the *itinerary*, the child's touring *group* membership, and *rules or norms.* Adults and children mingle together as they co-construct meaning at dark sites. For example, Roche and Quinn (2017) argue that social interactions with their peers and tour guides inform young tourists' preconceptions of dark tourism. Similarly, Burns observed this co-creation process taking place in her study of youth visiting dark sites:

> Over the course of the trip, students created a community of learners, sharing their learning and working together to make meaning of the experience…students also learned to "read" the space they occupied and identified and complied with implicit behavioral norms.
>
> (2018: iv–v)

Indeed, other situational aspects of a visit include the behavioral *rules and norms*, whether conveyed implicitly or explicitly by signage, staff, or other visitors. In a recent example, public comments revealed strong convictions about young tourists' demeanor at memorials. These comments appeared after a photograph of young children clambering on a memorial statue went viral on a social media site (Price and Kerr, 2017).

A young tourist's *itinerary* undoubtedly influences their experience. Optimally, young tourists receive instruction designed by a site's education staff (see Marcus et al., 2012). However, thousands travel on overnight school excursions with packed itineraries that may limit their visit time to less than one hour, thereby reducing their access to on-site educational programming (see Burns, 2018; Price, 2018). As a school trip director explained: "We do so many things, but it's just a bird's-eye view. You know… we see a piece of this, a snippet of that" (C. Greene, 2015, personal communication). In such instances, young tourists may receive minimal interpretation, leaving them to comprehend exhibits about death and dying on their own.

In summary, multiple sources – including archives, observations, interviews, and literature reviews – allowed us to construct an evolving conceptualization of young tourists' encounters with death and dying. We now turn to the implications of this new conceptual framework.

Implications for Future Research

As Robinson and Picard observed, we "need an understanding of tourists as persons and how they encounter, receive, respond, and react to the effective change in conditions which tourism ultimately entails. This is where things become complex" (2012: 23). Given the rise in dark tourism publications, one might conclude that dark tourism research has already identified these complexities. However, when one scrutinizes this literature, one sees that children "as persons" are largely overlooked. Adult theories and research methods cannot explain their encounters because the developing cognitive, emotional, physical, and interpersonal attributes of children and youth do not resemble those of adults.

Moreover, teachers and parents often choose itineraries and, in so doing, convey their own knowledge, beliefs, and norms. Once younger visitors arrive at a dark tourism site, they often encounter adult-oriented interpretation, exhibits, and perspectives. Adults at a site may communicate strong beliefs about how younger tourists should behave. Yet, young tourists may best experience a site through play, touch, make-believe, or other "forbidden" behaviors.

For reasons such as these, young tourists' encounters require novel conceptualizations and data collection approaches that recognize children as persons. To this end, our framework seeks to accelerate and deepen our understanding of young tourists' experiences at dark tourism sites. As Dallari and Mariotti conclude, "tourism for children and adolescents continues to be a field where it is difficult to obtain reliable and trustworthy data… Undoubtedly, much still needs to be done to produce a valid and universal methodology to tackle this phenomenon" (2016: 21).

How might research methodology be adapted to explore the intersecting influences cited in the framework? Firstly, researchers hesitant to work directly with young tourists can access dark tourism site archives, which include children's art, letters, tributes, and post-visit comments (see Kerr et al., 2017b for specific methods). As evidenced in this paper, these collections shed light on meaning-making, recollections, evaluations, and emotional expressions. Content analysis of artifacts (guided by the concepts articulated here) does not always require formal human subjects protection review, increasing its appeal to some.

Secondly, we acknowledge the importance of selecting methods suitable for the child's age, as highlighted by Khoo-Lattimore (2015) and others. This approach would take into account children's different understandings of death concepts. As illustrated in this paper, tourism researchers should adopt also *multiple* measures to optimally engage young tourists in sharing their experiences (Israfilova and Khoo-Lattimore, 2018; Kerr and Price, 2018).

Thirdly, our framework includes influences that are present before, during, and even after a trip (Falk and Dierking, 2018). Capturing these influences necessitates *phased* data collection across time. Based on our own experience, we concur with Roche and Quinn (2017) that such a phased approach also allows children to become more comfortable with adult researchers. Finally, we endorse the approaches of those working in childhood studies, who recognize young tourists as capable of exploring their own experiences. Our own team now includes youth members who co-design measures, collect data, and interpret findings.

Implications for Practice

The framework presented here has direct utility for site managers and interpreters who seek to engage young tourists. An example appears at the Flight

93 National Memorial commemorating victims hijacked on September 11, 2001. There, interpreters struggled to explain the terrorist attack to young tourists. By inviting youth to visit the site and give feedback on preliminary drafts of interpretive materials, the staff designed a child-friendly "Junior Ranger Program" that now engages young tourists in age-appropriate, multi-sensory activities. In another example, the staff considered young children's vulnerability to visual images. The result was the decision to limit video footage to one highly placed monitor and to move grim artifacts out of young children's immediate eye-level.

A more recent example took place at the National Pentagon 9/11 Memorial. This site receives 750,000 visitors per year, most of whom are on school trips. Using the framework, 13-year-old researchers conducted pre- and post-visit interviews and surveys to assess school trip visitors' preparation and their perceptions of the site and interpretation. Following their trip, the youth analyzed their data and presented recommendations on how to improve the youth visitor experience.

Without such research–practice partnerships, destination managers, interpreters, and even vendors may miss opportunities to engage younger tourists. Moreover, young tourists' negative experiences can influence their future motivations to avoid particular sites. Uninformed about their childhood experiences, the dark tourism industry may lose them as adults.

Conclusion

Importantly, a child's understanding of death and dying develops as the child matures. This less-than-fully developed understanding, coupled with emotional immaturity, often frames young tourists' experiences within dark tourism. Indeed, Aaron (whose letter to his parents introduced this paper) welcomed his father home from a research visit to Auschwitz-Birkenau. After hearing an explanation of Auschwitz – the people murdered and the nature of atrocities committed there – Aaron (then six years old) simply replied, "Daddy, can you hear them scream?" Of course, dark tourism is filled with the "screams" of the significant dead. How we present or represent the tragic dead to young tourists remains problematic. Consequently, the framework and its parameters beckon scholars to explore complex factors that influence, inform, and have consequences for young tourists' visits to dark tourism destinations. Indeed, the task for future cultural tourism and heritage scholarship is to utilize the framework or parts thereof, and reconnoiter research avenues illuminated by our agenda.

Hindered by the general lack of research on young tourists at dark tourism sites, we recognize the inherent risk in articulating a conceptual model. Nevertheless, we heed Veal's (2017: 76) advice that "a conceptual framework need not be a straitjacket; it can be a flexible, evolving device." It is our hope that those who embark on research with young tourists will adapt and refine this model to include differences among them and among the sites that they visit.

This framework proposes influences on a young tourist's experience at a dark tourism destination. It simultaneously calls on scholars to identify how best to pursue their questions, not with traditional adult tourism research methods, but rather with innovations suitable for engaging young tourists, perhaps even as researchers themselves. In doing so, the curious will find a rich landscape of new methods and discoveries that not only benefit the tourism field but also enrich these long-neglected childhood encounters.

Acknowledgments

We are deeply grateful to Cecilia Greene, Scott Marsh, Greg Wittig, and their students.

Note

1 This chapter originally appeared as Kerr, M.M., Stone, P.R. and Price, R. H., 2021. Young tourists' experiences at dark tourism sites: towards a conceptual framework. *Tourist Studies*, 21(2), pp. 198–218.

References

Bianucci R, Soldini M, Di Vella G, et al. (2015) The body worlds exhibits and juvenile understandings of death: Do we educate children to science or to voyeurism? *Clinical Therapeutics* 166(4): e264–268.

Bosangit C, Hibbert S and McCabe S (2015) "If I was going to die I should at least be having fun": Travel blogs, meaning and tourist experience. *Annals of Tourism Research* 55: 1–14.

Bowman MS and Pezzullo PC (2010) What's so "dark" about "dark tourism"? Death, tours, and performance. *Tourist Studies* 9(3): 187–202.

Burns L (2018) *Overnight school trips: An overlooked phenomenon.* PhD Thesis, University of Pittsburgh, US.

Canosa A, Graham A and Wilson E (2019) Progressing a child-centred research agenda in tourism studies. *Tourism Analysis* 24(1): 95–100.

Canosa A, Moyle BD and Wray M (2016) Can anybody hear me? A critical analysis of young residents' voices in tourism studies. *Tourism Analysis* 21(2–3): 325–337.

Caplan P (2007) "Never again": Genocide memorials in Rwanda. *Anthropology Today* 23(1): 20–22.

Carr N (2011) *Children's and Families' Holiday Experiences.* Abington: Taylor & Francis.

Chronis, A (2012) Tourists as Story-Builders: Narrative Construction at a Heritage Museum. *Journal of Travel & Tourism Marketing* 29(5–11): 444–59.

Cicero L and Teichert T (2018) Children's influence in museum visits: Antecedents and consequences. *Museum Management and Curatorship* 33(2): 146–157.

Coles TE, Hall CM and Duval DT (2009) Post-disciplinary tourism. In: Tribe J (ed) *Philosophical Issues in Tourism.* Bristol: Channel View Publications, pp.80–100.

Cooper C and Latham J (1988) English educational tourism. *Tourism Management* 9(4): 331–334.

Croom AR, Squitiero C and Kerr MM (2018) Something so sad can be so beautiful: A qualitative study of adolescent experiences at a 9/11 memorial. *Visitor Studies* 21(2): 157–174.

Cui R, Cheng M, Xin S, et al. (2019) International tourists' dark tourism experiences in China: The case of the memorial of the victims of the Nanjing Massacre. *Current Issues in Tourism* 23(12): 1–19.

Cullingford C (1995) Children's attitudes to holidays overseas. *Tourism Management* 16(2): 121–127.

Dallari F and Mariotti A (2016) Children in Tourism: a fresh perspective? The experience in Italy, from summer camps to the Seninter project. *Tourism Review* (10).

Elmi B, Bartoli E, Fioretti C, et al. (2020) Children's representation about travel: A comparison between what children remember and what children desire. *Tourism Management Perspectives* 33: 1–11.

Falk JH and Dierking LD (2018) *Learning from Museums* (2nd ed). Lanham: Rowman & Littlefield.

Frost W and Laing JH (2017) Children, families and heritage. *Journal of Heritage Tourism* 12(1): 1–6.

Graburn NH (1983) Editor's page. *Annals of Tourism Research* 10(1): 1–3.

Gutiérrez IT, Miller PJ, Rosengren KS, et al. (2014) III. Affective dimensions of death: Children's books, questions, and understandings. *Monographs of the Society for Research in Child Development* 79(1): 43–61.

Hartmann, R (2014) Dark tourism, thanatourism, and dissonance in heritage tourism management: New directions in contemporary tourism research. *Journal of Heritage Tourism* 9(2): 166–182.

Hirsch CR and Holmes EA (2007) Mental imagery in anxiety disorders. *Psychiatry* 6(4): 161–165.

Hunter SB and Smith DE (2008) Predictors of children's understandings of death: Age, cognitive ability, death experience and maternal communicative competence. *OMEGA-Journal of Death and Dying* 57(2): 143–162.

Israfilova F and Khoo-Lattimore C (2018) Sad and violent but I enjoyed it: Children's engagement with dark tourism as an educational tool. *Tourism and Hospitality Research*. Epub ahead of print 4 August 2018. DOI: 10.1177/1467358418782736.

Jackson M and Colwell J (2001) Talking to children about death. *Mortality* 6: 321–325.

Ji J, Anderson D, Wu X, et al. (2014) Chinese family groups' museum visit motivations: A comparative study of Beijing and Vancouver. *Curator: The Museum Journal* 57(1): 81–96.

Kellett M (2005) Children as active researchers: A new research paradigm for the 21st century? Available at: http://oro.open.ac.uk/7539/ (accessed February 04 2020).

Kellett M (2010) Small shoes, big steps! Empowering children as active researchers. *American Journal of Community Psychology* 46(1): 195–203.

Kerr MM, Dugan SE and Frese KM (2016, July/August) Using children's artifacts to avoid interpretive missteps. *Legacy: The Magazine of the National Association for Interpretation*.

Kerr MM, Fried SE, Price RH, et al. (2017a) Rural children's responses to the Flight 93 crash on September 11, 2001. *Journal of Rural Mental Health* 41(3): 176–188.

Kerr MM and Price RH (2016) Overlooked encounters: Young tourists' experiences at dark sites. *Journal of Heritage Tourism* 11(2): 177–185.

Kerr MM and Price RH (2018) "I know the plane crashed": Children's perspectives in dark tourism. In: Stone P, Hartmann R, Seaton T, et al. (eds) *The Palgrave Handbook of Dark Tourism Studies*. London: Palgrave Macmillan, pp.553–583.

Kerr MM, Price RH, Savine CD, et al. (2017b) Interpreting terrorism: Learning from children's visitor comments. *Journal of Interpretation Research* 22(1). Available at: https://www.interpnet.com/NAI/nai/_publications/JIR_v21n1_Kerr.aspx.

Khoo-Lattimore C (2015) Kids on board: Methodological challenges, concerns, and clarifications when including young children's voices in tourism research. *Current Issues in Tourism* 18(9): 845–858.

King-McKenzie E (2011) Death and dying in the curriculum of public schools: Is there a place? *Journal of Emerging Knowledge on Emerging Markets* 3: 511–520.

Knudsen BT (2011) Thanatourism: Witnessing difficult pasts. *Tourist Studies* 11(1): 55–72.

Koocher GP, O'Malley JE, Foster D et al. (1976) Death anxiety in normal children and adolescents. *Psychopathology* 9(3–4): 220–229.

Kucan L, Cho BY and Han H (2017) Introducing the historical thinking practice of contextualizing to middle school students. *The Social Studies* 108(5): 210–218.

Laing J and Frost W (2013) *Commemorative Events: Memory, Identities, Conflict.* New York: Routledge.

Larsen S and Jenssen D (2004) The school trip: Travelling with, not to or from. *Scandinavian Journal of Hospitality and Tourism* 4(1): 43–57.

Lazar A and Torney-Purta J (1991) The development of the subconcepts of death in young children: A short-term longitudinal study. *Child Development* 62(6): 1321–1333.

Lennon JJ and Foley M (2000) *Dark Tourism: The Attraction of Death and Disaster.* London: Continuum.

Light D (2017) Progress in dark tourism and thanatourism research: An uneasy relationship with heritage tourism. *Tourism Management* 61: 275–301.

Lovorn MG (2012) Historiography in the methods course: Training preservice history teachers to evaluate local historical commemorations. *The History Teacher* 45(4): 569–579.

McCabe S (2005) "Who is a tourist?" A critical review. *Tourist Studies.*

Mak MHJ (2011) Chinese secondary students' knowledge of, and attitudes toward, death, dying, and life education: A qualitative study. *Illness, Crisis & Loss* 19(4): 309–327.

Marcus AS, Stoddard JD and Woodward WW (2012) *Teaching History with Museums: Strategies for K-12 Social Studies.* New York: Routledge.

Martini A and Buda DM (2020) Dark tourism and affect: Framing places of death and disaster. *Current Issues in Tourism* 23(6): 679–692.

Matteucci X and Gnoth J (2017) Elaborating on grounded theory in tourism research. *Annals of Tourism Research* 65: 49–59.

Minnaert L (2014) Social tourism participation: The role of tourism inexperience and uncertainty. *Tourism Management* 40: 282–289.

Mooney S, Schänzel H and Poulston J (2017) Illuminating the blind spots. *Hospitality and Society* 7(2): 105–201.

Moscardo G, Ballantyne R and Hughes K (2007) *Designing Interpretive Signs: Principles in Practice.* Golden, CO: Fulcrum Publishing.

Nguyen SP and Gelman SA (2002) Four and 6-year olds' biological concept of death: The case of plants. *British Journal of Developmental Psychology* 20(4): 495–513.

Nguyen S and Rosengren K (2004) Parental reports of children's biological knowledge and misconceptions. *International Journal of Behavioral Development* 28(5): 411–420.

Noppe IC and Noppe LD (1997) Evolving meanings of death during early, middle, and later adolescence. *Death Studies* 21(3): 253–275.

Park H (2010) Heritage tourism: Emotional journeys into nationhood. *Annals of Tourism Research* 37(1): 116–135.

Patterson AR (2007) "Dad look, she's sleeping": Parent–child conversations about human remains. *Visitor Studies* 10(1): 55–72.

Penfold-Mounce R (2018) *Death, the Dead and Popular Culture*. Bingley: Emerald Publishing.

Perkins KD and Mackey B (2008) Supporting grieving children in early childhood programs. *Dimensions of Early Childhood* 36(3): 13–19.

Pinter A and Zandian S (2014) "I don't ever want to leave this room": Benefits of researching "with" children. *ELT Journal* 68(1): 64–74.

Poria Y, Atzaba-Poria N and Barrett M (2005) The relationship between children's geographical knowledge and travel experience: An exploratory study. *Tourism Geographies* 7(4): 389–397.

Poria Y and Timothy DJ (2014) Where are the children in tourism research? *Annals of Tourism Research* 47: 93–95.

Potts TJ (2012) "Dark tourism" and the "kitschification" of 9/11. *Tourist Studies* 12(3): 232–249.

Price RH (2018) *Expectations and revelations: Children discuss conducting research during a multi-day school excursion*. PhD Thesis, University of Pittsburgh.

Price RH and Kerr MM (2017) "'I thought it was cool how we were part of research': Engaging children as co-researchers on their school excursions." *The Qualitative Report 8th Annual Conference*, Fort Lauderdale, FL, 12–14 January.

Price RH and Kerr MM (2018) Child's play at war memorials: Insights from a social media debate. *Journal of Heritage Tourism* 13(2): 167–180.

Prout A and James A (1997) A new paradigm for the sociology of childhood? Provenance, promise and problems. In: James A and Prout A (eds) *Constructing and Reconstructing Childhood: Contemporary Issues in the Sociological Study of Childhood*. London: Farmer Press, pp.7–32.

Roberts C (2018) Educating the (dark) masses: Dark tourism and sensemaking. In: Stone P, Hartmann R, Seaton T, Sharpley R et al. (eds) *The Palgrave Handbook of Dark Tourism Studies*. London: Palgrave Macmillan, pp.603–637.

Robinson M and Picard D (eds) (2012) *Emotion in Motion: Tourism, Affect and Transformation*. Abington: Routledge.

Roche D and Quinn B (2017) Heritage sites and schoolchildren: Insights from the Battle of the Boyne. *Journal of Heritage Tourism* 12(1): 7–20.

Rosengren KS, Gutiérrez IT and Schein SS (2014) IV. Cognitive dimensions of death in context. *Monographs of the Society for Research in Child Development* 79(1): 62–82.

Ryan C (1992) The child as visitor. *World Travel and Tourism Review* 2: 135–139.

Schänzel H and Carr N (2015) Special issue on children, families and leisure – first of two issues. *Annals of Leisure Research* 18(2): 171–174.

Schänzel H and Carr N (2016) *Children, Families and Leisure*. New York: Routledge.

Schänzel H, Yeoman I and Backer E (2012) *Family Tourism: Multidisciplinary Perspectives*. Bristol: Channel View Publications.

Shaffer SE (2016) *Engaging Young Children in Museums*. New York: Routledge.

Sharpley R and Stone PR (2009) (Re)presenting the macabre: Interpretation, kitschification and authenticity. In: Sharpley R and Stone P (eds) *The Darker Side of Travel: The Theory and Practice of Dark Tourism*. Bristol: Channel View Publications, pp. 109–128.

Shavanddasht M and Schänzel H (2019) Measuring adolescents' tourism satisfaction: The role of mood and perceived parental style. *Tourism and Hospitality Research* 19(3): 308–320.

Simon N (2009) Review of an exhibition: "The Power of Children." Available at: http://www.exhibitfiles.org/the_power_of_children (accessed March 07 2019).

Small J (2008) The absence of childhood in tourism studies. *Annals of Tourism Research* 35(3): 772–789.

Sofka CJ (1997) Social support "internetworks," caskets for sale, and more: Thanatology and the information superhighway. *Death Studies* 21(6): 553–574.

Sofka CJ (2009) Adolescents, technology, and the internet: Coping with loss in the digital world. In: Balk D and Corr C (eds) *Adolescent Encounters with Death, Bereavement, and Coping*. New York: Springer, pp. 155–173.

Speece MW and Brent SB (1984) Children's understanding of death: A review of three components of a death concept. *Child Development* 55(5): 1671–1686.

Stevens Q and Franck KA (2015) *Memorials as Spaces of Engagement: Design, Use and Meaning*. New York: Routledge.

Stone PR (2006) A dark tourism spectrum: Towards a typology of death and macabre related tourist sites, attractions and exhibitions. *Tourism: An Interdisciplinary. International Journal* 54(2): 145–160.

Stone PR (2011) Dark tourism: Towards a new post-disciplinary research agenda. *International Journal of Tourism Anthropology* 1(3/4): 318–332.

Stone PR (2012) Dark tourism and significant other death: Towards a model of mortality mediation. *Annals of Tourism Research* 39(3): 1565–1587.

Stone PR, Hartmann R, Seaton T, et al. (eds) (2018) *The Palgrave Handbook of Dark Tourism Studies*. London: Palgrave Macmillan.

Stone PR and Sharpley R (2008) Consuming dark tourism: A thanatological perspective. *Annals of Tourism Research* 35(2): 574–595.

Sutcliffe K and Kim S (2014) Understanding children's engagement with interpretation at a cultural heritage museum. *Journal of Heritage Tourism* 9(4): 332–348.

Swarbrooke J and Horner S (1999) *Consumer Behaviour in Tourism*. Oxford: Butterworth-Heinemann.

Tagg S and Seaton AV (1994) How different are Scottish family holidays from English? In: Seaton A (ed) *Tourism: The State of the Art*. Chichester: John Wiley & Sons, pp.540–548.

Urry J and Larsen J (2011) *The Tourist Gaze 3.0*. Thousand Oaks: Sage.

Veal AJ (2017) *Research Methods for Leisure and Tourism*. London: Pearson.

Virgili S, Delacour H, Bornarel F, et al. (2018) "From the Flames to the Light": 100 years of the commodification of the dark tourist site around the Verdun battlefield. *Annals of Tourism Research* 68: 61–72.

Walby K and Piché J (2015) Staged authenticity in penal history sites across Canada. *Knudsens* 15(3): 231–247.

Walter T (2004) Body worlds: Clinical detachment and anatomical awe. *Sociology of Health & Illness* 26(4): 464–488.

Walter T (2009) Dark tourism: Mediating between the dead and the living. In: Sharpley R and Stone P (eds) *The Darker Side of Travel: The Theory and Practice of Dark Tourism*. Bristol: Channel View Publications, pp.39–55.

Woodhead M (2008) Childhood studies: Past, present and future. In: Kehily M (ed) *An Introduction to Childhood Studies*. London: Open University Press, pp.17–31.

Wu MY and Wall G (2017) Visiting heritage museums with children: Chinese parents' motivations. *Journal of Heritage Tourism* 12(1): 36–51.

Part II

Children as Tourists

Development in Context

3   The Youngest Tourists

Early Childhood Considerations and Challenges

Sue Dockett

Introduction

Prevailing assumptions about the innocence, vulnerability, and limited cognitive and socio-emotional development of young children have contributed to them being largely absent from research about tourism (Poria & Timothy 2014), and dark tourism in particular (Kerr & Price 2018). While young children do visit such sites (Kerr & Price 2016), they are rarely tourists on their own. Instead, they are usually accompanied by adults to tourism sites, and it is these adult voices that predominate in dark tourism research.

Recent conceptualizations question notions of childhood innocence, instead emphasizing children's developing competencies, particularly as they engage with people they know and trust. Coupled with recognition of children's rights to express views about issues that affect them, there have been increasing calls to recognize children as competent contributors to conversations about their lives and experiences (James & Prout 2014). Contemporary understandings of children's development recognize the importance of the family context as a site for the co-construction of meaning as children bring their own interpretations to shared experiences (Vygotsky 1978). These changes in how children and their development are conceptualized lead to the realization that they are actively involved in making sense of their experiences, and that their interpretations may well be different from those of the adults around them. Further, in supportive contexts and using a variety of strategies, there is clear evidence that young children are capable of sharing their perspectives (Christensen & James 2017).

It is through interactions within families and other adults and children, as well as the sites themselves, that young children construct meaning. The importance of recognizing young children's agency and engaging in conversations about emotionally provocative sites, and responding to their questions and concerns, provide the basis for the chapter. Methodologies that support such interactions are noted and the significance of including young children in studies of dark tourism is emphasized.

DOI: 10.4324/9781003032199-5

Young children's understanding of death

In contrast to prior assumptions about their development, a wave of recent research reveals greater sophistication in young children's social, emotional, and cognitive understandings. For example, the limitations outlined in Piagetian research have been replaced by awareness of the ways in which young children co-construct meanings through their actions and interactions in a variety of socio-cultural contexts.

Current explorations have concluded that some elements of understanding of death occur much earlier (Rosengren et al. 2014) than age 10 (Piaget 1929). While understanding of the universality, finality, non-functionality, and causality of death is not completed in early childhood, there is evidence that such biological understandings are developing (Menendez et al. 2020; Panagiotaki et al. 2018). Alongside these biological understandings, children's immersion in their socio-cultural contexts supports their exposure to religious or spiritual explanations of death (Rosengren et al. 2014).

Whether or not adults purposefully engage in conversations about it, children are likely to encounter death through their own experiences, through stories, and through media. For example, extensive media coverage of the death of Princess Diana in 1997 provoked an ongoing conversation between Harriet (aged 2 years and 10 months) and her mother, culminating in Harriet's question, 'Can Diana have another turn? If you die before you finish?' (Dockett 2000). While Menendez et al. (2020, p. 58) note that 'by age 6, most children seem to have a fairly sophisticated understanding of death', Harriet's sustained interest suggests that, at least for some children, this understanding builds from a much earlier time. Other researchers confirm early efforts to understand this complex concept (Renaud et al. 2015).

The influence of children's socio-cultural contexts is reflected in different understandings of, rituals around, and beliefs about, death (Rosengren et al. 2014). In exploring how children learn about death, Menendez et al. (2020) note the importance of parental conversations and adults' willingness to address children's questions—particularly those relating to the non-observable and abstract elements of death. Some researchers report a mismatch between children's questions and adult responses, as adults seek to shield children from information they think may be disturbing or too complex and incorporate spiritual and/or religious explanations with biological information (Gaab et al. 2013; Gutiérrez et al. 2019).

Recognizing children's interactions with more experienced others as a key driver of development is a cornerstone of cultural–historical theory. Drawing on Vygotsky's (1978) emphasis of the individual as a social being, cultural–historical theory (Rogoff 2003) stresses the role of interactions as adults—or more experienced others—mediate what is important culturally and assist children as they co-construct culturally relevant interpretations of experiences. The focus on learning as a social endeavor takes us away from the age-stage-based theories of development, such as those associated with Piaget. It also facilitates recognition of the knowledge and understanding children do have and are developing as strengths, rather than limitations.

Cultural–historical theory posits that, as children interact with others who have more experience and expertise, they are guided and supported in their learning. The responsiveness of those others is important. This was noted as Harriet and her mother engaged in dialogue about death, with Diana's death regarded as a legitimate topic for discussion and inquiry. Other adults may have avoided the topic, considering death a sensitive issue and children as vulnerable to distress (Powell et al. 2018).

Young children may ask about issues adults have not considered, or that adults regard as sensitive or inappropriate. However, Powell et al. (2018), note that depending on the context in which interactions occur and the relationships in which these are embedded, any topic has the potential to be perceived as sensitive. How adults respond to children depends on their images of children, particularly their views of young children's competence, and their willingness to listen to young children.

The image of the young child

Adult images of children are often complex, influenced by social and cultural contexts, experiences, and expectations. Interactions with children are influenced by images of children as vulnerable, innocent, immature, capable, confident, or any combination of these. As with children's understandings of death, recent research employing multiple methodologies has led researchers to question prevailing images of children and their competence.

Perspectives drawn from the study of sociologies of childhood have contributed to contemporary images of children as competent social actors, who actively engage in the co-construction of knowledge and identity as they interact with others (James & Prout 2014). These perspectives are encapsulated in the Mosaic Approach (Clark 2017, p. 20), which envisions young children as:

- experts in their own lives
- skillful communicators
- rights holders
- meaning-makers

These perspectives recognize that those engaged in the experiences, namely the children, are in the best position to comment on those experiences. While adults can observe and share interactions, they cannot 'be' the child. To know what the experience embodies for children, we need to rely on them to share their perspectives. While some young children will do so using language, others may use gestures, glances, or other bodily movements. They may refer to artifacts or even choose not to share their experiences at all (Pálmadóttir & Einarsdóttir 2016).

Careful documentation of children's perspectives can reveal not only their competence in a range of situations but also their perspectives of experiences

(Carr & Lee 2019; Dockett et al. 2019). Documentation also provides opportunities for adults to realize that children are skillful communicators. Examples from Reggio Emilia settings, which promote the hundred languages of children (Edwards et al. 2011), have influenced early childhood practice worldwide. They highlight the role of adults to recognize the many ways in which young children communicate and the importance of adults engaging in active listening (Rinaldi 2006).

The notions of children as experts on their own lives and as skillful communicators frame the provisions for children's rights in the Convention on the Rights of the Child (United Nations 1989). Those assert children's rights to express their views on matters affecting them, and to do this in ways that make sense for them. Further, the Convention promotes children's access to information and appropriate guidance as they exercise these rights (Dockett & Perry 2019) and their rights to engage in cultural activities. Importantly, these rights are not age-defined, nor are they divisible—such that some articles can be considered more important than others (Lundy & McEvoy 2012). Actualization of these rights requires that adults listen to children; take the interests, experiences, and expectations of young children seriously; and be prepared to provide information and guidance.

Acknowledging children as meaning-makers recognizes their active role in constructing knowledge and building understanding. This perspective positions learning as a social endeavor, situated within specific social and cultural contexts and occurring within relationships (Rogoff 2003). Further, the engagement of children and more experienced others sets the scene for children's thinking to be shared, stretched, and extended through processes of guided participation. Guided participation is collaborative, as children actively engage with others to explore a task, situation, or understanding that they would not be able to resolve on their own.

Research with young children

Researchers across many fields and disciplines have recognized the importance of engaging with young children in research about their everyday lives (Christensen & James 2017; Groundwater-Smith et al. 2015) and the impact this may have in enhancing children's experiences, contributing to their exercise of rights, building understandings, promoting respectful dialogue, and influencing social policy and service delivery (Davis 2009). At the same time, researchers also have noted the challenges inherent in such research.

These challenges include perceptions about what constitutes appropriate topics of investigation for young children, particularly around issues considered by adults to be sensitive, and where children are considered vulnerable (Powell et al. 2018). Other challenges are noted in identifying and applying suitable methods and gaining approval for research with young children from institutional ethics bodies and families and/or communities (Canosa & Graham 2016; Kerr & Price 2016; Khoo-Lattimore, 2015).

Five key characteristics of participatory rights-based research have been identified and offer guidance to navigate these challenges (Beazley et al. 2011, p. 161):

1 Research is genuinely respectful of children, regarding them as research partners. This respect extends to the nature of participation, which must be meaningful for children.
2 Research is grounded in ethical practice and children's participation is bound by the same ethical principles as research with adults, notably voluntary participation and avoidance of exploitation.
3 Research is systematic and valid. Data are generated systematically in ways that can be justified and replicated.
4 Analysis is rigorous, using techniques that are appropriate.
5 Research prioritizes local knowledge and expertise, focusing on children's own experiences and opinions within a specific context.

To these, Dockett and Perry (2014, p. 678) have added: 'Reporting and dissemination respects children's rights and aims to make a positive difference to their lives'.

The principles of rights-based research are encapsulated in the Ethical Research Involving Children (ERIC) framework (Graham et al. 2013) which provides guidance and support for researchers as they recognize the centrality of ethics for all involved in research with children. These principles include respect for the dignity of children, engagement in just and equitable research, consideration of the benefits and avoidance of harm for children involved in research, children's informed and ongoing consent, and reflexive engagement with research processes.

Underpinning rights-based research exploring young children's perspectives is a commitment to engaging with children in meaningful, relevant investigations of the things that matter for them, and to use this as the basis for improving their experiences. This commitment is evident in the Mosaic Approach.

The Mosaic Approach

The Mosaic Approach (Clark 2017; Clark & Moss 2001) uses multiple methods to collate the perspectives and experiences of young children. The Mosaic Approach provides a framework for building a mosaic—a composite picture—of data contributed by young children themselves, as well as those who engage with them. The aim is to explore everyday life and to create spaces where children are listened to within relationships that embody respect for them, their perspectives, and their competence.

Implementation of the Mosaic Approach produces a range of artifacts that are then pieced together. For example, the mosaic created to reflect 22-month-old Toni's experiences of being in a nursery (Clark 2017, p. 57) consisted of researcher observation, interviews with Toni's key educator, a parent interview, photographs taken by Toni's older sibling, and a record of Toni's engagement in role play. These data provided the basis for dialogue, reflection, and interpretation. Analysis of the data generated a picture of Toni's everyday life in the nursery—her preferences about where to be and with whom, as well as what she liked to do and what comforted as well as challenged her. The expectation from this case study was that the mosaic would be elaborated as Toni's experiences and communication skills expanded, and as she contributed more of the data.

An example of a more elaborate mosaic is provided for Gary (age 3) (Clark 2017, p. 60), where data were generated through the methods of drawing, child conferencing, observation, parent interview, photographs taken by Gary, a tour accompanied by recorded commentary, a map made by Gary and friends, and an interview with Gary's kindergarten educator. While the data produced by multiple methods are an important element of the Mosaic Approach, the theoretical framework provides the impetus for the ethical generation of data and reflexive dialogue, interpretation, and analysis involving the children and those who interact with them.

Methods and methodologies

Research involving young children often attends to the methods used to generate data. The methodological focus of participatory rights-based research, the Mosaic Approach, and ERIC framework reminds us that it is the methodology (the paradigm researchers bring to any inquiry) that guides the ways we think about social phenomena. It also informs the reasons that we regard the inquiry as important and worthy of study (Bessell 2009; Dockett, Einarsdóttir & Perry 2011a). The methodological framework influences the choice of methods (the practical tools or techniques) adopted for a study. Any particular approach can be employed in multiple ways and influenced by quite different methodologies. In seeking to engage in ethical, rights-based, and participatory research with young children, it is critical that we keep sight of '*why we choose* particular methods and, more importantly, *how we use them*' (Bessell 2009, p. 17). Not only does ethical, rights-based research require the use of appropriate methods, but it also requires an ethical framework for the selection and application of these methods.

Young children as tourists

An essential element of this ethical framework is that it recognizes the competence of children and their ability to reflect on their experiences, when others who are responsive active listeners provide them with opportunities to do so. To date, there has been limited evidence of this in tourism research. Notable recent exceptions have explored children's comment cards at terrorism sites (Kerr et al. 2017) as well as responses to children playing at a war memorial (Price & Kerr 2018), and have offered responses to methodological challenges (Khoo-Lattimore 2015). However, young children remain largely invisible in tourism research generally, and dark tourism research particularly. That level of invisibility continues in the face of 'the compelling reasons to undertake such research: to inform interpretation of emotionally provocative sites for children, to understand and mitigate children's psychological distress at dark sites, and to advance theoretical work on children as tourists' (Kerr & Price 2016, p. 177).

Despite this absence, young children do visit tourist sites. They also influence family decisions about tourism, both by contributing to destination decisions (Carr 2011) and through adult consideration of access to the specialist services required for their care and/or supervision (Thornton et al. 1997). Some research seeks children's perspectives as part of a whole-family approach to studying tourism experiences (Schanzel & Smith 2014). However, young children's perspectives can be masked by this overall approach.

Young children at dark tourism sites

Dark tourism sites invoke sadness and often intense emotional reactions as people reflect on tragedy and/or trauma (Cui et al. 2019). Young children visit sites such as cemeteries, memorials, battlefields, prisons, and museums (Figure 3.1). In Australia, for example, children and their families routinely visit sites of historical and modern tragedy (Wilson 2011).

The usual reasons cited for people to visit dark tourism sites are commemoration, pilgrimage, entertainment, and education (Lemelin et al. 2013). It is not clear if the same motivations drive families with young children to visit these sites. However, studies with older children note the anticipated educational value of such visits, as well as the opportunities for families to create memories from shared experiences (Israfilova & Khoo-Lattimore 2019).

There are many Australian sites where Indigenous tragedy and trauma has occurred. Some sites have 'keeping places' where people can record and share their stories of colonization, conflict, and displacement. Young children not only visit these sites but also share in the stories that surround them. Oral traditions remain central to Indigenous culture, supported by the intergenerational sharing of stories. These sites contribute to the reinterpretation of colonial narratives (Carrigan 2014). They are important for Indigenous families and communities because they preserve and convey narratives of suffering and survival. They are also important for Indigenous and non-Indigenous families and communities who seek to understand intergenerational trauma (Menzies 2019). Visits to these sites help to connect the past and the present and afford opportunities to mourn and honor and to celebrate continued survival in the face of severe difficulties and injustices. They also prompt discussion around narratives that are often contested and complex since 'personal, familial or national heritage is socially constructed, re-presented and performed' (Kidron 2013, p. 176).

While dark tourism sites are visited by young children, there are mixed views as to the appropriateness of this. Reporting visits to the site of the Nanjing Massacre in China, Cui et al. (2019) note that some adult visitors supported the educational potential of visits by young children, while others questioned their presence, expressing concern that the children were too young to understand the events, and would emerge with an incomplete

Figure 3.1 Children often accompany families on visits to cemeteries.

understanding of the event and its significance. One adult visitor commented on children's behavior at the site, indicating that young children running around and playing was disrespectful. That situation was similar to the social media debate described by Price and Kerr (2018). Kerr and Price (2016) also report concerns from some adults that making sites appropriate for children has the potential to 'water down' the history and significance of the sites. These studies highlight adult reactions to children at dark tourism sites and reflect different images of children and the ways in which they learn.

Young children as museum visitors

While we know little about how young children experience dark tourism sites, we do know quite a lot about how they engage with more general museum spaces, such as art galleries and natural history museums. We know, for example, that young children engage with many aspects of these museums—not just those that are designated as 'child-friendly' (Wallis 2018). They find museums most interesting when they are with others who guide and support their interactions, and when they can build on previous experiences of things that interest them (Dockett, Main & Kelly 2011b).

We also know that young children explore places by moving through them. Construing young children as wayfarers, Hackett (2016, p. 169) described how children's 'movement through place create[ed] embodied, tacit ways of knowing and experiencing the world' and 'enabled children to learn about places and routes, and led to the development of traditions, in which collective meanings and actions were attached to particular locations'. In that study, young children's embodied interactions with the museum space were integral to their meaning-making.

Many museums have specifically designed experiences for children, such as activity trails or interactive displays. One example is seen in the Children's Peace Monument situated in the Hiroshima Peace Park, which acknowledges the child victims of the atomic bombing of Hiroshima. Thousands of paper cranes made by children around the world are located outside the monument. They indicate children's contribution to the Peace Park and serve as an entry point for discussions about the events at Hiroshima. The Peace Bell offers another point of connection for young children (Figures 3.2 and 3.3).

While one purpose of these activities has been to promote learning (Andre et al. 2016), a complementary body of research notes that a child-rights perspective of museum visits acknowledges the importance of children engaging

Figure 3.2 The Children's Peace Monument Situated in the Hiroshima Peace Park acknowledges the child victims of the atomic bombing of Hiroshima.

Figure 3.3 Visitors are encouraged to ring The Peace Bell in Hiroshima Peace Park to promote world peace.

in social and cultural activities as an 'experience rather than *learning*' (Johanson & Glow 2012, p. 29). In their Australian study of young children's museum experiences, Piscitelli and Anderson (2001) report that most of the preschool- and early-school-age children they surveyed regarded the museum as a place for happy interactions as well as opportunities for learning. The children described the special objects in the museum as important, and also commented on the atmosphere, describing museums as places where people 'slowed down, spoke quietly, and took their time to look at objects' (Wong & Piscitelli 2019, p. 420).

The growing body of research around young children's museum encounters indicates that these can be 'sophisticated, multi-layered and personal' as they engage with the 'social, personal, intellectual, surprising, atmospheric, and physical aspects of museums' (Wallis 2018, p. 363). While dark tourism sites are different contexts, it is quite possible that the same conclusions could be made about young children visiting these sites.

Conclusion

Young children visit dark tourism sites, yet little is known about their experiences and perspectives of these visits. Even when children's perspectives are considered, these are rarely those of young children. This has been attributed to beliefs about young children's limited developmental competence and related concerns about their susceptibility to emotional distress; methodological complexity; and ethical challenges associated with researching with young children (Canosa & Graham, 2016). Each of these potential barriers is significant.

There is a fine line between recognizing children's vulnerability and their rights to protection and balancing this with their rights to participate in matters that affect them (Murray et al. 2019). Researching with children is methodologically complex and does call for specialist skills and knowledge (Kerr & Price 2018). Ethical commitments at both institutional and personal levels are also crucial (Graham et al. 2013).

Recent waves of research draw attention to the social and cultural contexts of young children and the ways in which they construct meaning. Many adults seek to shield young children from difficult, challenging, or emotionally charged situations. However, others advance arguments that honest and open communication conveyed in a safe context, where children can ask questions and discuss their concerns, provide the basis for meaning-making and building understanding (Rosengren et al., 2014). This research also notes that young children will encounter difficult or distressing experiences such as death, regardless of whether adults choose to discuss these, and that open communication contributes to children's coping abilities (Martincekova et al. 2018).

Recent research has been complemented by the fields of sociology of childhood and children's rights, which affirm children as social actors who influence and are influenced by their social contexts (Christensen & Prout 2005). Children are co-constructors of meaning through their many and varied interactions (Rogoff 2003); experts on their own lives (Clark 2017); and, with appropriate support, competent to share this expertise (James & Prout 2014).

From these perspectives, we can draw on ethical research practices such as those espoused in the Mosaic Approach, participatory rights-based research, and the ERIC framework. These recognize not only children's competence in supportive and responsive environments, but also the multiple means by which they may choose to share their experiences and expectations. It is important that we pursue such research as a means to recognize children's rights and the obligations of adults to actualize these.

References

Andre, L., Durkson, T., & Volman, M. (2016) 'Museums as avenues of learning for children: A decade of research', *Learning Environments Research*. DOI: 10.1007/s10984-016-9222-9.

Beazley, H., Bessell, S., Ennew, J., & Waterson, R. (2011) 'How are human rights of children related to research methodology?', in A. Invernizzi, & J. Williams (eds.), *The Human Rights of Children: From Visions to Implementation* (pp. 159–178). London: Taylor and Francis.

Bessell, S. (2009) 'Research with children: Thinking about method and methodology', in ARACY/NSW Commission for Children and Young People, *Involving Children and Young People in Research* (pp. 17–27). Sydney. https://www.aracy.org.au/publications-resources/command/download_file/id/108/filename/Involving_children_and_young_people_in_research.pdf

Canosa, A., & Graham, A. (2016) 'Ethical tourism research involving children', *Annals of Tourism Research*, 61: 219–221.

Carr, N. (2011) *Children's and Families Holiday Experience*. London: Routledge.

Carr, M., & Lee, W. (2019) *Learning Stories in Practice Education*. London: SAGE.

Carrigan, A. (2014) 'Dark tourism and postcolonial studies: Critical intersections', *Postcolonial Studies*, 17(3): 236–250. DOI: 10.1080/13688790.2014.993425.

Christensen, P., & James, A. (Eds.) (2017) *Research with Children: Perspectives and Practices* (3rd ed.). London: Routledge.

Christensen, P., & Prout, A. (2005) 'Anthropological and sociological perspectives on the study of children', in S. Greene, & D. Hogan (eds.), *Researching Children's Experience* (pp. 42–60). London: Sage.

Clark, A. (2017) *Listening to Young Children. A Guide to Understanding and Using the Mosaic Approach* (Expanded 3rd ed.). London: Jessica Kingsley.

Clark, A., & Moss, P. (2001) *Listening to Young Children: The Mosaic Approach*. London: National Children's Foundation and Joseph Rowntree Foundation.

Cui, R., Cheng, M., Xin, S., Hua, C., & Yao, Y. (2019) 'International tourists' dark tourism experiences in China: The case of the memorial of the victims of the Nanjing Massacre', *Current Issues in Tourism*. DOI: 10.1080/13683500.2019.1707172.

Davis, J. (2009) 'Involving children', in K. Tisdall, J. Davis, &M. Gallagher (eds.), *Researching with Children and Young People: Research Design, Methods and Analysis* (pp. 154–193). London: Sage.

Dockett, S. (2000) 'Child-initiated curriculum and images of children', in W. Schiller (ed.), *Thinking Through the Arts* (pp. 204–211). Amsterdam: Harwood.

Dockett, S., Einarsdóttir, J., & Perry, B. (2011a) 'Balancing methodologies and methods in research with young children', in D. Harcourt, B. Perry, & T. Waller (eds.), *Researching Young Children's Perspectives: Debating the Ethics and Dilemmas of Educational Research with Children* (pp. 68–81). London: Routledge.

Dockett, S., Einarsdóttir, J., & Perry, B. (Eds.) (2019) *Listening to Children's Advice about Starting School and School Age Care*. London: Routledge.

Dockett, S., Main, S., & Kelly, L. (2011b) 'Consulting young children: Experiences from a museum', *Visitor Studies*, 14(1):13–33. DOI: 10.1080/10645578.2011.557626.

Dockett, S., & Perry, B. (2014) 'Participatory rights-based research: Learning from young children's perspectives in research that affects their lives', in O. Saracho (ed.), *Handbook of Research Methods in Early Childhood Education: Review of Research Methodologies, Volume II* (pp. 675–710). Charlotte, NC: Information Age.

Dockett, S., & Perry, B. (2019) 'What do children expect out of research participation?', in J. Murray, B. Swadener, & K. Smith (eds.), *The Routledge International Handbook of Young Children's Rights* (pp. 460–471). London: Routledge.

Edwards, C., Gandini, L., & Foreman, G. (Eds.) (2011) *The Hundred Languages of Children: The Reggio Emilia Approach to Early Childhood Education* (3rd ed.). Santa Barbara: Praeger.

Gaab, E., Owens, G., & MacLeod, R. (2013) 'Caregivers' estimations of their children's perceptions of death', *Death Studies*, 37:693–703. DOI:10.10890/07481187.2012.692454.

Graham, A., Powell, M., Taylor, N., Anderson, D., & Fitzgerald, R. (2013) *Ethical Research Involving Children*. Florence: UNICEF Office of Research – Innocenti. https://childethics.com/

Groundwater-Smith, S., Dockett, S., & Bottrell, D. (2015) *Participatory Research with Children and Young People*. London: SAGE.

Gutiérrez, I., Menendez, D., Jiang, M., Hernandez, I., Miller, P., & Rosengren, K. (2019) 'Embracing death: Mexican parent and child perspectives on death', *Child Development*, 91(2): e491–e511. DOI: 10.1111/cdev.13263.

Hackett, A. (2016) 'Young children as wayfarers: Learning about place by moving through it', *Children and Society*, 30: 169–179. DOI: 10.1111/chso.12130.

Israfilova, F., & Khoo-Lattimore, C. (2019) 'Sad and violent but I enjoy it: Children's engagement with dark tourism as an educational tool', *Tourism and Hospitality Research*, 19(4): 478–487. DOI: 10.1177/1467358418782736.

James, A., & Prout, A. (2014) *Constructing and Reconstructing Childhood: Contemporary Issues in the Sociological Study of Children* (3rd ed.). London: Routledge.

Johanson, K., & Glow, H. (2012) 'It's not enough for the work of art to be great: Children and young people as museum visitors', *Journal of Audience & Reception Studies*, 9(1): 26–42.

Kerr, M. M., & Price, R. H. (2016) 'Overlooked encounters: Young tourists' experiences at dark sites', *Journal of Heritage Tourism*, 11(2): 177–185. DOI: 10.1080/1743873X.2015.1075543.

Kerr, M. M., & Price, R. H. (2018) '"I know the plane crashed": Children's perspectives in dark tourism', in P. R. Stone, R. Hartmann, T. Seaton, R. Sharpley, & L. White (eds.), *The Palgrave Handbook of Dark Tourism Studies* (pp. 553–583). London: Palgrave Macmillan.

Kerr, M. M., Price, R. H., Savine, C. D., Ifft, K., & McMullen, M. A. (2017) 'Interpreting terrorism: Learning from children's visitor comments', *Journal of Interpretation Research*, 22(1): 83–100.

Khoo-Lattimore, C. (2015) 'Kids on board: Methodological challenges, concerns and clarifications when including young children's voices in tourism research', *Current Issues in Tourism*, 18(9): 845–858. DOI: 10.1080/13683500.2015.1049129.

Kidron, C. (2013) 'Being there together: Dark family tourism and the emotive experience of co-presence in the Holocaust past', *Annals of Tourism Research*, 41: 175–194. DOI: 10.1016/j.annals.2012.12.009.

Lemelin, R., Thompson-Carr, A., Johnston, E., & Dawson, J. (2013) 'Indigenous people: Discussing the forgotten dimension of dark tourism and battlefield tourism', in D. Muller, L. Lundmark, & R. Lemelin (eds.), *New Issues in Polar Tourism: Communities, Environments, Politics* (pp. 205–215). Dordrecht: Springer.

Lundy, L., & McEvoy, L. (2012) 'Childhood, the United Nations convention on the rights of the child, and research: What constitutes a 'rights-based' approach?', in M. Freeman (ed.), *Law and Childhood Studies: Current Legal Issues*, Volume 14 (pp. 75–91). Oxford University Press.

Martincekova, L., Jiang, M., Adams, J., Menendez, D., Hernandez, I., Barber, G., & Rosengren, K. (2018) 'Do you remember being told what happened to grandma? The role of early socialization on later coping with death', *Death Studies*, 44: 78–88. DOI: 10.1080/07481187.2018.1522386.

Menendez, D., Hernandez, I., & Rosengren, K. (2020) 'Children's emerging understanding of death', *Child Development Perspectives*, 14(1): 55–60. DOI: 10.1111/cdep.12357.

Menzies, K. (2019) 'Understanding the Australian Aboriginal experience of collective, historical and intergenerational trauma', *International Social Work*, 62(6): 1522–1534. DOI: 10.1177/0020872819870585.

Murray, J., Swadener, B., & Smith, K. (Eds.) (2019) *The Routledge International Handbook of Young Children's Rights*. London: Routledge.

Pálmadóttir, H., & Einarsdóttir, J. (2016) 'Video observations of children's perspectives on their lived experiences: Challenges in the relations between the researcher and children', *European Early Childhood Education Research Journal*, 24(5): 721–733. DOI: 10.1080/1350293X.2015.1062662.

Panagiotaki, G., Hopkins, M., Nobes, G., Ward, E., & Griffiths, D. (2018) 'Children's and adults understanding of death: Cognitive, parental, and experiential influences', *Journal of Experimental Child Psychology*, 166: 96–115. DOI: 10.1016/j.jecp.2017.01.014.

Piaget, J. (1929) *The Child's Conception of the World*. London: Kegan Paul.

Piscitelli, B., & Anderson, D. (2001) 'Young children's perspectives of museum settings and experiences', *Museum Management and Curatorship*, 19(3): 269–282. DOI: 10.1080/09647770100401903.

Poria, Y., & Timothy, D. (2014) 'Where are the children in tourism research?', *Annals of Tourism Research*, 47: 93–95. DOI: 10.1016/j.annals.2014.05.009.

Powell, M., McArthur, M., Chalmers, J., Graham, A., Moore, T., Spriggs, M., & Taplin, S. (2018) 'Sensitive topics in social research involving children', *International Journal of Social Research Methodology*, 21(6): 647–660. DOI: 10.1080/13645579.2018.1462882.

Price, R. H., & Kerr, M. M. (2018) 'Child's play at war memorials: Insights from a social media debate', *Journal of Heritage Tourism*, 13(2): 167–180. DOI: 10.1080/1743873X.2016.1277732.

Renaud, S., Engarhos, P., Schleifer, M., & Talwar, V. (2015) 'Children's earliest experiences with death: Circumstances, conventions, explanations, and parental satisfaction', *Infant and Child Development*, 24: 157–174. DOI: 10.1002/icd.1889.

Rinaldi, C. (2006) *In Dialogue with Reggio Emilia: Listening, Researching, and Learning*. London: Routledge.

Rogoff, B. (2003) *The Cultural Nature of Human Development*. New York: Oxford University Press.

Rosengren, K., Miller, P., Gutierrez, I., Chow, P., Schein, S., & Anderson, K. (2014) 'Children's understanding of death: Toward a contextualised and integrated account', *Monographs of the Society for Research in Child Development* (No. 312), 79, 1–162. DOI: 10.1111/mono.v79.1.

Schanzel, H., & Smith, K. (2014) 'The socialization of families away from home: Group dynamics and family functioning on holiday', *Leisure Sciences: An Interdisciplinary Journal*, 36(2): 126–143. DOI: 10.1080/01490400.2013.857624.

Thornton, P., Shaw, G., & Williams, A. (1997) 'Tourist group holiday decision-making and behavior: The influence of children', *Tourism Management*, 18(5): 287–297.

United Nations (1989) *The United Nations Convention on the Rights of the Child*. New York: United Nations.

Vygotsky, L. (1978) *Mind in Society: The Development of Higher Psychological Processes*. Cambridge, MA: Harvard University Press.

Wallis, N. (2018) 'Titian, tapestries and toilets: What do preschoolers and their families value in a museum visit?', *Museum & Society*, 16(3): 352–368.

Wilson, J. (2011) 'Australian prison tourism: A question of narrative integrity', *History Compass*, 9(8): 562–571. DOI: 10.1111/j.1478-0542.2011.00789x.

Wong, K., & Piscitelli, B. (2019) 'Children's voices: What do young children say about museums in Hong Kong?', *Museum Management and Curatorship*, 34(4): 419–432. DOI: 10.1080/09647775.2019.1599994.

4 School-Aged Tourists

Pre-Adolescent and Adolescent Considerations and Challenges

Timothy M. Wagner

Introduction

Attending to the cognitive, social, and emotional developmental context of pre-adolescent and adolescent visitors to dark tourism sites is critical. The intersection of typical development and young people's experiences at dark tourism sites requires nuanced consideration and is necessary to illuminate the nature of pre-adolescence and adolescence in a tourism context. Research that examines those experiences exists but is often underrepresented in the literature (Kerr and Price, 2016; Roche and Quinn, 2016).

The manner in which pre-adolescents and adolescents plan for and experience dark tourism sites is multifaceted. For instance, pre-adolescents and adolescents desire a high degree of independence and autonomy. However, they may also be greatly influenced by peers and social contexts. Research also suggests that adolescents play an increasingly significant role in trip decision-making and planning. They are motivated to travel for a variety of reasons, and ultimately seek out specific tourist experiences based on their developmental stage (Khoo-Lattimore, 2015; Larsen and Jenssen, 2004).

With those points in mind, this essay attempts to (1) provide a framework for understanding typical pre-adolescent and adolescent development, (2) explain the ways that dark tourism sites relate to the domains of development, and (3) offer evidence-based recommendations for most appropriately guiding school-aged tourists through a dark heritage experience. In particular, school-based trips, tours, and student exchanges will be considered. These offer opportunities for understanding pre-adolescents' and adolescents' experiences with dark tourism, from both individual and group perspectives.

Introducing Pre-Adolescent and Adolescent Development

The field of human development can be traced back to Sigmund Freud's early 20th-century studies of psychosocial behavior (Gilmore and Meersand, 2014). Since that time, many developmentalists have taken up the work of grounding their theories of human development in empirical research.

DOI: 10.4324/9781003032199-6

Interestingly, conceptions of development overlap academic disciplines. For example, research in the disciplines of sociology, anthropology, biology, and psychology have all contributed to our understanding of development. This essay, however, primarily calls upon developmental psychologists to describe pre-adolescent and adolescent development.

To further understand the complexities of this developmental stage, it is important to quantify the age ranges of pre-adolescence and adolescence. Generally, adolescence is seen as 'the bridge' between childhood and adulthood (Milevsky, 2015). It is further suggested that pre-adolescence occurs between the ages of 11 and 14, while adolescence (or late-adolescence) takes place between ages 15 and 19 (Martin and Fabes, 2009).

The aforementioned age ranges are largely defined by the onset of significant physical maturation milestones (e.g., menarche, adrenarche) and the conclusion of formal education (Gilmore and Meersand, 2014; Steinbeis, Crone, Blakemore, and Kadosh, 2017). Although the onset of pre-adolescence and adolescence may occur earlier or later in some young people, the 11–14 and 15–19 age ranges represent typical development.

Alongside the physical changes that young people experience at these ages, cognitive and social–emotional changes also take place. The cognitive and social–emotional maturation that occurs during these age ranges is qualitative and substantive (Taylor, Barker, Heavey, and McHale, 2013). While physical development is important, this essay will focus on the domains of cognitive and social–emotional development. Those domains are particularly relevant because of their inherent link to pre-adolescents' and adolescents' participation in visits to dark heritage sites. Before outlining a conceptual framework, it is important to identify some of the controversies in the field of pre-adolescent and adolescent development.

Controversies in the Field

Three controversies in the literature are particularly salient. The first is whether development happens in distinct stages or in a continuous manner, the next is whether nature or nurture most impacts development, and the third is whether the concept of adolescence itself is merely a cultural invention or is a true developmental period.

Theories of Continuity versus Stage Theories and Nature versus Nurture

In the 5th century BC, Plato wrote about development and suggested that it is based on many factors and occurs with continuity over time (Martin and Fabes, 2009). In continuous development, growth across developmental domains is seen to be steady and gradual. One might visualize a graph with a smooth and linear upward progression while thinking about continuous development. Theorists who study from a continuous developmental perspective suggest that discontinuities may also occur. These discontinuities may be the result of an environmental disruption or a developmental delay

(Bornstein, 1984). Although there may be pauses or discontinuities in the steadiness of development, the manner of growth continues to be rooted in a continuous model of development.

On the other hand, a significant number of theories, including Jean Piaget's Stages of Cognitive Development and Erik Erikson's Stages of Psychosocial Development, contend that development takes place in distinct stages. In these theories of development, each stage builds upon the previous stage. Age range benchmarks outline when typically developing individuals progress to a subsequent stage (Lerner, 2002). While each developmental stage occurs in a determined order, atypical development is said to occur when a young person remains at a particular developmental stage well past their typically developing peers. The accomplishment of the specific criteria of each developmental stage over time (be they cognitive, social–emotional, or physical) indicates progress toward the successive stage (Ayer and Bornstein, 2005).

Because stage theorists mark developmental milestones using age ranges, and because so much of the most recent research in the child development literature has been conducted by stage theorists, this essay will utilize stage theorists' frameworks. Consequently, a young person's development will be considered to occur in stages.

Over time, two separate ways of thinking about how a young person develops have emerged in the literature. This debate is known as the 'nature versus nurture' controversy (Collier, 2008). Because the primary way that young people's development will be addressed via dark heritage experiences is sociocultural and environmental, theories surrounding nurture will be relied upon to frame this essay.

Cultural Invention versus Developmental Reality

Mid-20th-century literature suggests that adolescence may not be a universal and inevitable developmental reality for all young people (Dasen, 2000). Accordingly, some anthropological and sociological research points out that young people in certain cultures experience developmental characteristics between ages 11 and 19 quite differently than same-age peers in other cultures (Mead, 1928). It is this body of literature that indicates that adolescence may, in fact, be a cultural invention.

In contrast, more contemporary family studies and work in psychoanalysis indicate that adolescence is not a cultural construct, but rather a clear subset of developmental milestones within a lifespan (Gilmore and Meersand, 2014). These studies, many of which are based on caregiver self-reports and psychoanalysts' observations and interviews, contend that adolescents do have a particular set of biological and behavioral shifts. Those include periods of turmoil and relational upheaval, as they work to establish their identities.

As a framework for understanding how young people engage in dark heritage experiences, and to support the contention that adolescence is a developmental reality, the specific work of developmental stage theorists Jean Piaget and Erik Erikson is considered. Later in this essay, this conceptual framework will be called upon to explore how dark tourism and pre-adolescent and adolescent development intersect.

Conceptual Frameworks for Pre-Adolescent and Adolescent Development

In the early 20th century, developmentalist G. Stanley Hall first wrote about adolescence and cited the often-difficult cognitive and social–emotional challenges that adolescents face. Since that time, the period of 'sturm und drang' (storm and urges) that describes adolescence has been regularly studied from both a cognitive and social–emotional perspective (Bergman and Scott, 2011). Brain maturation, sensation-seeking, and moodiness are among the difficulties that have been cited as characteristic features of adolescence (Gilmore and Meersand, 2014).

To better understand the nuances associated with pre-adolescent and adolescent development, two developmental frameworks will be shared. The first, Jean Piaget's Stages of Cognitive Development, will venture to explain the intellectual growth and development of young people ages 11–19. Erik Erikson's Stages of Psychosocial Development will then be reviewed, with particular emphasis on those stages linked to pre-adolescent and adolescent development.

Cognitive Development

Cognitive development is the process of developing the intellectual means to adapt to one's environment (Langer and Killen, 1998). Furthermore, cognitive development consists of establishing mental processes used to make sense of information, grow in awareness, solve problems, and gain knowledge (Martin and Fabes, 2009).

Cognitive developmentalist Jean Piaget was particularly interested in 'the process of coming to know' and established a series of incremental stages associated with an individual's cognitive development (Huitt and Hummel, 2003, p. 1). As a developmentalist whose work centered primarily on young people, Piaget organized his framework for cognitive development into four stages: Sensorimotor, preoperational, concrete operational, and formal operational. Particular age ranges characterize each of those stages of cognitive development, along with a set of features associated with an individual's intellectual development during that time.

Generally speaking, therefore, movement from concrete operational to formal operational development occurs during pre-adolescence and adolescence. Although there is a return to egocentrism at the start of this stage, young people also assimilate concepts such as justice, equality, and fairness into their way of seeing the world (Zigler and Gilman, 1998). Conceptual ideas such as these represent an adolescent's cognitive growth from concrete to abstract thinking. For instance, a young person is better able to solve for the variable *x* in an Algebra problem during this stage. Similarly, during this stage dark heritage site visitors will more keenly grapple with ideas such as mortality, integrity, or grief. Conceptual abstract thinking evolves during this stage of cognitive development, as does the ability to apply knowledge, skills, and conceptual understanding to new and unique situations (Ahmad, Ch, Batool, Sittar, and Malik, 2016). Finally, and of important note, some developmentalists believe that the formal operational stage of cognitive development continues well into adulthood (Martin and Fabes, 2009).

Next, with an idea of how pre-adolescents' and adolescents' cognitive reasoning looks (largely linked to evolving abstract thinking skills), we once again call upon a longstanding developmental framework. We turn our attention to key tenets of young people's social–emotional development.

Social–Emotional Development

Researchers in the field define social–emotional development as the processes related to interactions with other people. Included in this domain is the study of relationships, emotions, personality, and moral development (Martin and Fabes, 2009). Erik Erikson was a lifespan developmentalist who, although recognizing that biological factors were at play when it came to development, grounded his theories in how one's environment influences personal changes over time. Erikson referred to the conditions of young people's environments as *cultural contexts* (Côté, 2005). Furthermore, Erikson noted that an environmental transaction occurs during development. Individuals influence their environment, while the environment simultaneously influences individuals (Côté, 2005).

With Erikson's work on Stages of Psychosocial Development as a backdrop for describing social–emotional development, *identity versus role confusion* is acknowledged as a stage for specific consideration for pre-adolescent and adolescent development. It is during this stage that young people must navigate the development of an identity or otherwise experience the crisis of sustained role confusion until the crisis is resolved.

Answering the existential question 'Who am I?' is challenging work for young people. Adolescents are exploring this question not only relative to their social identities, but also as it relates to a prospective occupation or family status (Shaffer and Kipp, 2010). When this stage is resolved, individuals have a grounded sense of self and are able to describe their own belief systems. In addition to gaining a strong sense of personal self-identity, individuals who have successfully navigated this stage can also understand the perspective of others with greater ease. In so doing they begin to recognize how personal identity integrates across environments (McLeod, 2013).

While it is noted that a majority of individuals successfully navigate this stage from a social–emotional perspective, unresolved *role confusion* has potentially debilitating outcomes (Côté, 2005; McLeod 2013). For instance, individuals who experience role confusion may begin to experiment with different lifestyles because they do not have a clear concept of where and how they fit into society. If a negative identity emerges, we see the oft-cited rebellious nature of an adolescent surface or, in more dire situations, depression and suicidality may develop (McLeod, 2013; Shaffer and Kipp, 2010).

At the intersection of Erikson's *identity versus role confusion* stage (and its accompanying developmental features) is a series of relevant contexts in which this stage of development has been studied. James Côté (2005, p. 409) identified four contexts in which Erikson's work has been studied and applied:

- Cultural differences in child-rearing
- How the structuring of children's play reproduces cultural ideologies
- The interplay between individual biography and historical events
- The contribution of youth to cultural renewal

The aforementioned contexts will provide insight into how a dark heritage experience relates to social–emotional development, particularly in pre-adolescents and adolescents as they form an identity. Accordingly, we now return to both Piaget's and Erikson's developmental frameworks to uncover the connection between dark tourism and development.

Considering Development and Dark Tourism

The intersection of typical development and young people's experiences at dark tourism sites requires nuanced consideration that is often underrepresented in the literature (Kerr and Price, 2016; Roche and Quinn, 2016). The underrepresentation of young people's voices and experiences in heritage studies is, in part, related to the absence of researchers' knowledge of human development and accompanying research methods appropriate for young people (Irwin and Johnson, 2005; Wu et al., 2019). The discreet differences between how young children and adolescents, for instance, participate in dark heritage site visits have also been noted as a key consideration (Wu et al., 2019).

While there is a lack of research into how young people experience dark heritage sites, we do have substantial understanding of how pre-adolescents and adolescents develop. Consequently, this section ventures to explicitly connect (1) features of pre-adolescent and adolescent development and (2) what the literature reveals about dark heritage sites and young people. Israfilova and Khoo-Lattimore (2018) suggest four distinct dark site visit outcomes for young people: Increasing knowledge, shaping personality, challenging emotions, and entertaining.

The Intersection of Pre-Adolescent and Adolescent Development and Dark Tourism

Table 4.1 outlines a variety of developmental features, from both cognitive and social–emotional perspectives. The first column's content derives from the developmental framework of either Piaget or Erikson. The second column references dark tourism and/or heritage tourism studies that illustrate each developmental feature.

Table 4.1 The Intersection of Development and Dark Tourism Research

Pre-Adolescent/Adolescent Developmental Features (Corresponding Domain)	Examples of Developmental References Appearing in Dark Tourism and Heritage Tourism Literature
Abstract conceptualization (cognitive development)	Adolescents are better able to understand their physical situation and have an evolving knowledge of geography. In short, they can understand how abstract concepts such as political geography are relevant when visiting dark tourism sites (Poria, Atzaba-Poria, and Barrett, 2005). Dark tourism invites visitors to understand the tangible as well as intangible and socio-cultural elements of a site (Tinson, Saren, and Roth, 2015).
Assimilation of concepts such as justice, equity, etc. (cognitive development)	Not only do dark tourism sites offer an opportunity to memorialize an event, but they often provoke deeper questions of ethics, for instance (Winter, 2009). Pre-adolescents and adolescents have an emerging ability to engage in complex and conceptual thinking.
Continued egocentric thinking (cognitive development)	Because dark tourism experiences are designed to be highly personal and inspire the individual to make meaning of their experience, a visit will likely align with some level of egocentric thinking (Bosangit, Hibbert, and McCabe, 2015).
Identity development (social–emotional development)	School-aged tourists at dark sites note the opportunity to enhance their self-identities through trips – moreso than they express interest in the topic/content of the experience (Tinson et al., 2015). New identities may be forged during visits to dark tourism sites (Roberson, 2007). How young people see themselves and define personal identity will, in turn, influence their experiences at dark tourism sites (Levy, 2014).
A search for autonomy and emancipation (social–emotional development)	Young people have the ability to influence parental and family decision-making when it comes to trip locations and activities (Khoo-Lattimore, 2015). Adolescents have more influence than younger children in family tourism decisions and are typically more motivated by personally pleasing (fun and physical) activities (Khoo-Lattimore, 2015).
Belief and value system emergence (social–emotional development)	Studies suggest that values and ideals are established and affirmed when pre-adolescents and adolescents visit dark tourism sites and engage with concepts such as death (Tinson et al., 2015).

(*Continued*)

Table 4.1 (Continued)

Pre-Adolescent/Adolescent Developmental Features (Corresponding Domain)	Examples of Developmental References Appearing in Dark Tourism and Heritage Tourism Literature
Recognition and identification of one's role in a group or culture (social–emotional development)	Young people's ethnic identities involve both how they view themselves and how others view their ethnicity (Levy, 2014). Dark tourism sites that involve individuals from a young person's own ethnic group further this recognition and identification process. Pre-adolescent and adolescent tourists who identify with a group (e.g., ethnic, religious), and visit dark heritages sites where that group is reflected, reinforce connections by speaking about themselves as members of that group (Israfiloza and Khoo-Lattimore, 2018). Students' racial, ethnic, national, and religious backgrounds influence how they interpret history (Levy, 2014).
An increased ability to gain perspective (social–emotional development)	Interactions with history (e.g., visits to dark tourism sites) prompt young people to not only question what they are learning, but also begin to grasp the perspective of the individuals and groups whose experiences are being shared (Levy, 2014).
The importance and influence of peers (social–emotional development)	Young people indicate that social connection is an important factor in identifying why travel is important. Social belonging and social engagement are both strong influencers in their travel experiences (Hilbrecht, Shaw, Delamere, and Havitz, 2008; Schanzel, 2012; Small, 2008). Young people are more motivated by social factors than by any other reason for travel (Larsen and Jenssen, 2004).

As shown in Table 4.1, a growing body of research reveals a connection between typical development and dark tourism. Consequently, the next section addresses how students learn and details evidence-based recommendations for guiding young people through dark heritage sites.

Implications of the Intersection of Development and Dark Heritage Experiences

Visitor experiences at dark heritage sites tend to be predominantly interpretive in nature, prioritizing meaning-making rather than fact, figure, and content memorization (Light, 2017; Roche and Quinn, 2016). Unfortunately, much of the current research focuses on young people's content acquisition, rather than meaning-making (Sutcliffe and Kim, 2014).

However, many of the content-focused pedagogical strategies of a traditional classroom would not be appropriate for a young person's visit to a dark heritage site. Accordingly, the opportunity to *experience* a site, rather than be *educated* about a site, becomes critical. Adults require developmental knowledge to meaningfully guide young people through dark tourism sites. What follows is an overview of several evidence-based recommendations for those who guide students through dark heritage experiences. There are deliberate activities that young people might experience to maximize their time at dark heritage sites. These activities position those who guide pre-adolescents and adolescents to engage them in interpretation, versus a more routine pedagogy. Table 4.2 reveals practical opportunities to meaningfully ground young people's experiences at dark heritage sites in a developmental context, by considering their cognitive and social–emotional development as an 'in the moment' experience and opportunity for future growth.

As outlined in Table 4.2, translating evidence-based recommendations to experiential practices offers a guide to meaningfully engage pre-adolescent and adolescent travelers. Cutler and Carmichael (2010) underscore how experience, and the many ways it is described, informs tourism:

> The study of experience has one of the most significant areas of tourism research. Extant literature has examined phases and modes of experiences, the role of authenticity, its relationship with self-identity, and dimensions of specific types of tourist experiences. It has also examined the roles of narratives, sacredness and spirituality, skill, information and learning, place and mobility, social relationships, the imagery, influential elements of experience, and conceptual research.
>
> (in Bosangit, Hibbert, and McCabe, 2015, p. 3)

To enhance this definition of experience, consideration of pre-adolescent and adolescent development deserves a relevant place in dark heritage visits. The nature of dark heritage visits offers a profound opportunity to influence how pre-adolescents and adolescents view themselves and their world.

Conclusion

When pre-adolescents and adolescents visit dark tourist sites, their own developmental context informs the experience. Accordingly, it is possible for young people to move from a mere site visit to a profound, perspective-broadening experience when guides have an understanding of typical development. Future research in this area might consider (1) pre-adolescents' and adolescents' voices and feedback in response to the application of evidence-based recommendations and (2) a review of dark tourist site programs and materials using the principles considered in this essay. An opportunity for both individual and collective transformation exists when developmental research, dark tourism practices, and accounts of lived experiences are coordinated.

Table 4.2 Evidence-Based Recommendations for Guiding Students through Dark Heritage Experiences

Evidence-Based Recommendation	*Practical Application*
Build background knowledge for pre-adolescents and adolescents.	In advance of a dark heritage site visit, use lectures, readings, maps, primary source documents, and films/documentaries to orient young people to the site. Much of this 'pre-work' may come in the form of traditional pedagogy.
Engage pre-adolescent and adolescent visitors in experiences that are multi-sensory (Schorch, 2014).	Look for opportunities for students to use all of their senses on a trip. What can students touch or how can students move/stand, for instance, to bring them into authentic, period-based interaction?
Invite young people into targeted sociocultural conversations.	Ask students to articulate their personal views on topics such as politics, war, or equity. Then, use the dark heritage site to compare and contrast young people's perceptions of their own cultural understandings and how the site reflects and communicates components of culture. Strategies such as discussion and debate are useful.
Offer experiential learning opportunities. Experiential learning provides young people with concrete experiences, time for reflection, and a chance to generalize conclusions into a unique context (Kolb, 2015).	While visiting dark heritage sites, give students a chance to experience things firsthand (Larsen and Jenssen, 2004). Use interactive strategies to move students through heritage sites. In addition to simply visiting a dark heritage site, look for ways for students to *do* something at that site. For instance, is there an interactive display or an activity (e.g., making a paper crane at the Hiroshima Peace Memorial Park) that can involve young people?
Build empathy and broaden perspective. Pre-adolescents and adolescents are naturally reflecting on personal identity.	Some dark heritage sites might ask visitors to follow the life of a person from the time period, engaging in history through the lens of a real person. Use a case study or individual profile/biography to connect students to the site on an interpersonal level.
As students learn beyond school walls, activate reflection through contemporary technology tools such as blogs or social media 'stories.'	Young tourists typically have access to web-based and social media tools. Leverage students' use of these tools by structuring opportunities for reflection and meaning-making via contemporary technology.

(*Continued*)

Table 4.2 (Continued)

Evidence-Based Recommendation	Practical Application
Encourage students to question if and how their own identities and/or engagements in the world were influenced by the site (Burnett, 2001).	Following a visit, take time for synthesis. Asking questions such as how students felt, what they enjoyed, and what they learned helps to inspire personal reflection and extend the site's influence to young people's own lives.

References

Ahmad, S., Ch, A.H., Batool, A., Sittar, K., and Malik, M. (2016) 'Play and cognitive development: Formal operational perspective of Piaget's theory', *Journal of Education and Practice*, 7(28): 72–79.

Bergman, M.M. and Scott, J. (2011) 'Young adolescents' wellbeing and health risk behaviours: Gender and socioeconomic differences', *Journal of Adolescence*, 24: 183–197.

Bornstein, M.H. (1984) 'Developmental psychology and the problem of artistic change', *The Journal of Aesthetics and Art Criticism*: 55, 42131–42145, https://doi.org/10.1016/j.annals.2015.08.001

Bosangit, C., Hibbert, S., and McCabe, S. (2015) '"If I was going to die I should at least be having fun": Travel blogs, meaning and tourist experience', *Annals of Tourism Research*, 55: 1–14.

Burnett, K.A. (2001) 'Heritage, authenticity and history', in S. Drummond and I. Yeoman (eds.), *Quality Issues in Heritage Visitor Attractions* (pp. 39–53). Oxford: Butterworth-Heinemann.

Collier, W.G. (2008) 'Nature versus nurture', in Stephen F. Davis and William Buskist (eds.), *21st Century Psychology: A Reference Handbook*. (Vol. 2). Thousand Oaks: SAGE Publications, Inc., 2008. SAGE Knowledge, pp. 11–72.

Côté, J.E. (2005) 'Erikson's theory', in C. Fisher and R. Lerner (eds.), *Encyclopedia of Applied Developmental Science* (pp. 406–409). Thousand Oaks, CA: SAGE Publications, Inc.

Cutler, S. and Carmichael, B. (2010) 'The dimensions of tourist experience', in M. Morgan, P. Lugosi, and J.R. Brent Richie (eds.), *The Tourism and Leisure Experience. Consumer and Management Perspective* (pp. 3–260). Bristol: Channel View Publications.

Dasen, P.R. (2000) 'Rapid social change and the turmoil of adolescence: A cross-cultural perspective', *International Journal of Group Tensions*, 29: 17–49.

Gilmore, K. and Meersand, P. (2014) *The Little Book of Child and Adolescent Development*. Oxford: Oxford University Press.

Hilbrecht, M., Shaw, S.M., Delamere, F.M., and Havitz, M.E. (2008) 'Experiences, perspectives, and meanings of family vacations for children', *Leisure/Loisir*, 32(2): 541–571.

Huitt, W., and Hummel, J. (2003) Piaget's theory of cognitive development. Educational Psychology Interactive. Valdosta, GA: Valdosta State University. Retrieved [date] from http://www.edpsycinteractive.org/topics/cognition/piaget.html

Khoo-Lattimore, C. (2015) 'Kids on board: Methodological challenges, concerns and clarifications when including young children's voices in tourism research', *Current Issues in Tourism*, 18(9), pp. 845–858.

Kolb, D.A. (2015) *Experiential Learning: Experience as the Source of Learning and Development* (2nd ed.). Upper Saddle Ridge, NJ: Pearson Education, Inc.

Larsen, S. and Jenssen, D. (2004) 'The school trip: Travelling with, not to or from', *Scandinavian Journal of Hospitality and Tourism*, 4(1): 43–57, DOI: 10.1080/15022250410006273.

Lerner, R.M. (2002) *Concepts and Theories of Human Development* (3rd ed.). Mahwah, NJ: Erlbaum.

Levy, S.A. (2014) Heritage, History and, Identity. Teachers College Record, 116.

Light, D. (2017) 'Progress in dark tourism and than a tourism research: An uneasy relationship with heritage tourism', *Tourism Management*, 61: 275–301.

McLeod, S. (2013) 'Erik Erikson', *Simply Psychology*. [online] Available at: https://docuri.com/download/erik-erikson-psychosocial-stages-simply-psychology_59c1e23ff581710b286a64c6_pdf

Mead, M. (1928) 'The role of the individual in Samoan culture', *The Journal of the Royal Anthropological Institute of Great Britain and Ireland*, 58: 481.

Milevsky, P.L. (2015) *Understanding Adolescents for Helping Professionals*. New York: Springer Publishing Company.

Poria, Y., Atzaba-Poria, N., and Barrett, M. (2005) 'Research note: The relationship between children's geographical knowledge and travel experience: An exploratory study', *Tourism Geographies*, 7(4): 389–397.

Roberson, S. (2007) 'Geographies of the self in nineteenth century women's travel writing', in M. Bruckner and H. Hsu (eds.), *American Literary Geographies: Spatial Practice and Cultural Production* (pp 281–295). Delaware: University of Delaware Press.

Schanzel, H. (2012) 'The inclusion of fathers, children and the whole family group in tourism research on families', in H. Schanzel, I. Yeoman, and E. Backer (eds.), *Family Tourism: Multidisciplinary Perspective* (pp. 67–80). Bristol: Channel View.

Schorch, P. (2014) 'Cultural feelings and the making of meaning', *International Journal of Heritage Studies*, 20(1): 22–35.

Shaffer, D. and Kipp, K. (2010) *Developmental Psychology*. Belmont, CA: Wadsworth Cengage Learning.

Small, J. (2008) 'An absence of childhood in tourism studies', *Annals of Tourism Research*, 35(3): 772–789.

Steinbeis, N., Crone, E., Blakemore, S.-J., and Kadosh, K.C. (2017) 'Development holds the key to understanding the interplay of nature versus nurture in shaping the individual', *Developmental Cognitive Neuroscience*, 25: 1–4. Elsevier Science.

Taylor, S. J., Barker, L. A., Heavey, L., and McHale, S. (2013) The typical developmental trajectory of social and executive functions in late adolescence and early adulthood. *Developmental Psychology*, 49, 1253–1265. DOI: 10.1037/a0029871.

Tinson, J.S., Saren, M.A.J., and Roth, B.E. (2015) 'Exploring the role of dark tourism in thecreation of national identity of young Americans', *Journal of Marketing Management*, 31(7): 856–880.

Winter, C. (2009) 'Tourism, social memory and the Great War', *Annals of Tourism Research*, 36(4): 607–626.

Wu, M.-Y., Wall, G., Zu, Y., and Ying, T. (2019) 'Chinese children's family tourism experiences', *Tourism Management Perspectives*, 29: 166–175.

5 Development of Death Concepts

Childhood and Adolescence – Considerations for Tourist Experience and Research

Andrea Croom and Gopika Rajanikanth

Introduction

Decades of research have shown that death concepts begin to develop in early childhood and proceed through predictable stages of development during middle childhood and adolescence. Understanding the development of death concepts will help readers to (a) recognize how children of different ages navigate and make sense of dark tourism sites, (b) design materials and signage to help caregivers, educators, and docents facilitate exploration of these sites, and (c) inform research methodologies. We begin by reviewing what researchers have learned about children's conceptualizations of death.

History of Children's Understanding of Death

In the early 20th century, popular opinion was that "normal" children did not think about or even recognize the reality of death because this was a morbid topic. Death would often occur in the home and parents did not necessarily shield children from death experiences, but rather did not consider that it would have an impact on them. The experience of children in relation to death was not considered an important factor even in times when death was prominent in society (e.g., war, pandemics). However, pioneering research in the 1930s and 1940s conducted by Paul Schilder, David Wechsler, Sylvia Anthony, and Maria Nagy showed that children as young as 3 have begun to conceptualize the meaning of death. Rather than leading to meaningful conversations with children about death, this started the practice of shielding children from death by not taking them to funerals, not informing them of losses, and often not even telling them about their own terminal diagnoses (see Parsons et al., 2007; Walter, 2009). Beginning in the late 1960s, interest grew in trying to understand children's development of death concepts (see Kenyon, 2001 for a review of this literature). The research clearly showed that the development of death concepts begins at a young age; however, this did not immediately alter the way that health care professionals, parents, and educators thought about and communicated with children about death.

DOI: 10.4324/9781003032199-7

In the 21st century, few would argue that young children are not aware of death-related themes. Yet, even now, conversations about death and dying are difficult for parents, educators, and health care professionals to broach with young children and adolescents. Death is often discussed with children and adolescents in the context of a personal loss, suicide or accidental death of a student, or in response to highly publicized mass casualty events, such as school shootings or terrorist attacks (King-McKenzie, 2011). Dark tourism affords children and adolescents another opportunity to experience encounters with death (Kerr, Stone, & Price, 2021).

Accordingly, the first purpose of this chapter is to introduce the development of death concepts, across childhood and adolescence. Understanding the development of death concepts will allow readers to recognize how young tourists may make meaning of dark tourism sites. This background is critical to designing appropriate research measures to understand children's comprehension of death represented at dark tourism sites. Moreover, this chapter provides background information needed to design materials and signage. Lastly, readers may gain insights as to how they can help caregivers, educators and docents facilitate exploration of these sites.

How Do Children Learn about Death?

Freud believed that there was an innate death instinct called *thanatos* (Kahn & Liefooghe, 2014). However, most theorists believe that death is something that children learn through experience and from observing others, often in a disorganized and impromptu way (Hunter & Smith, 2008). Experiences, also known as "teachable moments," occur throughout childhood and open the door for parents to discuss the feelings and biology that surround death. These events include the loss of a grandparent or a pet, a violent crime in the neighborhood, or even something as simple as stepping on a bug. Following these moments, children are observant of their parents' responses, and they use their parents' reactions as a cue to how they are supposed to respond to this thing called "death."

Even children who have never had personal experiences with death are nevertheless exposed to these concepts through peers, teachers, religious affiliations, and the media (Despelder & Strickland, 2005). From early on, young children are exposed to death concepts through children's stories, fairy tales, nursery rhymes, and lullabies. One study examined two hundred popular nursery rhymes and found that over half examined death-related themes or characters being mistreated (Despelder & Strickland, 2005). Death themes play a central role in older children's movies, books, and video games, often becoming more vivid and explicit. Visits to dark tourism sites are yet another way that children and adolescents may be exposed to the concept of death, with these sites often representing mass casualty events (e.g., 9/11 Memorials, Holocaust museums, battlefields, and cemeteries).

Over the years, research reports mixed results on whether prior death experience has an effect on children's understanding of death concepts (Cotton &

Range, 1990; Longbottom & Slaughter, 2018; Menendez, Hernandez, & Rosengren, 2020; Rosengren, Gutiérrez, & Schein, 2014). Some research has found that, in general, individuals with fewer personal death experiences have less well-developed death concepts. However, other research suggests that children with terminal or chronic illnesses do not differ from healthy children in their understanding of death concepts (Kenyon, 2001).

How Death Concepts Develop during Childhood: A Theoretical Overview

Nagy (1948) found that children go through three developmental stages between the ages of three and ten in order to reach a mature understanding of death. The first stage (age 3–5) is when children understand death as being "less alive." In this stage, children view the dead as living on inside of their coffin with the same needs and desires as a person who is not dead; and they believe that if the circumstances were correct then the dead person could return to life. In the second stage (age 5–9), children understand that death is final once it occurs, however, they believe that death is an entity (e.g., personified like the grim reaper) that can be tricked or avoided. Also in this stage, death seems like something distant from the child and they are unaware or hesitant to acknowledge that someday they and the people they love will die (Swain, 1979). In the third and final stage (age 9 and older), children have reached a point where they recognize that death is a biological process that is universal, final, and inevitable.

In the 1950s, Erikson developed his stages of psychosocial development in which each stage involves a crisis that must be dealt with in order to successfully pass through the remaining stages. Although this theory was not based on the development of death concepts, it is interesting to see how the death of someone in a close interpersonal relationship might impact a child at each stage of development. In infancy, Erikson is most concerned that the death of a parent or other important person might lead to stress in the family and could cause the infant to see the world as unpredictable ("someone was there and now they're gone"). During the toddler years if a death occurs, the child may regress back to a dependent state instead of being independent; because when they left to explore the environment someone that they cared about disappeared ("I'll never leave you again"). Early childhood (age 3–6) is when children have many defiant thoughts and quick turning emotions. If they have thoughts where they wish that their parent, a pet, or a friend was dead then they may feel guilty afterward and maybe scared that their thoughts have the power to make death happen. Middle childhood (age 6–11) is when children begin to compare themselves to peers. At this stage, children who have had more death experiences (e.g., a parent, sibling, or grandparent) may feel that they are different from and cannot relate to others. Finally, during adolescence children have dreams and goals that are within their grasp, but they realize that death could put a stop to all of that. For some adolescents, the idea of personal mortality becomes more real, especially if there is a death in the school or a similar event, which can lead to a more rapid "growing up" period. However, some adolescents develop a sense of invulnerability where they feel like negative consequences such as death can be avoided (Despelder & Strickland, 2005; Noppe & Noppe, 1991).

The most frequently cited developmental theorist in the death concept literature is Piaget. Some research has found that measures of cognitive ability based on concrete operational tasks (e.g., liquid conservation) are a better predictor of a child's death concept score than either their intelligence quotient (IQ) or their chronological age (Cotton & Range, 1990; Kenyon, 2001; White, Elsom, & Prawat, 1978). For the majority of the sensorimotor period (age 0–2), until object permanence develops, the child is unable to hold a person in their head or represent them if they are not physically present. Therefore, if someone dies it is the same kind of phenomenon as if a parent leaves the room, they simply vanish (Despelder & Strickland, 2005). During the preoperational period (age 2–7) children begin to use language and symbols to represent objects and so when asked they can begin to describe how they conceptualize death. The concrete operational period (age 7–12) is believed to be the most critical turning point in the development of children's death concepts. At this stage, children's thinking becomes more flexible and they are able to understand the concept of universal rules, which helps them to realize that everyone must die including themselves. Also, the ability to successfully navigate and make sense of conservation tasks allows them to understand the concept of irreversibility. Finally, during the formal operational period (starting at age 11–12) children can begin to think about death in more abstract and metaphysical ways, which can contribute to death concepts becoming increasingly "fuzzy" (Koocher, 1973).

Primary Death Concepts

There are four generally accepted concepts that make up a mature understanding of death (Cotton & Range, 1990; Poling & Evans, 2004). Researchers have employed many measures and techniques to investigate how children conceptualize and reason about death-related concepts. Validated measures, such as the Development of Death Concept Questionnaire (DDCQ) and the Derry Death Concept Scale, assess death concepts using a series of close-ended questions (Kenyon, 2001). Other measures, such as the Concepts of Death Interview (CDI), rely on semi-structured interview questions to compare how children view inanimate objects (e.g., Is a car living?), animals (e.g., Zip, the dog, has died – will Zip ever be alive again?), and humans (e.g., Mr. Zong has died, where is he?; Jay, Green, Johnson, Caldwell, & Nitschke, 1987). Additional measures include comparing stuffed animals and taxidermy animals, phenomenographic study of drawings, and narrative stories (Tamm & Granqvist, 1995; White, Elsom, & Prawat, 1978). Clinically, there are often indicators in drawings and language that can tell you how a child or adolescent conceptualizes death, and it is these indicators that we will focus on below.

Non-functionality

This is the concept that all bodily functions cease after death. Grasping this concept requires a basic understanding of biological or "life" concepts such as recognizing organs and their functions (e.g., these are my lungs, they help me breathe) and being able to differentiate between things that are alive (animals, plants, humans) and inanimate objects. If children understand that death is the opposite of life then once they learn that the body functions to support life, they can infer that when the body stops functioning there is death (Slaughter & Lyons, 2003).

Lazar and Torney-Purta (1991) suggest that children first understand non-functionality regarding more obvious biological functions (e.g., the dead cannot eat or speak) and, as they grow older, they start understanding the less obvious dysfunctions (e.g., the dead cannot drink or hear). Lack of understanding can be seen in children leaving food items for the deceased, talking about being dead as sleeping, or attributing emotions to the deceased (e.g., referring to the deceased as sad or angry).

Finality or Irreversibility. This is the idea that once a living organism has died, it cannot come back to life again. Sometimes talking with a child about a loss can feel like a broken record when this concept has not yet been grasped (e.g., "I know Grandpa is dead, but he'll be at my baseball game next week, right?").

Media can offer a confusing view of death, especially cartoons and video games, which do not give clear evidence for finality or irreversibility because nobody ever really gets hurt and they can always come back to life (Rushforth, 1999). Graham, Yuhas, and Roman (2018) compared the differences about the portrayal of death between ten Disney movies released between 1937 and 2003 and eight Disney movies released between 2003 and 2016. This study incorporated Cox, Garrett, and Grahams' earlier 2005 study which analyzed the portrayal of death in movies released between 1937 and 2005. In the older movies, deaths of antagonists were always permanent, whereas protagonists' deaths were frequently reversible, or the character returned in an altered form (e.g., Snow White coming back to life after a kiss from Prince Charming). Disney and Pixar have shifted away from showing death as being reversible as evidenced by only one movie during 2003–2016 with any form of reversible death. This has led Graham et al. (2018) to conclude that Disney has shifted towards a more realistic and implicit portrayal of death.

Universality or Inevitability. This concept states that all living things must die and that this death is inevitable, unavoidable, and unpredictable. *Personal mortality* is related to the concept of universality, but it involves applying a universal concept to the self (the idea that someday *I* will die; Despelder & Strickland, 2005). This concept can develop at a separate time such that a child would answer that everything dies one day but would deny that they are going to die one day.

The realization of personal mortality can lead to *death anxiety*. Although adults attempt to spare children from death anxiety, in other ways they indirectly teach children death anxiety through cautions (e.g., "Slow down on the steps," "Wash your hands before you eat," "Chew your food slowly," or "Look both ways before you cross the street"). In all of these safety rules, we are telling our children how to avoid death. Too much death anxiety is unhealthy because it can be paralyzing, while no death anxiety is dangerous because individuals do not take precautions. Moderate death anxiety was evolutionarily advantageous; our cautious ancestors survived to pass on their genes.

Causality

Causality refers to an understanding of the causes that lead to death (e.g., illnesses, accidents, intentional actions, and weather-related disasters). The depth of this understanding can vary at different ages. Researchers have also investigated additional ideas that children hold about death, such as the notion of imminent justice – the idea that death, illness, or other tragic events occur because a person did something warranting punishment. Children who score high on this concept will believe that "mean" people play a role in causing their own death, while "nice" people die by accidental means or because it is their time (White, Elsom, & Prawat, 1978).

This concept may be disrupted when children visit dark tourism sites, which often commemorate those who did not "deserve" to die. The following conversation was captured during the first author's qualitative research (see Patterson, 2007 for study details) of parents and children exploring *The Mysterious Bog People Exhibit*, which features well-preserved mummies, many of whom suffered violent deaths.

M: *I think the hair is real. And look, there's her face. And she had a cord wrapped around her neck so how do you think she died?*
B: *Of choking.*
M: *Choking, strangling.*
B: *Who choked her?*
M: *Someone.*
B: *Who would choke a sixteen-year-old girl?*
M: *That's a good question.*

Other Conceptualizations of Death Concepts

In 1995, Tamm and Granqvist asked 431 children to make a drawing of what comes to their mind when they hear the word "death" and then to explain their drawings. This phenomenographic study found that children view death in one of three ways: (1) Biological, (2) psychological, and (3) metaphysical. Biological depictions of death include drawing the moment that a death occurs, showing a violent death, or showing a person in the state of death (e.g., at a morgue or in their coffin). Psychological depictions of death are when children show the emotions associated with death (e.g., grief or sadness), mental images of fearing death, or the emptiness associated with death.

Metaphysical depictions of death consist of religious or cultural symbols for death, personifying death (e.g., the grim reaper), or perceptions of heaven and hell.

Recognizing these different ways to conceptualize death is vital when asked the difficult question from a child of "why did this person/animal/loved one die?" Our first inclination is often to give a causal answer. To name the specific thing that led a loved one to pass away (e.g., grandpa had cancer, your mom got in a car accident). The child might get frustrated and say, "I know that, but why did they die?" so then we can move in different directions. If this is a biological question, then we might explain how the heart stops pumping or what happens when a body loses too much blood. If we see this as a psychological/emotional question, then we might acknowledge the sadness or anger that they're feeling. Sometimes the question of why is responded to with a spiritual explanation or response (e.g., your mom went to be an angel).

Perhaps the most honest answer is a combination of all three – "your mom got in a car accident, she lost a lot of blood and so her body stopped working. It hurts to lose someone we love so much. And what I believe is that your mom is still with us even though we can't see her anymore." And other times, even with the best of intentions, we might miss the mark completely as happened to one of the first author's clients when interacting with her 4-year-old son at a birthday party for his deceased grandfather. He asked her whether or not Grandpa would be attending the party. She gave what she thought was a well-thought-out response addressing biological, psychological, and metaphysical aspects of the grandfather's death, along the lines of "I miss Grandpa too, but he is dead so although we believe he's here with us in spirit and we like to remember him, he won't actually be at the party," to which her son replied, "Great, I just wanted to know if I could have his piece of cake!".

How Death Concepts Progress throughout Childhood and Adolescence

It would be nice if research allowed us to paint a pretty picture that tells us at what specific age a child can expect to "achieve" a mature understanding of a specific death concept; however, as with most things in developmental research, it's not that simple. There are many factors that can impact the development of death concepts including, but not limited to, the child's personality, direct education, media exposure, and personal life experiences. Although the timeline for the development of these concepts remains disputed, we will highlight some considerations and observations about each age group.

Toddlers (up to Age 3)

Research of death concept development typically begins with 3-year-olds as the youngest cohort (Renaud, Engarhos, Schleifer, & Talwar, 2015; Slaughter & Lyons, 2003). However, the first author distinctly remembers going on a walk with her oldest daughter when she was 20 months old and out of the blue, she began to wail and point to a dead squirrel on the sidewalk. She cried inconsolably for several minutes even after they moved past the deceased body. Moreover, the first author's middle daughter at 3 years old found a dead bumble bee in their driveway and immediately wanted to give it attention "the bumble bee is broken, Mommy, fix it." She returned to the bumble bee day after day (until her mother finally removed it) with ongoing concern about what might have happened to the bumble bee, frequently generating narratives about potential causes (e.g., "the big ant stepped on the bee and smushed it"). Personally, the first author grew up hearing stories about how for more than 2 years she was terrified to ride in white cars (luckily the one her family owned was blue) following her grandfather's death when she was 2 years old. Her grandfather was taken away in a white hearse and she could only imagine that getting into a white car would result in her being taken away and never coming back. These stories reflect how toddlers may not be able to articulate to researchers their understanding of death. Nevertheless, they are certainly able to recognize death as something undesirable and emotional.

Young Children (Age 4–6)

As vocabulary improves and life experiences abound, children begin to form a more complex understanding of death. When asked about causality of death, young children like kindergartners might give examples like violence or accidents (Kenyon, 2001). Their emerging understanding of death often leads young children to ask questions about death which are often met with uncomfortable responses from parents and educators. Young children are also more likely to express themselves through artwork rather than writing, so exploring artwork can reveal areas of concern, as well as be a useful tool for processing death experiences.

School-Age Children (Age 6–10)

During this age range is when we see most of the mature death concepts described above fully develop (Cotton & Range, 1990; O'Halloran & Altmaier, 1996; Poling & Evans, 2004). When asked about causality of death, school-age children may give examples of natural causes or heart attacks (Kenyon, 2001). Considering events such as the 9/11 attacks, we could imagine how a child's understanding of cause could become more complex across time going from recognizing that a plane crashed to describing how "bad guys" crashed the plane and ultimately to a more complex understanding of international events contributing to terrorism. Chapter 3 of this text offers further examples of children's developing death concepts.

During early childhood, parents often curate their children's experiences; however, as children venture into school, they become more exposed to media, peer influence, and non-shared experiences with their parents which means that parents may not be aware of or reliably recall which experiences their children have or have not had. This is illustrated in another exchange between a mother and son recorded at *The Mysterious Bog People Exhibit.*

M: *You've been to funerals right? Where people are in the…*
B: *I've never been to a funeral.*
M: *You've never been to a funeral?*
B: *No.*
M: *[pauses and looks at her son with disbelief]*
B: *No.*
M: *You've never been to a funeral?*
B: *Not that I can remember. If I was there as a baby then. I've never been to a funeral.*
M: *Like my Uncle Doug?*
B: *No.*
M: *Daddy's Uncle Jack?*
B: *No.*
M: *You weren't at those funerals?*
B: *No. I never went to a funeral.*
M: *Okay, well when people, um, when you see a person in a coffin they're usually dressed up kind of fancy, right?*
B: *Yeah, in their best.*
M: *So that's dressed up in their finest cloak for the ritual of passing.*

Adolescents (11–18)

Personal mortality is one of the last concepts to develop, resulting in the common phenomenon of adolescent invincibility (Kenyon, 2001; Speece & Brent, 1992; Zajanckauskaite, 2018). When discussing causality of death, younger adolescents may give spiritual examples like being called by God. Interestingly, adolescents begin to show a gradual *decrease* in their scores on ratings of a mature death concept, which continues into young adulthood. This decrease is thought to reflect the idea that as children mature, their death concepts become more abstract. Adolescents begin to incorporate religious beliefs and medical knowledge, which often is contrary to the more concrete notions of irreversibility and non-functionality (Despelder & Strickland, 2005; Kenyon, 2001). Biological depictions are most prevalent in children who are 12 and younger, while metaphysical depictions become more common in adolescence when children begin to have a more complex system of religious and philosophical thoughts.

For example, Israfilova and Khoo-Lattimore (2018) interviewed Azerbaijani students after their school trip to the Guba Genocide Memorial Complex, located in Azerbaijan. The two predominant cultural identities of the students were Muslim and Soviet culture. Students who identified as Muslim tended to have more religious views about death and the dark tourism site they were visiting. Those who identified with the Soviet culture had a more secular view. When asked about the victims of the genocide, one student said, "I believe the spirits of those who were killed in this dark event are in the heaven because they were innocent" (p. 71).

Adolescents are also better able to recognize and communicate about dual emotions evoked by death experiences and visits to dark tourism sites, as illustrated by the following verbatim excerpts of teenage visitors to the Flight 93 National Memorial in Shanksville, Pennsylvania, which honors the victims of a hijacked plane that crashed on September 11, 2001 (Croom, Squitiero, & Kerr, 2018):

- "Overall, it was depressing, but learning about the sad parts of our history is necessary." [2015, female].
- "I really liked it because it brings back all those loved ones who died on flight 93 it was really emotional. But I loved it." [2015, male]
- "I think it was cool we got the chance to visit were Flight 93 occurred, but also very sad because it was a very upsetting day for our nation. The people that lost there [sic] lives will never be forgotten and did a great part in our history and will never be forgotten." [2015, female]

Considering Death Concepts at Dark Tourism Sites – Visitor Experience and Research

As the importance of teaching children about death concepts is becoming clearer, society is creating new ways to teach parents and children how to communicate about death and dying. Some schools, especially in Europe, have begun formal training classes about death concepts for children that are conducted in the same format as sex education classes (Higgins, 1999; Jackson & Colwell, 2001). These formalized programs, as well as a number of books that have been published about talking to children about death, have paved the way for death to become a topic of open communication instead of something taboo. Children have responded positively to these programs and the vast majority have expressed interest in having open and honest discussions about death concepts as illustrated above.

As death becomes less taboo, dark tourism sites and museum exhibits focused on death-related themes, such as *Endings: An Exhibit about Death and Loss*, *Body Worlds*, and *The Mysterious Bog People* (Despelder & Strickland, 2005; Patterson, 2007; Walter, 2004), are becoming more popular destinations. Families and schools are more likely to incorporate these destinations into their itinerary bringing a higher frequency of young visitors to sites that are often designed for adults and often include graphic and visceral information and experiences (e.g., ability to hear recorded phone messages between victims and their loved ones from the moments before the Flight 93 crash on 9/11). Considering the experience of young visitors at these exhibits and sites will aid in their processing of death-related concepts as well as decrease the discomfort of parents, educators, and docents who facilitate the experience for young visitors. Open communication about death-related topics is vital because otherwise, children will be left with their emotions and fears about death without the tools to cope with and process those emotions.

Based on the first author's research and clinical experiences, as well as a review of the literature, we offer the following recommendations to facilitate exploration of these sites. First, it is important to offer a wide variety of interactive materials including opportunities for artwork, writing, and discussion so that children are able to access the materials that are best suited to their age range and personal preferences. Second, young visitors benefit from preparation prior to these visits, so that they have background information on the events that led to the site or exhibit, as well as knowledge of emotionally salient features of the site or exhibit (e.g., bones or skulls, crash site, burned artifacts). Young visitors also benefit from debriefing immediately following and several days after their visits as children process these experiences at different rates. These preparatory and debriefing conversations acknowledge to the child that these experiences are unique compared to other tourist sites that the families or schools may visit. Third, it could be helpful to provide educators and parents with prompts to discuss difficult concepts as well as helpful tips for communicating with children of different ages. Important skills to promote are how to ask questions in an open-ended format (e.g., "what questions do you have?" rather than "do you have any questions?"), how to answer only the question that has been asked and then prompt if they have additional questions, how to respond to and recognize emotion talk, and the importance of not using euphemisms (e.g., comparing death to being sick or asleep). It is also useful to encourage parents and educators to share their own emotions and reflections with the children and to discuss shared and non-shared personal experiences. This can help children integrate their personal experiences and normalize the emotions and responses that children are having. We want to caution parents and educators not to flood the young visitors with adult emotions (e.g., expecting them to help us process our emotions about the site, providing them with more information than they have asked for or want). Finally, it is important that young visitors have mechanisms to channel their emotions, which sites can provide by offering opportunities for memorialization (e.g., leaving a picture or item on a memorial wall or lighting a candle).

References

Cotton, C. R., & Range, L. M. (1990) Children's death concepts: Relationship to cognitive functioning, age, experience with death, fear of death, and hopelessness. *Journal of Clinical Child Psychology*, *19*(2): 123–127.

Cox, M., Garrett, E., & Graham, J. A. (2005) Death in Disney films: Implications for children's understanding of death. *OMEGA: The International Journal of Management Sciences*, *50*(4): 267–280.

Croom, A. R., Squitiero, C., & Kerr, M. M. (2018) Something so sad can be so beautiful: A qualitative study of adolescent experiences at a 9/11 memorial. *Visitor Studies*, *21*(2): 157–174.

DeSpelder, L. A., & Strickland, A. L. (2005) *The Last Dance: Encountering Death and Dying*. (7th edition). New York: McGraw-Hill Companies.

Graham, J. A., Yuhas, H., & Roman, J. L. (2018) Death and coping mechanisms in animated Disney movies: A content analysis of Disney films (1937–2003) and Disney/Pixar films (2003–2016). *Social Sciences*, 7(10): 199.

Higgins, S. (1999) Death education in the primary school. *International Journal of Children's Spirituality*, 4(1): 77–89.

Hunter, S. B., & Smith, D. E. (2008) Predictors of children's understandings of death: Age, cognitive ability, death experience and maternal communicative competence. *OMEGA-Journal of Death and Dying*, 57(2): 143–162.

Israfilova, F., & Khoo-Lattimore, C. (2018) Sad and violent but I enjoyed it: Children's engagement with dark tourism as an educational tool. *Tourism and Hospitality Research*. Epub ahead of print 4 August 2018. DOI: 10.1177/1467358418782736.

Jackson, M., & Colwell, J. (2001) Talking to children about death. *Mortality*, 6(3): 321–325.

Jay, S. M., Green, V., Johnson, S., Caldwell, S., & Nitschke, R. (1987) Differences in death concepts between children with cancer and physically healthy children. *Journal of Clinical Child Psychology*, 16(4): 301–306.

Kahn, S., & Liefooghe, A. (2014) Thanatos: Freudian manifestations of death at work. *Culture and Organization*, 20(1): 53–67.

Kenyon, B. L. (2001) Current research in children's conceptions of death: A critical review. *OMEGA: The International Journal of Management Sciences*, 43(1): 63–91.

Kerr, M. M., Stone, P. R., & Price, R. H. (2021) Young tourists' experiences at dark tourism sites: Towards a conceptual framework. *Tourist Studies*, 21(2): 198–218.

King-McKenzie, E. (2011) Death and dying in the curriculum of public schools: Is there a place? *Journal of Emerging Knowledge on Emerging Markets*, 3: 511–520.

Koocher, G. P. (1973) Childhood, death, and cognitive development. *Developmental Psychology*, 9(3): 369–375.

Lazar, A., & Torney-Purta, J. (1991) The development of the subconcepts of death in young children: A short-term longitudinal study. *Child Development*, 62(6): 1321–1333.

Longbottom, S., & Slaughter, V. (2018) Sources of children's knowledge about death and dying. *Philosophical Transactions of the Royal Society B: Biological Sciences*, 373(1754): 20170267.

Menendez, D., Hernandez, I. G., & Rosengren, K. S. (2020) Children's emerging understanding of death. *Child Development Perspectives*, 14(1): 55–60.

Nagy, M. H. (1948) The child's theories concerning death. *Journal of Genetic Psychology*, 73: 3–27.

Noppe, L. D., & Noppe, I. C. (1991) Dialectical themes in adolescent conceptions of death. *Journal of Adolescent Research*, 6(1): 28–42.

O'Halloran, C. M., & Altmaier, E. M. (1996) Awareness of death among children: Does a life-threatening illness alter the process of discovery? *Journal of Counseling and Development*, 74: 259–262.

Parsons, S. K., Saiki-Craighill, S., Mayer, D. K., Sullivan, A. M., Jeruss, S., Terrin, N., Tighiouart, H., Nakagawa, K., Iwata, Y., Hara, J., & Grier, H. E. (2007) Telling children and adolescents about their cancer diagnosis: Cross-cultural comparisons between pediatric oncologists in the US and Japan. *Psycho-Oncology: Journal of the Psychological, Social and Behavioral Dimensions of Cancer*, 16(1): 60–68.

Patterson, A. R. (2007) "Dad look, she's sleeping": Parent–Child conversations about human remains. *Visitor Studies*, *10*(1): 55–72.

Poling, D. A., & Evans, E. M. (2004) Are dinosaurs the rule or the exception? Developing concepts of death and extinction. *Cognitive Development*, *19*: 363–383.

Renaud, S. J., Engarhos, P., Schleifer, M., & Talwar, V. (2015) Children's earliest experiences with death: Circumstances, conversations, explanations, and parental satisfaction. *Infant and Child Development*, *24*(2): 157–174.

Rosengren, K. S., Gutiérrez, I. T., & Schein, S. S. (2014) IV. Cognitive dimensions of death in context. *Monographs of the Society for Research in Child Development*, *79*(1): 62–82.

Rushforth, H. (1999) Communicating with hospitalized children: Review and application of research pertaining to children's understanding of health and illness. *Journal of Child Psychology and Psychiatry*, *40*(5): 683–691.

Slaughter, V., & Lyons, M. (2003) Learning about life and death in early childhood. *Cognitive Psychology*, *46*: 1–30.

Speece, M. W., & Brent, S. B. (1992) The acquisition of a mature understanding of three components of the concept of death. *Death Studies*, *16*(3): 211–229.

Swain, H. L. (1979) Childhood views of death. *Death Education*, *2*(4): 341–358.

Tamm, M. E., & Granqvist, A. (1995) The meaning of death for children and adolescents: A phenomenographic study of drawings. *Death Studies*, *19*: 203–222.

Walter, T. (2004) Body worlds: Clinical detachment and anatomical awe. *Sociology of Health & Illness*, *26*(4): 464–488.

Walter, T. (2009) Dark tourism: Mediating between the dead and the living. In: Sharpley, R. and Stone, P. (eds) *The Darker Side of Travel: The Theory and Practice of Dark Tourism*. Bristol: Channel View Publications, pp. 39–55.

White, E., Elsom, B., & Prawat, R. (1978) Children's conceptions of death. *Child Development*, *49*(2): 307–310.

Zajanckauskaite, E. (2018) Decoding the adolescent brain. *The Lancet Child & Adolescent Health*, *2*(3): 169.

6 Young Tourists with Disabilities

Considerations and Challenges

Cristina Restrepo-Harner, Kristen Marsico, and Mary Margaret Kerr

When one considers the literature on children's development of understandings of death and dying, and those of children with intellectual disabilities, one understands that children with disabilities may possess different understandings of death and dying from those of their typically developing peers. Of interest to tourism managers is the fact that children with intellectual or emotional disabilities may have distressing or confusing experiences at a dark tourism site. That may lead their families to leave or altogether avoid these destinations. Therefore, this chapter will first review the small amount of literature on child tourists with disabilities or mental health conditions, highlighting issues especially relevant for those managing dark tourism sites. Also, case studies that reflect child and family travel experiences will be presented. Rarely studied online travel reviews from families at dark tourism sites worldwide are also used to illustrate industry challenges and solutions to address them.

> The quality I love most about my daughter is her inquisitiveness. Not only do her teachers and me see it, but everybody we interact with. When we are on a museum tour, she is the first person to raise their hand when the guide asks if there are any questions. Her mind, without fail, thinks of at least ten questions that never even crossed my mind. Her insight brings a whole new perspective.
>
> *– Mother of a 9-year-old child with autism*

This chapter invites the reader to consider the needs of children and young people with disabilities as they travel, especially to dark tourism sites. While tourism may improve the physical and mental health of many children and reinvigorate family relationships (Kim & Lehto, 2013; McAvoy, Rynders, Smith, Scholl, Newman, & Holman, 2003), recent studies expose the challenges facing families traveling with children with disabilities or children with special needs resulting from an illness or injury (Brewster & Coleyshaw, 2011; Freund et al., 2019; Williams & Aaker, 2002).

DOI: 10.4324/9781003032199-8

The first barrier facing young people with special needs is exclusion from travel altogether. After being refused by a travel agency, a young woman with cancer observed, "there you are trying to feel normal and do normal things that everyone else takes for granted and all of a sudden it's [cancer] thrown back in your face all over again" (Hunter-Jones, 2004, p. 255).

In a U.S. survey of nearly 900 parents, researchers uncovered prevalent patterns of exclusion from school trips owing to disabilities (Wynarczuk et al., 2020a). Nearly one-third of parents reported that their child was not allowed on a school trip (despite federal laws prohibiting their exclusion). Lack of planning emerged as a major barrier, as children's needs received only last-minute acknowledgment. Parents reported calls the night before a trip asking them to keep their children at home. This practice echoes the notion that persons with disabilities are "invisible" members of society (Aslam, 2013).

Once children do arrive at a destination, they may face ostracizing stares or comments from other tourists, because they cannot conform to presumed norms for behavior, communication, or emotional expression (Sedgley et al., 2017). This uncomfortable situation may be more prevalent at memorials, where many expect even young children to exhibit "respectful" behavior (Price and Kerr, 2018). Such encounters no doubt contribute to recent reports that travel can negatively influence family dynamics, with increased anxiety and tension for parents (Freund et al., 2019).

Moreover, children with disabilities may possess different understandings of death and dying from those of their typically developing peers, becoming distressed or confused (Kerr et al., 2021). In the event that disabilities result from illness or a life-altering accident, young tourists (and their families) may face grim reflections of their own conditions when they encounter representations of death and human suffering at a dark site (Stone et al., 2018). For example, the Thackray Museum of Medicine (UK) offers "a reconstruction of 1842 surgery, before anaesthetics were in use. Visitors watch as 11-year old Hannah undergoes amputation of her leg after it was crushed in a mill accident" (https://museu.ms/museum/details/485/thackray-medical-museum).

In order to create more welcoming environments, one must first understand what these young tourists and their families need. Therefore, we begin by clarifying disabilities and accompanying challenges, before moving to the specific context of dark tourism sites. We conclude with suggestions for destination managers and interpreters.

Types of Disabilities and Possible Accommodations

Children with disabilities face additional challenges when visiting tourism sites. While there are many types of disabilities, this chapter will primarily focus on six categories of disability: Physical, visual, hearing, mental/

behavioral, intellectual, and sensory/processing. Some child tourists will have one disability, while others will have more than one. A child's type of disability, as well as the severity of difficulties, may dictate barriers to travel and tourism. However, with guidance from advocates for inclusion and legal authorities, sites across the world now offer more accommodations to increase participation by tourists (United Nations World Tourism Organization, 2017). The section below provides a general overview of each of the six categories of disability, potential challenges, and accommodations that can make tourism and travel sites more accessible.

Physical disabilities – those which limit one's physical functioning, mobility, dexterity, or stamina – can make traveling and accessing tourism sites particularly challenging. Youth with physical disabilities may require a walker, wheelchair, or other devices to move around. Most tourism sites have legal obligations to provide and maintain accessibility to visitors with disabilities. This includes physical access to the site, such as providing accessible entrances (e.g., ramps and power-operated doors) and routes throughout the site, wide paths, and ticket lines that are wheelchair accessible (U.S. Department of Justice, n.d.). To make content more accessible for youth with physical disabilities, sites typically offer *some* content lowered to eye-level for those in wheelchairs. However, while some efforts are made to increase the accessibility of tourism sites to those with physical disabilities, it is common for the majority of the site to be inaccessible. For example, gift shops often have narrow aisles, and counters are typically designed for visitors who are standing (Poria et al., 2008). Notably, some individuals with physical disabilities have "invisible" difficulties which cannot be outwardly seen, such as chronic pain or fatigue.

Visual impairment refers to decreased sight abilities that may not be easily remedied by corrective measures such as glasses. Visual impairments may be partial or total (e.g., full blindness). Under accessibility guidelines and recommendations, public spaces should reduce bumping hazards such as crowded entryways or pathways that can be difficult for individuals with canes to navigate. Additional accommodations might include alternate formats of materials (e.g., Braille, large print, audio tours) and increased opportunities for tactile experiences.

Hearing impairments or hearing loss can be permanent or fluctuating and can be partial or total (e.g., deafness, hearing loss in one ear). Tourism sites can offer accommodations such as sign language interpreters, guided tours, subtitles on videos, and increased visual material to increase positive experiences of visitors with hearing impairments.

Mental health conditions create distress and/or disruption in daily functioning. Roughly one in five youth between the ages of 12 and 18 experiences a mental health disorder (National Alliance on Mental Illness, 2017). Children with mental and emotional challenges may be more easily overwhelmed at travel sites and are likely to be more sensitive to difficult materials. For example, youth with anxiety may avoid reminders of death and dying or become overly focused on thoughts of tragedy striking their own families.

Travel itself can cause anxiety. As McIntosh (2020) observed in her poignant account of travelers with epilepsy:

> For people with disabilities, emotions and anxiety are similarly found to amplify during travel due to the unfamiliarity of the tourism environment (Devile & Kastenholz, 2018; Lehto et al., 2018; Sedgley et al., 2017; Small, 2015). These emotions are compounded by the social discrimination of their disability (Darcy, 1998), yet remain in contrast to the more commonly reported positive emotional stimulation and hedonism of travel.
>
> (Krippendorf, 1987a, 1987b) (p.7)

Youth with other behavioral difficulties such as attention deficits or high levels of hyperactivity may have significant challenges sitting still, speaking quietly, and ignoring impulses to touch materials or yell out their thoughts. These challenges can be particularly disruptive at dark tourism sites, such as memorials, where respect is shown through quietness and stillness. Sites infrequently provide specific accommodations for individuals with mental health challenges, instead of focusing on physical disabilities and general accessibility measures (Freund et al., 2019). Although efforts such as using trigger warnings for individuals with sensitivities or post-traumatic stress disorder (PTSD) are becoming more common (Frost, 2017), they tend to serve adults.

Intellectual disabilities are those that limit a child's reasoning, learning, and problem-solving, as well as adaptive behaviors such as everyday social, communication, and daily living skills (American Psychiatric Association, 2013). For some youth, an intellectual disability is their primary diagnosis, while others have limited intellectual disabilities as a result or co-occurring symptom of another disability. Children with intellectual disabilities have cognitive abilities well below their biological age and can have difficulty understanding exhibits, tours, and materials specifically designed for most children. Depending on the extent of the intellectual disability, the child may not be able to speak. Material that does not rely on abstraction and is simplified, repeated multiple times, and provided in multiple formats (e.g., oral, visual, and graphically) is better understood by individuals with intellectual disabilities (Gillovic et al., 2018).

Lastly, sensory and processing disabilities such as the neurodevelopmental disorder, autism spectrum disorder, or traumatic brain injury can pose significant challenges. Children with these disabilities are often sensitive to input such as light, noise, and touch, leading to discomfort and interruption to daily functioning and emotional–behavioral patterns (Miller et al., 2007). Tourism sites can be over-stimulating for visitors with sensory and processing disabilities due to bright lights, loud sounds, and big crowds (Langa et al., 2013). Autism may also cause youth to react to disruptions in routines or a preferred activity, such as being told to abruptly move from one museum exhibit or activity to another, which may present challenges for youth with autism (American Psychiatric Association, 2013). Some aspects of tourism can make youth with sensory disabilities feel uncomfortable or unwelcome, such as feeling like security guards are watching them or the presentation of novel ("weird") exhibits (Langa et al., 2013). Recommendations include the creation of autism-friendly environments with well-trained staff and clear guidance regarding how to support children with autism (Amet, 2013).

While this chapter primarily focuses on the needs of child tourists with disabilities, it is important to recognize that these young visitors also bring many assets to tourism sites, as demonstrated in the opening quote. For example, youth with attention deficits are often energetic, enthusiastic, and excited to explore new environments. Youth with limited social skills are often truthful, passionate, and ask important questions. Young tourists who have vision or hearing impairments find ways to explore materials in novel ways and can provide excellent insight into their sensory experience of sites. There are endless strengths that children with disabilities bring with them to tourism sites, and sites should strive to not only provide accommodations to help these visitors access the materials, but to find ways to highlight these strengths.

Table 6.1 outlines some of the most prevalent difficulties in each of the six categories discussed above as well as methods for increasing accessibility of tourism sites.

While it is helpful to be aware of specific disabilities and accommodations that address common challenges, sites may instead opt to implement general design principles that are helpful for visitors with a variety of ability levels. Many of the accommodations above are components of "universal design," meaning designs that can be used and experienced by people of all ages and abilities to the greatest extent possible and are fully integrated into the environment (Story, 1998).

Universal Design Principles: A Brief Overview

Universal design methods promote equitable, flexible, and simple use of space and content and encourage engagement for all visitors, including those with and without disabilities (Rappolt-Schlichtmann & Daley, 2013; Stringer, 2014). These approaches include effective communication of information, minimal hazards, low physical effort, and appropriate size and space for individuals to interact with the materials or environment regardless of body size, posture, or mobility (Story, 1998).

In her text, "Programming for People with Special Needs: A Guide for Museums and Historic Sites," Stringer outlines seven key elements to the creation of programs for audiences with disabilities at historic sites and museums. Key elements include measures to take before, during, and after the visit serving as a model for sufficient planning and tourist engagement across all aspects of the tourist experience. Destination managers may use these key elements, aligned with Universal Design, to plan an inclusive experience for all tourists. Key components include (1) sensitivity and awareness training, (2) planning and communication, (3) timing, (4) engagement, (5) object-centered

Table 6.1 Overview of Disabilities, Challenges, and Accommodations

Disability	Possible Challenges	Tourism/Travel Accommodations
Physical disabilities and mobility impairments	– Walking, standing, or specific physical activities – Walkers, canes, wheelchairs, and other devices may require additional space – Wheelchair use can make it difficult to see exhibits	– Free wheelchair rentals – Accessibility map with entrances, parking, etc. – Clear, open entryways and pathways – Ramps and alternatives to stairs (e.g., elevators, escalators) – Content provided at multiple heights
Visual impairments	– Partial or complete inability to see – Reading small font or seeing small objects	– Braille content – Large font – Audio or guided tours – Increased tactile opportunities
Hearing impairments	– Partial or complete inability to hear – Sound discrimination in loud places	– Assistive listening devices – Closed captioning or subtitles – Sound amplification – Transcripts of audio
Behavioral difficulties *Mental health disability*	– Sitting still – Paying attention for extended periods of time – Impulsivity (e.g., touching items not meant to be touched, blurting out inappropriate comments) – Frequent movement – Tantrums – Anger, defiance (e.g., refusal to follow rules), arguing	– Areas where movement and standing are less disruptive – Areas for moving around and speaking loudly – Short learning activities and engaging, technology-based content – Clear rules and expectations provided in multiple formats (e.g., signs, oral reminders) – Room for movement and breaks
Emotional difficulties *Mental health disability*	– Intense fears (generalized or specific) – Regulating emotions and thoughts – Inappropriate emotions or behaviors	– Clear warnings of graphic exhibits – Easily accessible exits in disturbing or distressing areas – Rooms for calming down – Post-visit debriefs
Intellectual disability/delay	– Understanding complex or abstract ideas – Communication (receptive and expressive) – Tasks of daily living, including caring for one's self	– Simplified materials – Materials provided in multiple formats – Small group or individual workshops and hands-on activities

(Continued)

Table 6.1 (Continued)

Disability	Possible Challenges	Tourism/Travel Accommodations
Autism Spectrum Disorder *Sensory processing disability*	– Understanding emotional concepts – Changes in routine or moving from one activity to another – Communication (receptive and expressive) – Appropriate social interactions – Restricted interests – Repetitive behaviors – Sensitivity to noise, light, touch	– Stories that tell children what to expect before their visit (see https://www.sbbh.pitt.edu/families/travel-guides-families) – Sensory rooms with less input or quiet reflection rooms (e.g., 9/11 Memorial & Museum) – Exhibits with decreased noise and light – Quiet times to visit the site – Noise-reducing headphones

and inquiry-based design, (6) structure, and (7) flexibility. The following sections draw from Universal Design and Stringer's work to offer specific recommendations for children, young people, and dark tourism.

Sensitivity and awareness training expands to all personnel who support operations on site, including security guards, cashiers, and greeters (Stringer, 2014). All tourists are worthy of respect and have a right to participate in the tourism experience. Throughout the entire tourist experience, the staff members on site should be equipped with the tools and knowledge to contribute to an accessible environment. Sensitivity training is inclusive of learning about disabilities and the barriers tourists with disabilities may encounter, understanding preferred terminology (person-first or identity-first language), and general etiquette in supporting all tourists. For example, if a young tourist using a wheelchair is attempting to navigate a space, one should ask before offering assistance. Wheelchairs provide an individual with mobility and independence. A wheelchair is their space, so one should not touch their wheelchair unless asked. Of special interest to those in dark tourism sites, disability laws in many countries require that a certain proportion of artifacts be displayed at wheelchair level. However, one should be sensitive to the fact that these artifacts may be too grim for a very young tourist. It may be useful to check with their family member before showing them a particular exhibit.

The next component, effective planning, involves the tourist site communicating with tourists prior to the arrival. For example, if a school trip is scheduled to arrive, destination managers may contact the group prior to their arrival. By contacting the group proactively, they can identify teacher and student needs, as well as discuss specific experiences tourists may encounter. For example, if a tourist site has a simulation room consisting of loud noises and other overwhelming sensory input, this can be communicated with the group prior to their arrival and plans for appropriate accommodations may begin. Other methods to allow for effective planning may include an inquiry or contact form that tourists with disabilities may use prior to their visit, allowing for open communication between the tourist site and the visitor.

Following effective planning is effective timing. Effective timing notes that each interpretation segment should be no longer than thirty minutes to allow for questions and inquiry, but not prolonged such that visitors lose interest (Stringer, 2014). It is also important to consider that the most appropriate length of time in an exhibit will differ based on individual needs, so remaining flexible is essential in planning. To illustrate, a visitor at the Museum Auschwitz-Birkenau Państwowe Muzeum notes that the timing and pace of the tour was essential to a positive experience for a visitor with autism,

> This time I was returning with a young grandchild at her request, – she became interested in the war after the poppy exhibition at Tower Hill in London – with some trepidation (especially as she is autistic). We went on a guided tour that, again I was a bit worried about in advance as her attention span is very limited. However all went really well. The guide was informative and experienced and moved us along at a good pace.
> *Trip Advisor Review, Maria T. September 14, 2015 "thought provoking"*

When planning for effective timing, destination managers should consider the time of day and how this may affect young tourists. Peak hours may present more challenges, while slower times may allow for increased accommodations and fewer barriers for visitors with disabilities.

The fourth component focuses on engagement through differentiation and a holistic tourist experience. As Stringer (2014) notes in her research, many special education teachers emphasize that the overall experience, extending beyond the educational content, is highly significant. Teachers value broader life experiences for their students, including both social interactions: How are students engaging with the staff on site? Is the museum docent welcoming and open to converse with a student with disabilities? How is the student navigating the new environment and generalizing skills such as asking for help, asking where to find the restroom, or finding a space to take a break? For students with disabilities, the value of a visit goes well beyond the content presented in the museum.

Engagement leads to Stringer's fifth component, object-centered and inquiry-based experiences. It is recommended that tourism sites create multimodal, sensory-based experiences, incorporating primary sources, photographs, artifacts, and spaces as young tourists navigate the content. Exhibits that allow for touch provide opportunities for tourists to further explore and find meaning. When a multimodal approach is not taken into account, the tourist experience may be less informative, valuable, and accessible. As a parent observed in an online review of a dark tourism site,

> Not for small kids. My 8 yr old is interested in science, so we decided to explore the museum of medicine. Unfortunately, the museum turned out not to be designed for kids. It had not hands-on or interactive exhibits, and explorations are geared toward older students and adults. The museum is small and quiet, no tours or anything of this kind.
>
> *Trip Advisor, Julia B. July 16, 2013 National Museum of Health and*
> *Medicine, Silver Spring, Maryland, USA*

The sixth element focuses on framework and planning via structure. Structure creates a clear plan and sequence of events, which may ease anxiety for visitors in a new setting and experience. Moreover, structure allows for the accomplishment of educational goals and a clear path of what the young tourist will experience. Museums may provide an additional level of structure by providing an outline of suggested activities, as well as the timing for each activity, upon a tourist's arrival. The outline can be easily adapted to meet the needs of each tourist. For example, an itinerary template with a space for five to ten write-in activities, accompanied by a site map is a resource that leads to a more accessible and structured tourist experience. While structure is important, tourism sites must also be flexible and able to adapt to a group's interests or abilities.

Thus, flexibility becomes the final component in Stringer's seven key elements. Taking into account the reality of site funding, staffing limitations, and the changing nature of a destination manager's responsibilities, flexibility is key when supporting young tourists with disabilities. As Stringer (2014) notes, rather than recreating entire programs, staff can begin by adapting elements of the programs they already have in place to create a better, more inclusive tourist experience. For example, a no-cost accommodation is to ask children if they would like to revisit their favorite exhibit. Often, young children and children with developmental delays want to return and spend more time in exhibits they enjoy rather than moving on to see every exhibit at the site, as adults might want to do.

On the other hand, children may need to skip exhibits due to sensitive or frightening content. Consider the experience of a mother at the Mütter Museum,

> While I was busy oohing and aahing over the wax models of faces devoured by syphilis and ovarian cysts the size of Montana, my 11 year old daughter was cowering in the corner, hugging her knees to her chest, and rocking back and forth while repeating affirmations of health and safety. I think the fetuses in the jars put her over the edge.
>
> *–Trip Advisor, Julia G. "My Daughter is Still Having Nightmares"*
> *April 9, 2013*

Evaluating Site Accessibility

Tourism sites are encouraged to assess their accessibility for visitors with disabilities (Stringer, 2014). This allows them to better understand the needs of their visitors, plan for possible adaptations or accommodations, and advertise accessibility efforts. Site managers can use tools such as the School Trip Accommodations Tool-Site Form to examine mobility, communication, and sensory considerations for visitors with disabilities (Wynarczuk, et al., 2020b). For example, this form asks site managers to answer questions about the frequency of encountering stairs or loud noises and provide information about where these may be encountered and what, if any, accommodations are available. Similarly, the Self-Assessment Accessibility Survey includes questions for tourism sites about the experiences, prevalence, and needs of visitors and staff with disabilities (Art Beyond Sight/Art Education for the Blind, Inc., 2011).

Sites might also become credentialed to host these visitors. For example, the Autism Travel Network highlights the International Board of Credentialing and Continuing Education Standards (IBCCES). IBCCES offers varying degrees of certification that reflect an organization's competency to support individuals with autism or other cognitive disorders. Obtaining and advertising this certification can demonstrate to parents that sites have taken steps to make the physical space and content accessible to their children with disabilities.

Other forms of evaluation typical in tourism are visitor surveys and interviews, which should, but rarely do, include young tourists or tourists with disabilities. Chapter 17 offers guidance. Comment cards provide a platform for young tourists to share their voice and experience, while also informing the tourist site of adaptations they could create to support future tourists. Inclusive research and planning should include the voices and feedback of those with disabilities via thoughtful and flexible research models (Gillovic et al., 2018). Semi-structured interviews with adaptations for a variety of communication methods provide a starting point for tourism sites to gain feedback from the lived experiences of tourists with disabilities. Adapted materials may include visual response cards to facilitate communication (e.g., picture responses) or opportunities for the young tourist to draw their reflection, interpretation, or questions as they navigate the site. Effective tourism research should continue to improve and allow for the true inclusion of the feedback and voices of tourists with disabilities (Gillovic et al., 2018).

Lastly, families offer their own evaluations through online travel sites (Divaker and Kerr, in press). Parents and individuals with disabilities have taken to virtual platforms to plan and design meaningful trips with feedback from others who plan for similar barriers when traveling. Trip Advisor, a travel site with an international audience, hosts multiple parent forums focused on travelers with disabilities as well as open comment threads on individual dark tourist sites. Dark tourism destination managers can review comments on feedback forums. For example, wheelchairtravel.org hosts feedback from tourists on the physical accessibility of a site.

Let us illustrate this assessment approach. To examine tourist feedback for the prior section, we first identified a list of dark sites (Forbes, 2019), then conducted keyword searches for each site, using *disabilities/disability/disabled, family, autism/autistic, sensory*, and *children/child*. Among the 59 reviews we found, common themes included accommodations provided by the site, the level of accessibility, and the appropriateness of the material presented. Common challenges addressed by families of those with disabilities include lack of access to the entire site, exhibits that lead to sensory overload, and inappropriate or lack of modified content. We now turn to specific recommendations for addressing these challenges at dark tourism sites.

Supporting Young Tourists with Disabilities at Dark Tourism Sites

Children with disabilities may encounter unique challenges and barriers when visiting dark tourism sites. For many families, these challenges may determine the sites they choose to visit, as well as their experience at the site. For example, consider these comments:

- "I would NOT recommend taking children with a limited attention span to this museum; you'll want to see the museum, but will be frustrated that you can't because of them. Trust me, I found this out first hand, when I took my teenager with developmental disabilities, even though he's the one who wanted to go." – NYC 9/11 Memorial and Museum Trip Advisor Review by Joan K, December 2017.
- "Museum is incredibly informative, thorough, and well done. My 14 year old with intellectual delays was mesmerize, just like our 21 year old and my husband and I. There is wonderful mix of multi-media materials for all ages!" – NYC Memorial Trip Advisor Review by middB, January 2016.
- "Our daughter has an autism spectrum disorder and is sensitive to sound, some of the noises on the audio tour frightened her a little bit, but they are few and far between, and she still managed to cope. The views of San Francisco from the Island are amazing too!" – Alcatraz Federal Penitentiary Trip Advisor Review by HolidayHappy6, January 2014.
- "It is Accessible (we have a child in a wheelchair), but they make you work for it! Seriously – the entrance is around the back and in order to navigate the floors you have to find a guard to let you into a secret passage to gain access to the elevator." – Mütter Museum Trip Advisor Review by June_Boys, November 2016.
- "I will keep this short BC I don't want to take away from your experience here. BUT I took my 6 year old add nephew and 8 year old autistic niece. Crazy right?? Not at all. They were very intrigued and there was a lot to keep them occupied." – Oklahoma City Memorial and Museum Trip Advisor Review by Bellyfool, April 2015.

What follows are more detailed recommendations drawn from the broader literature to address challenges and support an accessible experience for all tourists at dark tourism sites. We begin with pre-visit resources.

Pre-Visit Resources

Before a young person with a disability visits a dark tourist site, preparation is essential (Wynarczuk et al., 2020a). Preparation eases children's anxiety about an unknown situation and allows families and teachers to determine the appropriateness of the site. Dark tourism sites can support meaningful visits by providing online information that meets international Web accessibility standards (see Web Accessibility Laws & Policies | Web Accessibility Initiative (WAI) | W3C). Useful content includes a comprehensive overview of the material covered, behavioral expectations, and spaces where children can go if they feel overwhelmed or need to move around. Examples include frequent exits at distressing sites or a standing section for watching videos or performances. For example, the Lisboa Story Centre in Lisbon, Portugal shares the story of the Earthquake of 1755 via a video and sensory simulation in graphic detail of the day's events. Prior to entering this part of the exhibit, an audio message is played via a self-guided headset informing visitors of the content and providing a window of time for visitors to skip the video and simulation.

Several dark tourism sites (e.g., Auschwitz-Birkenau State Museum, Srebrenica Genocide Memorial) offer virtual tours. These help families and teachers understand what the visit will entail. They also help children "practice" their visit and communicate what they would like to see or avoid. Virtual tours should include images of visitors with disabilities when possible. A tourism site may choose to show exhibits through the point of view of someone moving through the spaces and engaging with interactive elements. If shared on the destination's website, families and teachers can then use the video model to prepare children. Video modeling is an effective method for teaching an array of tasks and activities (Bellini & Akullian, 2007). Tourist sites can supplement these virtual tours by providing a travel guide or FAQ on their website, informed by gathering feedback from tourist comments (Kerr and Price, 2018).

Many young tourists and their families will have familiarity with "Social Stories™" (Gray, 2015; Williams and Wright, 2016), which present brief child-friendly information and steps the child can take to enjoy a visit. Stories written for a child walk them through familiar coping steps ("If I feel scared, I can hold Mommy's hand."), and remind them of the rules ("I need to keep my hands in my pockets so I don't bump anything." A destination manager may choose to develop a Social Story™ or similar resource featuring some of the more sensitive content a tourist may encounter while on site.

The Museum of London Docklands provides a photographic story, depicting what young tourists will encounter on their trip in the Museum Accessibility tab on their website. Within this story, tourists can expect photos and descriptions of what they will interact with throughout their tourist experience. The story offers suggestions for accommodations, as well. In the excerpt "Sailortown," tourists are advised to skip a specific area of the tour and to inform a host they would prefer a different route to avoid the gallery if an exhibit is not preferred.

On-Site Supports

Destination managers can enhance young tourists' experiences through adapted interpretation, such as those included in Table 6.1. Adapted materials of this nature are helpful for young children as well as youth with intellectual disabilities. Chapter 7 of this volume describes dark tourism interpretation examples and recommendations.

For exhibits displaying graphic images, video, or sound, site managers should consider easily accessible exits for distressed young tourists. Some sites may place all dark exhibits on one floor or in one area so that families can plan their tours in other spaces, if needed. For example, the Imperial War Museum becomes increasingly dark as visitors ascend the floors, with exhibits on the Holocaust and crimes against humanity on the upper floors. Additionally, to help families, site managers may provide notices about graphic or disturbing information and exhibits. When tour guides are available, they should tell visitors what to expect on the tour and warn them of possibly distressing content. For example, prior to entering an exhibit at The Museum of London Docklands, the wall reads "This area of the exhibition contains skeletal remains of babies and young children." A clear warning allows for visitors to reroute their tour, and allows caregivers the opportunity to provide verbal preparation and modify the experience for a younger visitor.

Hands-on learning activities are beneficial for many youth with disabilities as they provide additional opportunities to interact with and understand the

Sailortown

This gallery shows you what a street would have been like in London's Docklands in the 1840s. It can be dark and smelly in here.

Find out what animals were sold in animal emporiums 150 years ago. Look through the window of this shop and see if you can find models of a camel, a monkey, and a parrot.

You can walk through narrow, dark streets and listen to the sounds of 1840s' London. If you would rather not walk through Sailortown, let a host know and they will take you on a different route to avoid the gallery.

museumoflondon.org.uk

Figure 6.1 Excerpt from Museum of London Docklands Visitor Guidebook.

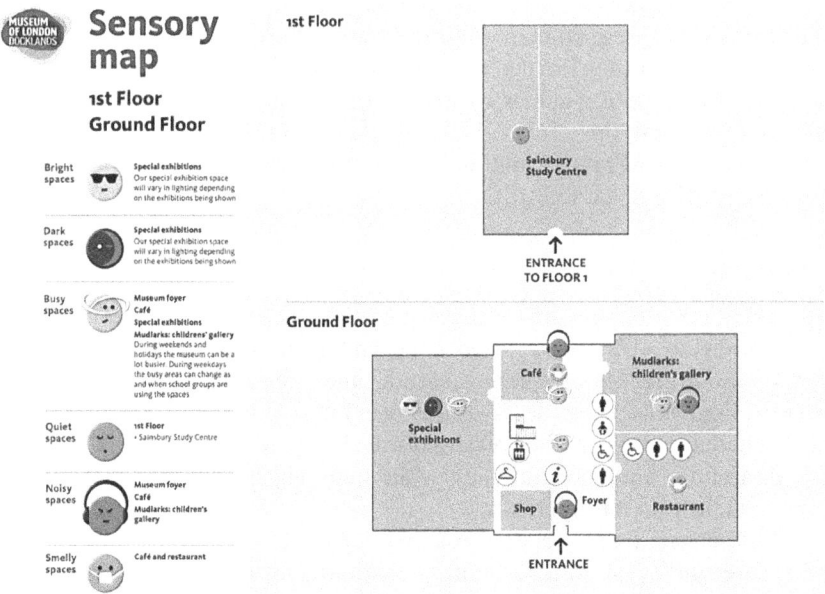

Figure 6.2 Museum of London Docklands Sensory Map.

material. Specifically, tactile experiences allow youth with sensory and intellectual disabilities to interact with museum materials in new ways. For example, the Mütter Museum in Philadelphia (U.S.) offers sensory-friendly events which allow children to touch certain objects, participate in a guided tour, or take a break in a quiet room. Site managers should carefully consider which materials can be and should be touched – both in regard to the fragility and safety of items as well as the developmental appropriateness of them. For example, it may not be appropriate for youth with disabilities to hold weapons, even if they have been modified for safety.

Post-Visit Supports

Young tourists' learning does not end when they leave. Destination managers can provide materials designed to facilitate the learning process following the visit. Materials provided to families and school groups may include written guides with reflection questions. Reflection questions can guide conversation beyond the walls of the exhibit. The Immigration Museum in Victoria, Australia offers post-visit reflection questions for teachers to pose to their class in an easily accessible format on their museum website (https://museumsvictoria.com.au/immigrationmuseum/). The suggested reflection questions include "At the museum I thought we would... and we....," "Before I went to the museum I... now I..." Additionally, the museum references a guide the students received while on site containing follow-up activities to continue the learning and reflection in their classrooms.

The Illinois Holocaust Museum and Education Center offers post-visit reflection activities and student programs centered around sharing their voice and applying what they learned to make the world a better place. Some dark tourism sites sell books, games, or toys that allow young visitors to continue reflecting on what they learned during their visit. After all, exploration through play is how many children learn and process new experiences (Kerr et al., 2021; Price & Kerr, 2018).

Conclusion

Although it is widely recognized that young tourists with disabilities face barriers worldwide, their needs have gone largely unnoticed in the tourism research literature. Nevertheless, museums and sites around the world have provided innovative examples to guide us in improving their tourism experiences. Therefore, this chapter reviewed the challenges facing young tourists with disabilities and their families, while highlighting accommodations to improve their tourism experiences.

As Shiraani and Carr (2021) observed, "While related fields (e.g., sports and leisure studies) show a substantial commitment to engage with disabled children, the same cannot be said for tourism…" (p. xx). Dark tourism research has rarely mentioned the experience of young tourists with disabilities, making this a priority for future research (Hunter-Jones, 2004). Future research should use an inclusive model, taking into account the overlooked voices of those with disabilities and designing studies to ensure that they can teach us what they know.

References

American Psychiatric Association, (2013). *Diagnostic and Statistical Manual of Mental Disorders* (5th ed.). Arlington, VA: Author.

Amet, L., (2013). Holiday, what holiday? Vacation experiences of children with autism and their families. *Autism*, 3(3): 123.

Art Beyond Sight/Art Education for the Blind, Inc., (2011). Self-Assessment Accessibility Survey. Retrieved from: http://www.artbeyondsight.org/mei/wp-content/uploads/Initial-Accessibility-Survey-2011.pdf

Aslam, A., (2013). *The State of the World's Children 2013: Children with Disabilities*. United Nations Children's Fund (UNICEF).

Bellini, S., & Akullian, J., (2007). A meta-analysis of video modeling and video self-modeling interventions for children and adolescents with autism spectrum disorders. *Exceptional Children*, 73(3): 264–287.

Brewster, S., & Coleyshaw, L., (2011). Participation or exclusion? Perspectives of pupils with autistic spectrum disorders on their participation in leisure activities. *British Journal of Learning Disabilities*, 39(4): 284–291.

Darcy, S., (1998). People with a disability and tourism: A bibliography. Online Bibliography, 7.

Devile, E., & Kastenholz, E., (2018). Accessible tourism experiences: The voice of people with visual disabilities. *Journal of Policy Research in Tourism, Leisure and Events*, 10(3): 265–285.

Forbes,(2019).https://www.forbes.com/sites/duncanmadden/2019/09/25/dark-tourism-eight-of-the-worlds-most-gruesome-tourist-attractions/?sh=6130e3613590

Freund, D., Cerdan Chiscano, M., Hernandez-Maskivker, G., Guix, M., Iñesta, A., & Castelló, M., (2019). Enhancing the hospitality customer experience of families with children on the autism spectrum disorder. *International Journal of Tourism Research*, 21(5): 606–614. https://doi.org/10.1002/jtr.2284

Frost, L., (2017). No safe spaces: Notes on the national september 11 museum. *Journal of Urban Cultural Studies*, 4(1): 221–239. https://doi.org/10.1386/jucs.4.1-2.221_1

Gillovic, B., McIntosh, A., Cockburn-Wootten, C., & Darcy, S., (2018). Having a voice in inclusive tourism research. *Annals of Tourism Research*, 71(C): 54–56

Gray, C., (2015). *The New Social Story Book: 15th Anniversary Edition*. Arlington, TX: Sensory World/Future Horizons.

Hunter-Jones, P., 2004. Young people, holiday-taking and cancer—an exploratory analysis. *Tourism Management*, 25(2): 249–258.

Kerr, M.M., & Price, R.H., (2018). "I know the plane crashed": Children's perspectives in dark tourism. In: Stone, P., Hartmann, R., Seaton, T., et al. (eds) *The Palgrave Handbook of Dark Tourism Studies*. London: Palgrave Macmillan, pp.553–583.

Kerr, M.M., Stone, P.R., & Price, R.H., (2021). Young tourists' experiences at dark tourism sites: Towards a conceptual framework. *Tourist Studies*, 21(2): 198–218.

Kim, S., & Lehto, X.Y., (2013). Travel by families with children possessing disabilities: Motives and activities. *Tourism Management*, 37: 13–24.

Krippendorf, J., (1987a). Ecological approach to tourism marketing. *Tourism Management*, 8(2): 174–176.

Krippendorf, E., (1987b). The dominance of American approaches in international relations. *Millennium*, 16(2): 207–214.

Langa, L.A., Monaco, P., Subramaniam, M., Jaeger, P.T., Shanahan, K., & Ziebarth, B., (2013). Improving the museum experiences of children with autism spectrum disorders and their families: An exploratory examination of their motivations and needs and using web-based resources to meet them. *Curator: The Museum Journal*, 56(3): 323–335. https://doi.org/10.1111/cura.12031

Lehto, X., Luo, W., Miao, L., & Ghiselli, R.F., (2018). Shared tourism experience of individuals with disabilities and their caregivers. *Journal of Destination Marketing & Management*, 8: 185–193.

McAvoy, L., Rynders, J., Smith, J., Scholl, K., Newman, J., & Holman, T., (2003). Inclusive outdoor adventure programming: A training manual. Unpublished manuscript. University of Minnesota, Minneapolis.

McIntosh, A.J., (2020). The hidden side of travel: Epilepsy and tourism. *Annals of Tourism Research*, 81: 102856.

Miller, L.J., Anzalone, M.E., Lane, S.J., Cermak, S.A., & Osten, E.T., (2007). Concept evolution in sensory integration: A proposed nosology for diagnosis. *American Journal of Occupational Therapy*, 61: 135–140.

National Alliance on Mental Illness, (2017). *Mental Health by the Numbers*. Retrieved from: https://www.nami.org/learn-more/mental-health-by-the-numbers.

Poria, Y., Reichel, A., & Brandt, Y., (2008). People with disabilities visit art museums: An exploratory study of obstacles and difficulties. *Journal of Heritage Tourism*, 4(2): 117–129.

Rappolt-Schlichtmann, G., & Daley, S.G., (2013). Providing access to engagement in learning: The Potential of universal design for learning in museum design. *Curator*, 56: 307–321.

Price, R.H., & Kerr, M.M., (2018). Child's play at war memorials: Insights from a social media debate. *Journal of Heritage Tourism*, 13(2): 167–180.

Sedgley, D., Pritchard, A., Morgan, N., & Hanna, P., (2017). Tourism and autism: Journeys of mixed emotions. *Annals of Tourism Research*, 66: 14–25.

Shiraani, F., & Carr, N. (2021). Disabled children are not voiceless beings. *Annals of Tourism Research*, 103257.

Small, J., (2015). Interconnecting mobilities on tour: Tourists with vision impairment partnered with sighted tourists. *Tourism Geographies*, 17(1): 76–90.

Stone, P.R., Hartmann, R., Seaton, T., et al. (eds) (2018). *The Palgrave Handbook of Dark Tourism Studies*. London: Palgrave Macmillan.

Story, M.F., (1998). Maximizing usability: The principles of universal design. *Assistive Technology*, 10(1): 4–12.

Stringer, K., (2014). *Programming for People with Special Needs: A Guide for Museums and Historic Sites*. Lanhan, MD: Rowman & Littlefield.

U.S. Department of Justice, (n.d.). *Marinating Accessibility in Museums*. Retrieved from https://www.ada.gov/business/museum_access.htm

Williams, C., & Wright, B., (2016). *A Guide to Writing Social StoriesTM: Step-by-Step Guidelines for Parents and Professionals*. Jessica Kingsley Publishers.

Williams, P., & Aaker, J.L., (2002). Can mixed emotions peacefully coexist?. *Journal of Consumer Research*, 28(4): 636–649.

World Tourism Organization, (2017). UNWTO Annual Report 2016, UNWTO, Madrid. https://doi.org/10.18111/9789284418725

Wynarczuk, K., Schaefer, A., Bongiovanni, J., Monaco, N., & Kellar, B. (2020a). Participation of students with disabilities in school trips: Parents' experiences and perspectives. https://doi.org/10.13140/RG.2.2.13718.70720.

Wynarczuk, K., Sanders, E., Potter, A.M., Kahn, L., Curley, B., & Kresge, L. (2020b). Parents' perceived utility of the School Trips Accommodation Tool.

Part III

Dark Tourism and Interpretation

A Child's Perspective

7 Interpretation for children

Turning horror and hurt into healing and hope

Roy Ballantyne, Jan Packer, Karen Hughes, and Tobias Broughton

Introduction

Dark tourism sites are places of remembrance, memorial, education, and entertainment. They often serve as cultural and historical landmarks with the potential to contribute to the development of national and global identity and the promotion of human rights and social justice. Such sites are ideally placed to teach visitors about the horrors of the past and instill hope for the future. But how do we do this in a way that doesn't elicit distress among younger audiences? Should we shield younger audiences from the horrors of war and other such potentially distressing topics, or is there a way to safely harness their transformative power?

To date, the impact of dark tourism sites on young visitors has received little scholarly attention. Kerr and Price (2018) call for further research not only *on* children but *with* children as collaborators. They argue that such research will reveal new perspectives that can enrich interpretive design and practice. They claim that research is needed to understand young visitors' experiences at dark tourism sites because (a) children, especially under age six, have an incomplete understanding of death and may be unable to grasp the meaning of the site; (b) children have limited agency in choosing their tourism destinations; (c) children approach sites from the perspective of play and active exploration, and thus may engage in ways that seem disrespectful or inappropriate to adults; and (d) children may experience a greater level of distress in response to dark sites.

Research and discussion about strategies to engender positive responses among younger audiences is thus both warranted and timely. Accordingly, this chapter will explore the approaches and techniques used by dark tourism sites to help visitors understand and relate to difficult events and issues. It will also discuss how interpretation can be used to create positive experiences for younger audiences.

Interpretation at dark tourism sites

Tilden (1957, p. 9) defined interpretation as 'an educational activity which aims to reveal meanings and relationships through the use of original objects, by firsthand experience, and by illustrative media.' Interpretation is more

DOI: 10.4324/9781003032199-10

than the communication of factual information – it seeks to reveal to visitors the significance of a site. At its core, interpretation has four key elements: It presents information in a way that is enjoyable, relevant to the target audience, organized into digestible and logical 'chunks,' and based on an underlying theme (Ham 1992, 2016; Tilden 1957). Moscardo, Ballantyne and Hughes (2007) also stipulate that interpretation should engage the target audience and respect and consider visitors' individual differences. There are many ways to interpret events that occurred at dark tourism sites, including stories, poems, illustrations, displays, models, audio-visual presentations, signs, role-plays, recordings, augmented reality and/or virtual reality, and immersive experiences.

Interpretation at dark heritage sites is often intentionally designed to arouse people's feelings and emotions, an approach that Uzzell (1989) termed 'hot' interpretation. Hot interpretation is specifically designed to 'tap into' personal emotions in a bid to harness emotional power and thereby instill values of reconciliation, peace, and understanding (Ballantyne 1998; Ballantyne & Uzzell 1999; Uzzell & Ballantyne 1998). This approach rose to prominence in the late 1980s in response to the realization that visitors' encounters with past events are rarely only cognitive. Places and issues that relate to personal values, beliefs, and memories are highly likely to also elicit emotional responses (Uzzell 1989). In this respect, Ballantyne (1998) remarks that interpreters need to take responsibility for the impact that their material has on visitors. Ballantyne (1998, p. 2) stated

> We should not see ourselves as 'hired hands' – people skilled in the techniques of presenting the messages of other people and organisations … (it is not) acceptable to rebuff criticism of our work with the Nuremberg defence, 'we were only following orders.'

It is essential, therefore, for interpreters to be cognizant that the process of deliberately provoking an emotional response from visitors may elicit feelings of distress or other challenging emotions. That is particularly true among younger audiences. Despite this concern, there is widespread acceptance that children will and should visit dark tourism sites (Kerr & Price 2016). Commenting on the importance of engaging children in knowing and understanding the difficult history of the Holocaust, Lewin (1988, p. 4) explains, 'The Holocaust is a legacy that we can use to help our children recognize the depth of human baseness, learn the height to which the human spirit can soar, and respond to that difference.'

Interpreting dark tourism sites for children

Tilden (1957, p. 9) cautioned against providing 'watered down' versions of information for children, stating: 'Interpretation addressed to children (say up to the age of twelve) should not be a dilution of the presentation to adults, but should follow a fundamentally different approach. To be at its best it will require a separate program.' While this mantra has been repeated numerous times in the intervening 60 years, researchers have yet to clearly articulate what the approach should entail. Interpreters need to avoid sensationalizing the content at one extreme, and overly sanitizing it at the other. Arnold-de Simine (2013, p. 46) argues that both sides of this dilemma are equally concerning.

On the one hand, 'images of suffering might brutalize spectators and normalize atrocities' (Arnold-de Simine 2013, p. 46). As explained by the curator of the Liverpool Slavery Museum,

> Some wanted us to construct an emotive but authentic hold to walk through, with manacled bodies covered in excrement, groans, smells–the full works ... but we did not want something that frightened people (particularly children) and we did not want to sensationalise.
>
> (Tibbles, cited by Arnold-de Simine 2012, p. 34)

On the other hand, there is a danger that presenting a 'comfortable, rose-tinted or safe, sanitised version of past events' has the potential to 'trivialise, commoditise, distort, soften or depersonalise the event commemorated or represented' (Sharpley & Stone 2009, p. 115). By trivializing history, interpretation could inculcate within the public 'a reactionary, superficial and romantic view of the past' (Uzzell 1998, p. 1). For example, Sturken (2007) notes that attempts to produce feelings of comfort through the sale of souvenirs such as teddy bears in firefighter helmets at the Ground Zero site in New York act to prescribe and contain emotions, collapse history into simple narratives, and strip the site of its larger political meaning.

In relation to the display of weapons and objects of death and torture at the Imperial War Museum's (London) Holocaust Exhibition, Popescu (2020, p. 234) asks, 'Is it necessary to display such potentially emotionally harmful objects? Since such objects are bound to provoke a visceral reaction, what are the moral responsibilities of the museum curators?' Her answer lies in the framing of the presentation 'so that visitors are encouraged to think about ethical questions.' She advocates using exhibition design strategies to create transformative encounters that go beyond 'shock value' and provoke 'deeper consideration of the content and how to respond to it' (p. 238).

So how do we walk the fine line between horror and hope? How can we ensure that dark tourism sites are interpreted for children in a way that engages, inspires, and heals? Such understanding is essential to inform the development of learning aims, the curation of experiences, and the framing of sensitive messaging to provide children with positive learning experiences. With the rise of terrorism, mass shootings, extreme weather events, and the recent global pandemic, the number of sites memorializing traumatic events is likely to rise. That will potentially render dark tourism experiences even more influential in children's cognitive, emotional, and moral development. Increasingly, such experiences may shape their actions, values, and beliefs as they become adults with agency. Understanding the impact that visiting dark tourism sites has on children is therefore critically important, especially in relation to how learning from the past can prepare them for the future.

Impacts of dark tourism on children

Studies regarding the impact of dark tourism on young people have largely focused on adolescents visiting dark tourism sites on school field trips (Croom, Squitiero, & Kerr 2018; Israfilova & Khoo-Lattimore 2019; Popescu 2020). These studies show that students experience a variety of emotions including sadness, nonchalance, empathy, pride, gratitude, excitement, and inspiration. Croom et al. (2018, p. 167) conclude that '[t]he adolescents could tolerate experiences that brought them face-to-face with human suffering and death.' Further, they noted that students respected the solemnity of the site and were able to reflect on issues of bravery and heroism. In Israfilova and Khoo-Lattimore's (2019) study, although students found the excursion emotionally challenging, the visit had positive impacts on their knowledge, awareness of human rights, character development, sense of identity, and for some, inspired an interest in politics. Popescu's (2020) study further illustrates the capacity of older students (aged 16–18) to think critically and reflectively about their experiences.

In a study focusing on younger children (aged 8–10) at the South Australian Maritime Museum, Sutcliffe and Kim (2014) investigated the impact of interpretation on school children's engagement with the 'Life on Board' exhibition, which focuses on the conditions early European migrants faced when traveling to Australia by boat in the early 19th century. They found that stories of death and dying had a particularly powerful effect on children's learning. They also noted that opportunities for children to interact and engage with each other through play were important in conveying cultural lessons. In particular, children who participated in social or exploratory play developed an understanding of the discomfort and dangers of the migrant experience.

Research at the District Six Museum in South Africa, where visitors experience a hot interpretive display of Apartheid-era 'forced removals,' clearly identifies the distress these exhibits engender in past residents of the area (Ballantyne 2003; Ballantyne & Uzzell 1993). In this regard, photographs of the area, old street signs, maps, and the personal stories of guides trigger personal memories and strong emotional responses. A major focal point of the exhibit is the 'memory cloth' upon which past residents and visitors are encouraged to write comments. Reading these comments reduces many visitors to tears 'as they confront words that hint at the real impact and hurt of the removals upon community members' (Ballantyne 2003, p. 287). School students who visit the District Six Museum, some of whom are the children or grandchildren of former District Six residents, express feelings of anger, resentment, and overall distress regarding the forced removals and displacement of past residents of the area. Notwithstanding this, many school students report learning positive lessons for the future from what they experienced: 'I have learned about my roots and heritage, where my ancestors lived, their hardships, struggles and it makes me realize that people can live in harmony, despite religion, colour or class.' In those instances, hot interpretive exhibits and experiences aid students' personal reflection and introspection, thus promoting understanding and reconciliation rather than distress.

The above studies provide some reassurance that difficult emotions do not need to be avoided in interpreting dark tourism sites for young visitors. Given appropriate support, the children and young people in these studies were able to engage respectfully with suffering and death and make meaning from their experience. Encouraging children to empathize with and immerse themselves in the topics, stories, and issues being discussed, and then providing them with opportunities to express their feelings, helps them process the 'difficult' information being presented.

Interpreting difficult topics for children: Examples from the field

This section presents further examples of interpretive techniques that are designed to engage children in difficult topics, and the learning objectives these techniques aim to achieve.

The 'Remember the Children: Daniel's Story' exhibition in the United States Holocaust Memorial Museum in Washington, DC, is designed especially for children. It tells the story of 'Daniel' and his family in Nazi Germany. (The character Daniel was created specifically for the purposes of the exhibition, based on the wartime writings of young people, historical imagery, family photographs, and survivors' memories.) The exhibition invites visitors to 'enter realistic environments where they can touch, listen to, and engage in Daniel's world as it changes during the Holocaust' (United States Holocaust Memorial Museum 2020). The exhibition uses narration, personal diaries, and walk-through environments to present an accurate and sensitive account of the impact of the Holocaust upon children and families. At the end of the exhibition, children are invited to express their feelings or write down their thoughts. The exhibition aims to present this history to children, but to also 'encourage them to relate the story to events in the world today' and 'stimulate children to understand and respect people who are different and to examine critically the tendency to tease and exclude' (Lewin 1988, p. 4).

The original traveling exhibition provided discussion points and guiding questions for teachers and parents to sensitize students to stereotyping and prejudice ('How are people similar and different?'; 'How do you react to newcomers in your community'); put-downs ('Have you ever put someone down because they were different?'; 'How do put-downs hurt people?'); discrimination ('How can discrimination become a habit?'; 'How does prejudice hurt people?'); hostility and persecution ('What conditions in your community can lead to violence against a particular type or group of people?') (United States Holocaust Memorial Council 1988, p. 8). This approach was deliberately chosen to help children internalize and reflect upon the exhibit content.

Another Holocaust museum, the Yad Layeled Museum in Israel, was established as a memorial site for children who died in the hostilities. The site is considered both 'a memorial site for the child victims' as well as 'an educational institution for today's children' (Heidecker 2016, p. 274). The goal is to both commemorate the victims and allow young visitors to have agency in the space. Importantly, interpreters recognized that a museum intended primarily for children (aged 10 and above) 'demanded a different concept of what a museum should be' (p. 281). Thus, the museum is designed to create a positive, non-threatening experience that arouses curiosity and empathy. Three-dimensional installations provide young visitors with an opportunity to experience, touch, and explore the world of children who lived during the Holocaust. The focus is on creating an emotional experience and telling children's stories via visual and audio means, rather than overreliance on text. Throughout their visit, young visitors are encouraged to actively explore and express their feelings. At the end, they are invited to write letters or paint a picture to capture their emotions and reactions.

Also commemorating World War II is the Anne Frank House Museum in Amsterdam, NL. When it opened, the *New York Times* reported that Otto Frank (Anne's father) wanted the site to be not only a memorial to Anne and her suffering, but also 'a building in which the ideals of Anne will find their realization' (Allar 2013, p. 191). A focus on education has been central to the museum's continual growth, and the museum website provides links to educational resources created by and for the Anne Frank House. Students can take digital lessons, where they learn about Anne's story, the history of Jewish persecution, and World War II. A digital game called 'Fair Play' focuses more broadly on discrimination and gives players the choice of how to deal with discrimination in a number of different social, sporting, and educational environments. 'Stories That Move' provides interactive online lesson material about discrimination and encourages students to reflect on diversity. In this way, young audiences are encouraged to view issues such as conflict and discrimination from a range of perspectives and to reflect upon their current and future behavior.

The Flight 93 National Memorial in Somerset County, Pennsylvania interprets a terrorist event (the September 11 hijacks and subsequent attacks) by focusing on the victims as protagonists and heroes in the narrative. Located at the site of the crash of United Airlines Flight 93, the park presents the story of the 40 passengers who decided to fight the terrorists and force the plane to crash, ending the lives of all on board. The story is framed as one of hope, courage, and unity and praises those who chose to make the ultimate sacrifice to protect others. The site's 'Junior Ranger Handbook' guides children through the learning experience and, on its final page, asks children to draw a picture or write letters of gratitude 'to the heroes of Flight 93.' The images, words, and objects left by children as tributes during a 16-year period have been collected and displayed as part of the 'Through Their Eyes,' exhibit at the Memorial's Learning Center (National Park Service 2007, p. 16).

This focus on the more positive elements of conflict and tragedy has also been adopted by The Australian War Memorial, in Canberra, AU. A 'Discovery Zone' invites children to practice sending Morse code in a replica World War I trench, see what life was like in Australian homes during World War II, and climb inside an Iroquois helicopter. The museum also has a diverse range of online resources to help children explore themes associated with conflict. For instance, the theme of 'Art in War' has a module titled 'Art in the Aftermath' with a section that explores how peace was commemorated after World War I (Australian War Memorial 2020). Children are asked to look at Joseph Fennimore's painting, 'The Signing of Treaty of Peace at Versailles, 28 June 1919,' and describe the mood of the people illustrated, suggest reasons for the choice of location for the signing, explore the contents of the treaty and its significance for Australia, and imagine the reactions of people living in Britain and Germany. Children are also asked to imagine being a German or English citizen and to write a newspaper article about their response to the treaty conditions. These exercises are designed to immerse younger audiences in the issues and events of the day, to familiarize them with what led to the events in question, and to encourage them to see the conflict and its repercussions from a variety of perspectives or viewpoints. Again, the attention is on exploring issues and stories from a variety of perspectives rather than focusing too much on conflict and death.

Another dark tourism site in Australia that also uses stories to engage younger visitors is the Port Arthur Historic Site, a former convict settlement established in Tasmania in 1830. The site offers a range of family and school activities based on the stories of convicts who were incarcerated in the prison. These include a 'Hidden Stories' activity book with puzzles and quizzes and a 'Hidden Mystery Tour' that challenges school children to piece together stories and events using convicts' records and artifacts. There is also a 'Breaking the Code' activity where children learn about the challenges of communication and how to operate a semaphore station using replica equipment. For children over the age of 10, a nightly ghost tour tells stories of the site's inmates and ghostly encounters. Convicts' stories are the foundation of all these interpretive experiences. While the hardships are not glossed over, they are balanced with stories of survival, success, and hope.

The Canterbury Cathedral in the UK uses play-based experiences to interpret dark history. While groups of school children take great delight in dressing up and re-enacting the violent murder of Archbishop Thomas Becket in the cathedral in 1170, staff must take precautions. They need to ensure that these activities remain respectful and do not interfere with the experiences of other visitors, particularly pilgrims. Keeping a tight rein on proceedings can be difficult at times because children tend to become engrossed in their role-play and, consequently, excitement and noise levels invariably rise. To temper proceedings, the team asks children to reflect on the events leading up to the murder; the emotions likely to be experienced by the main protagonists; the response of King Henry II; and the impact on Canterbury Cathedral and the town after Thomas was made a saint. Teachers report that integrating these discussions into the role-play helps children develop a real appreciation and understanding of why the site is important, and how the events shaped practices and politics even to this day (T. Heslop, personal communication, June 25, 2015).

Children's exhibitions and interpretations have also been installed at tourism sites that deal with environmental issues. As these generally deal with potential future disasters, rather than past atrocities, they may have an even greater capacity to disturb young visitors. The stated mission of the Climate Museum in New York City is 'to inspire action on the climate crisis with programming across the arts and sciences that deepens understanding, builds connections, and advances just solutions' (Climate Museum 2020). Many wide and varied initiatives are undertaken by the Climate Museum to engage students directly. These have included public art installations, walking tours, 'Ask a Scientist Day,' and the 'Climate Art for Congress' online program, which encourages students at all grade levels to submit illustrated notes to Congress. As noted on their website, 'K-12 students can't vote, but they have agency and a huge stake' (Climate Museum 2020).

The Eden Project in Cornwall, UK, also strives to educate visitors about halting the devastating impact of climate change. Massive biomes, built on the site of a former china clay pit, house a variety of ecosystems including the world's largest indoor rainforest. The site is tailored to both younger and older children, with features such as a play area that mimics a giant 'bug hotel' and 'discovery zones' that explore the invisible interconnections of the natural world.

The focus throughout the site is on discovering the wonders of nature and how the health of the ecosystem affects, and is affected by, each one of us. Exploration and play underlie much of what is offered, but on-site interpretive activities and online resources all urge children to introduce environmentally friendly practices into their everyday lives. Messaging is accompanied by questions that prompt children to think about their own role in the ecosystem and how their decisions impact the future health of the planet. While issues such as pollution are discussed in detail, tips for being part of the climate change solution are also presented. The site clearly attempts to balance information about environmental destruction with messages of hope and healing.

Interpreting dark tourism sites for children: Suggestions for practice

Many dark tourism sites encourage young visitors to regard themselves as agents of change. These 'new moral spaces' (Stone 2009) have the capacity to teach young people about empathy, to critically examine stereotypes, and to promote reconciliation (Ballantyne 2003). It is evident that through sensitive and supportive interpretation, children can be assisted through the process of applying lessons learned from history to their own time and place.

Drawing on principles developed by Ballantyne, Packer, and Bond (2012) and current approaches to accommodating children at dark tourism sites, five suggestions for designing children's interpretive experiences are presented.

1   *Personal stories should take a central place in children's interpretation.*
    Personal stories help visitors engage and connect with the experiences
    and feelings of others. Highlighting similarities between visitors and
    those whose story is being told helps audiences imagine themselves in the
    other's place, experience the other's feelings, and develop an understand-
    ing of their situation. The United States Holocaust Memorial Museum,
    Yad Layeled Museum, and Anne Frank House Museum all use video
    and audio commentaries, images, documents, and artifacts to tell the sto-
    ries of children who lived during World War II.

    Even if children do not feature strongly in the site's history, stories can
    still be used to engage younger visitors. As an example, past events could
    be described and then young visitors could be asked to consider how
    they would feel or react in similar circumstances. To be successful, these
    stories need to feature situations with which the children are familiar.
    Thus, a story of a convict shipped to Australia might focus on feelings of
    loss and loneliness and ask children to imagine how they would feel if cut
    off from their family and friends, and how they would go about making
    a new life for themselves. They could also be asked to consider their own
    personal strengths and describe what they as a convict would contribute
    to a new colony. In this way, the 'impersonal' history is transformed into
    a positive personal experience. Play activities can also be used to help
    children understand events from the perspective of those who experi-
    enced them.

2   *Interpretation should help young visitors counter-balance negative reactions
    such as despair and sadness with messages of healing and hope.*
    Hot interpretation is designed to elicit emotions such as sadness, anger,
    guilt, embarrassment, or shame, depending on the relationship between
    the visitor and the event or issue presented. Ballantyne et al. (2012) argue
    that interpretation should be framed in such a way that these negative
    emotions are balanced by positive responses, such as feelings of hope
    and admiration. Achieving such a balance should help visitors deal with
    their feelings and move forward.

    Balancing negative with positive messages for the future is particu-
    larly important for younger audiences. Although it is good to encour-
    age children to empathize with characters involved in historic events,
    undue emphasis on illness, death, misery, and tragedy may lead to feel-
    ings of despair that younger visitors are ill-equipped to manage. Thus,
    wherever possible, interpretation should balance accounts of tragedy
    with positive messaging, such as lessons learned, hidden benefits for
    subsequent generations, and strategies to manage similar situations.
    This approach is used extensively at Port Arthur Heritage Site, where
    stories of incarceration and hardship are balanced by stories of con-
    victs completing their sentences and subsequently becoming respected
    citizens in the new colony. The underlying and clear message is that
    hardship and despair can build the resilience required to achieve a
    better life.

3   *Interpretation should aim to educate rather than persuade.*

Ballantyne et al. (2012) caution against relying too heavily on emotional accounts, as this can make visitors feel that they are being manipulated into agreeing with or adopting a particular viewpoint. Children can be especially vulnerable in this regard as they often lack the experience and maturity to realize that alternative interpretations of a site's history are possible. It is therefore important to express different viewpoints, recount events using a variety of voices, and provide balanced and reasoned explanations of why events happened.

As noted above, Canterbury Cathedral helps students re-enact the murder of Thomas Becket, but ensures this is done with sensitivity. This is accompanied by discussions of why the main protagonists acted as they did, how this affected the relationship between church and monarchy, and why the site became a place of pilgrimage whose importance continues to this day. These discussions help to answer the perennial 'why' questions, and in doing so, provide children with the details needed to form their own opinions and views.

4   *Young visitors should be given examples that resonate with their own experiences and opportunities to reflect upon how they would react.*

Hot interpretive exhibitions generally incorporate a time and space for visitors to reflect on their responses, and to make meaning that they can apply to their own lives and experiences. This might include creating a personal response through art, music, drama, or reflective writing. These responses often then become part of the experience for subsequent visitors.

The Yad Layeled Museum invites children to write a letter or paint a picture to express their feelings about the Holocaust. The Flight 93 National Memorial asks children to draw a picture or write letters of gratitude to their fallen heroes and has used this in their subsequent exhibits. The Climate Museum has an art program that encourages children to visually express their thoughts and feelings about the environment, and The Australian War Memorial asks children to imagine they have just witnessed a war and write a letter describing their feelings on the signing of a peace treaty. These reflective activities are all designed to help children mentally and emotionally process the events described at the site, and should be included in dark tourism sites wherever possible.

5   *Interpretation should help children bridge the gap between past events and future actions.*

Achieving a balance between past atrocities and possible future actions can help visitors make sense of distressing events and learn how to make better decisions in their own lives. Activities that help younger audiences to see a brighter future are likely to be especially powerful and memorable. The Eden Project is particularly successful in this regard. They aim to build resilience and impart skills to their young audiences to address future problems. This is done by offering activities that range from building nests for woodland creatures to recycling and reusing products in their homes. The message underpinning all their interpretation is that even children can make a difference to the health of the planet. As visitors leave the site, they are asked to consider what changes they will implement to make the world a better place.

Conclusion

Internationally, dark tourism sites attract millions of visitors each year, including many children and adolescents. Yet little is known about how young people experience and respond to the potentially distressing emotional content presented at such sites, which often present 'evocative recreations of traumatic experiences that are designed to unsettle and disturb visitors' (Arnold-de Simine 2012, p. 23).

It is widely acknowledged that a gap exists in research around children's voices in tourism, and dark tourism is no exception (Kerr, Stone, & Price 2021). Yet there is an increasing need to hear from young people about their visitor experience, as difficult heritage sites continue to grow in popularity with both school excursion groups and families. There are certainly methodological challenges that arise when researching child respondents (Khoo-Lattimore 2015). Nevertheless, there is a pressing need to better understand the experiences of children at different development levels and to use that understanding to inform the design, implementation, and evaluation of hot interpretive exhibitions.

Children are future agents for social and environmental change. Therefore, it is important that their perspectives and reactions are taken into account and that their visits to dark tourism sites result in a positive outcome. We provide some evidence and examples of how this might be achieved, but systematic research is required to ascertain the impacts of these interpretive approaches on young audiences. That will allow us to better understand how to design dark tourism experiences that have a beneficial and lasting impact on young people. Young tourists should exit hot interpretive exhibits in a positive rather than distressed frame of mind, enthusiastically empowered with the knowledge of how and what they can do to address the mistakes of the past.

References

Allar, K.P. (2013) 'Holocaust tourism in a post-Holocaust Europe: Anne Frank and Auschwitz', in L. White & E. Frew (eds.), *Dark tourism and place identity: Managing and interpreting dark places* (pp. 189–201). London: Routledge.

Arnold-de Simine, S. (2012) 'The "moving" image: Empathy and projection in the International Slavery Museum, Liverpool', *Journal of Educational Memory, Media, & Society*, 4(2): 23–40.

Arnold-de Simine, S. (2013) *Mediating memory in the museum: Trauma, empathy, nostalgia*. New York: Palgrave Macmillan.

Australian War Memorial (2020) *Peace at last*, viewed 23 December 2020, https://www.awm.gov.au/learn/schools/resources/art-in-the-aftermath/peace-at-last

Ballantyne, R. (1998) 'Problems and prospects for heritage and environmental interpretation in the new millennium: An introduction', in D. Uzzell & R. Ballantyne (eds.), *Contemporary issues in heritage and environmental interpretation: Problems and prospects* (pp.1–10). London: The Stationery Office.

Ballantyne, R. (2003) 'Interpreting Apartheid: Visitors' perceptions of the District Six Museum', *Curator: The Museum Journal*, 46(3): 279–292.

Ballantyne, R., Packer, J., & Bond, N. (2012) 'Interpreting shared and contested histories: The Broken Links exhibition', *Curator: The Museum Journal*, 55(2): 153–166.

Ballantyne, R., & Uzzell, D. (1993) 'Environmental mediation and hot interpretation - a case study of District Six, Cape Town', *Journal of Environmental Education*, 24(3): 4–7.

Ballantyne, R., & Uzzell, D. (1999) 'International trends in heritage and environmental interpretation: Future directions for Australian research and practice', *Journal of Interpretation Research*, 4(1): 59–75.

Climate Museum (2020) viewed 23 December 2020, https://climatemuseum.org

Croom, A.R., Squitiero, C., & Kerr, M.M. (2018) 'Something so sad can be so beautiful: A qualitative study of adolescent experiences at a 9/11 memorial', *Visitor Studies*, 21(2): 157–174.

Ham, S. (1992) *Environmental interpretation*. Golden, CO: Fulcrum Publishing.

Ham, S. (2016) *Interpretation: Making a difference on purpose*. Golden, CO: Fulcrum Publishing.

Heidecker, N. (2016) '*Yad Layeled* at the Ghetto Fighters' House: A museum about children in the Holocaust or a museum for children about the Holocaust?', *Dapim: Studies on the Holocaust*, 30(3): 274–281.

Israfilova, F., & Khoo-Lattimore, C. (2019) 'Sad and violent but I enjoy it: Children's engagement with dark tourism as an educational tool', *Tourism and Hospitality Research*, 19(4): 478–487.

Kerr, M.M., & Price, R.H. (2016) 'Overlooked encounters: Young tourists' experiences at dark sites', *Journal of Heritage Tourism*, 11(2): 177–185.

Kerr, M.M., & Price, R.H. (2018) '"I know the plane crashed": Children's perspectives in dark tourism', in P. Stone, R. Hartmann, T. Seaton, R. Sharpley, & L. White (eds.), *Palgrave handbook of dark tourism studies* (pp. 553–583). London: Palgrave MacMillan.

Kerr, M.M., Stone, P.R., & Price, R.H. (2021) 'Young tourists' experiences at dark tourism sites: Toward a conceptual framework', *Tourist Studies*.

Khoo-Lattimore, C. (2015) 'Kids on board: Methodological challenges, concerns and clarifications when including young children'svoices in tourism research', *Current Issues in Tourism*, 18(9): 845–858.

Lewin, A. (1988) 'A message from the president', in *The National Learning Center presents "Remember the children": An exhibit for children about the Holocaust*. Capital Children's Museum, Washington, D.C. [U.S. Holocaust Memorial Council], viewed 23 December 2020, http://hdl.handle.net/2027/umn.319510029597936

Moscardo, G., Ballantyne, R., & Hughes, K. (2007) *Designing interpretive signs: Principles in practice*. Golden, CO: Fulcrum Publishing.

National Park Service (2007) *Junior Ranger Journal for Older Children: Flight 93 National Memorial*, viewed 23 December 2020, http://npshistory.com/publications/flni/jr-ranger.pdf

Popescu, D.I. (2020) 'The potency of design in holocaust exhibitions: A case study of The Imperial War Museum's *Holocaust Exhibition* (2000)', *Museum & Society*, 18(2): 218–242.

Sharpley, R., & Stone, P.R. (2009) '(Re)presenting the macabre: Interpretation, Kitschification and authenticity', in R. Sharpley & P.R. Stone (eds.), *The darker side of travel: The theory and practice of dark tourism* (pp. 109–128). Bristol: Channel View.

Stone, P. (2009) 'Dark tourism: Morality and new moral spaces', in R. Sharpley & P.R. Stone (eds.), *The darker side of travel: The theory and practice of dark tourism* (pp. 56–72). Bristol: Channel View.

Sturken, M. (2007) *Tourists of history: Memory, kitsch, and consumerism from Oklahoma city to ground zero*(E-Duke books scholarly collection). Durham, NC: Duke University Press.

Sutcliffe, K., & Kim, S. (2014) 'Understanding Children's engagement with interpretation at a cultural heritage museum', *Journal of Heritage Tourism*, 9(4): 332–348.

Tilden, F. (1957) *Interpreting our heritage: Principles and practices for visitor services in parks, museums, and historic places.* Chapel Hill, NC: University of North Carolina Press.

United States Holocaust Memorial Council (1988) *The National Learning Center presents "Remember the children": An exhibit for children about the Holocaust.* Capital Children's Museum, Washington, D.C. [U.S. Holocaust Memorial Council], viewed 23 December 2020, http://hdl.handle.net/2027/umn.319510029597936

United States Holocaust Memorial Museum (2020) *Remember the Children: Daniel's Story, viewed 23 December 2020*, https://www.ushmm.org/information/exhibitions/museum-exhibitions/remember-the-children-daniels-story

Uzzell, D. (1989) 'The hot interpretation of war and conflict', in D.L. Uzzell (ed.), *Heritage interpretation volume 1: The natural and built environment* (pp. 33–47). London: Belhaven.

Uzzell, D.L. (1998) 'Interpreting our heritage: A theoretical interpretation', in D.L. Uzzell & R. Ballantyne (eds.), *Contemporary issues in heritage and environmental interpretation: Problems and prospects* (pp. 11–25). London: The Stationery Office.

Uzzell, D., & Ballantyne, R. (1998) 'Heritage that hurts: Interpretation in a post-modern world', in, D.L. Uzzell & R. Ballantyne (eds.), *Contemporary issues in heritage and environmental interpretation: Problems and prospects* (pp. 152–171). London: The Stationery Office.

8 Understanding Children's Visits to Difficult Heritage Sites

Children's Sense of Place

R. Scott Marsh

> Every mind must know the whole lesson for itself, must go over the whole ground. What it does not see, what it does not live, it will not know.
>
> (Ralph Waldo Emerson 1836)

Introduction

Late May brings warmer, longer days to the Flight 93 National Memorial in the Laurel Highlands of western Pennsylvania. Standing in front of one of the 40 disconnected, but seemingly solidly unified Wall of Names exhibits, a young visitor is fully aware of the sound of somber silence. A soft yet persistent breeze whispers through the hemlock forest beyond the large boulder marking the site of impact, yet the stillness creates power in itself. This cherished open and peaceful place is part of the story of what happened here. The plane that crashed on this spot was meant to send a violent, devastating message in a much more crowded place. That message instead became one of brave resistance and sacrifice, a lesson that is not lost on young visitors to the wall.

The resonance fostered by the significance of a place is something we experience through our senses and emotions. It is a physical yet also intangible personal encounter that we carry with us through our everyday memories and experiences. With children, finding meaning in the places they visit is even more significant. According to Lim and Barton (2010), children form contextual, intimate, and symbolic meaning through their interactive and adaptive engagement with specific places. When encountering new places, and especially dark sites, children will address their new challenges and experiences based on how they interpret the information provided by those places (Briggs, Stedman and Krasny, 2014). Therefore, the importance of this interaction must be taken into account when planning and preparing difficult heritage site visits that expose children to them.

A sense of place is not just physical or geographical, but also encompasses the connection and groundedness felt when there (Youngblood, 2004). How young tourists encounter a sense of place at difficult heritage sites should guide the utilization of those sites as a powerful teaching tool. To use sense of place effectively, teachers, interpreters, and parents must understand how the concept can be developed and instilled in young learners before, during,

DOI: 10.4324/9781003032199-11

and after visits. Establishing an emotional and multisensory approach to understanding place allows children to appropriately engage in difficult heritage sites, where interpretation can be demanding and stressful for them. It is important first to understand the concept of sense of place and how it is established and fostered within us during childhood.

What Is Sense of Place?

'Place' refers to a specific physical setting that is bordered, locatable, and able to be identified by name. More importantly, 'places deploy space' (Blair, Dickinson and Ott, 2010, p. 23). They have the power to invoke memories that are specific and analogous to the locations or unique spaces in which they occur. When we think back on places, we remember feelings and activities, not just the physical settings (Glassberg, 2001). Perhaps Vergeront (2013, p. 1) states it best when she writes,

> Place is...intangible, carrying the spirit of a physical setting that emanates from the shape and the feel of the land; from the ground and vistas and how they meet; and from the quality of the light and the blue of the sky.

Place is something we experience privately, intimately, and intensely through our senses, but it is also something we share with others. We return to places that hold meaning for us. We bring others along to experience the same emotions we feel while we are there. Places speak to us. They tell us stories and have the power to pull us into their resonance. Places constantly evolve with layers of memory and excitement that continuously allow us to reconnect and reflect within ourselves and with our history (Vergeront, 2013).

Our sense of place incorporates our emotional connections with the bonds, values, and meanings we associate with the spaces where we live, work, travel, and make our lives. Sense of place is continuously built and rebuilt in our minds, based on our enduring awareness of the cultural, historical, and spatial context of our space (Lengen and Kistemann, 2012). We remake these places by creating memories and attaching meaning to them.

Furthermore, our personal identities are based upon our understanding of the physical world in which we live. These identities differ, depending on how we experience our places, and these experiences are shaped by social characteristics, such as age, gender, race, class, and physical condition (Glassberg, 2001). This comprehension also includes memories, feelings, attitudes, values, preferences, behavioral concepts, and experiences, which make up the varieties and intricacies of the physical setting that define who we are (Glassberg, 2001; Lengen and Kistemann, 2012). These many attitudes, feelings, and memories create an experience known as 'place attachment.'

Place attachment is the deep emotional connection people develop toward specific places over time, by repeated positive interactions. Place attachment promotes the emergence, development, and continuation of the identity of a person, group, or culture, making it extremely applicable during adolescence (Dallago et al., 2009). Our emotional attachment to our places deepens our connection to the history and heritage that is associated with them. This connection does not just pertain to the places where we spend our daily lives, but it also applies to historical sites that are meaningful to us.

We, as visitors to historical places, must be allowed to connect via what we do, see, and feel. This connection ties us to what we already know, understand, and acknowledge (Filene, 2010). According to Filene (2010, p. 173), history, among other things, is 'empathy.' The ability to see humanity in the distant figures of the past and bring their experiences to life makes us feel that we know them. Coming to know and understand others, whether they are people who live down the street or across a millennium, requires that we educate our senses (Wineburg, 2001). Our understanding of historical places, and the people who made them so, creates a sense of history which reveals the importance of emotional attachments to certain places (Glassberg).

Our experiences with place, or seeing ourselves in that place, is what we long for when we travel to historic sites such as battlefields or spaces of public memory. Our reflections on concepts such as war and sacrifice at these sites reinforce our own identities (Willard, Lade and Frost, 2013). In fact, place-making, as a pursuit of public memory, becomes important in the formation of memory and the ways in which it can be embraced (Blair, Dickinson and Ott, 2010). Place becomes an object of preference because it has the power to 'represent, inspire, instruct, remind, admonish, exemplify and/or offer the opportunity for affiliation and public identification' (Blair, Dickinson and Ott, 2010, p. 26). The effective emotional and symbolic significance of landscapes helps us to understand how the presentation of these sites of memory, in both an individual and collective way, establishes emotional resonance. We, as visitors, bring our own viewpoints to historical sites (Bird, 2013). We make meaning of places through our own lens, which is influenced by our national culture, memory, and collective identity. Dehoorne and Jolliffe (2013, p. 156) remind us, 'national identity is increasingly connected to tourism, as is the identity of a place that differentiates it from other destinations.'

As we visit places of historical significance, the power of those sites often takes on a transcendent quality. Bowman (2010, p. 203) describes visitors' descriptions of the Mary Queen of Scots House [MQSH] in Jedburgh, Scotland, as 'drenched, soaked, oozing and dripping' in history. This eery experience allows visitors to approach artifacts and spaces on their own terms. Visitors can feel like they are reliving a moment from the past, almost as if they had been there and experienced it themselves (Blair, Dickinson and Ott, 2010). Tourists typically have moments where the sense of being a visitor is erased, and the experience of the site creates a 'sense of belonging' (Bird, 2013, p. 174).

Bowman describes this sense of belonging as hauntology, an expression coined by Derrida (1994) to illustrate the occasion of being in the presence of a ghost which is neither present nor absent, dead nor alive. This term also describes how a place 'recruits and mobilizes bodies to perform acts of remembrance' (Bowman, 2010, p. 208). During hauntology, the place and the artifacts associated with it leave a lasting mark on visitors. This impression, a result of the 'history lesson' they receive, intoxicates them and allows the place to captivate them, bringing their experience to life. Visitors to the MQSH in Bowman's study declared, 'To think that you're walking where she [Mary Stuart] walked, touching the same walls, being in the same room…' (Bowman, 2010, pp. 210–211). Similarly, in describing an unmarked Confederate burial site on Culp's Hill at Gettysburg National Military Park, Carmichael (2019, pp. 43–44) writes, 'I stared at the slight depression, and it felt like a portal into the past where history is imagined and felt.' In describing sitting alone in the room at Petersen House where Abraham Lincoln died, Clinton (2019, p. 179) expounds, 'Closing our eyes, we can almost taste the sour metallic smell of drying blood and feel the stillness within.' Such connections between people and their places are powerful, intense, and profound.

Children and Their 'Place'

Vigorous, meaningful learning has been connected to the places we have lived and explored from our earliest memories. From birth, children learn and engage with their surroundings (Vander Ark, Liebtag and McClennon, 2020). According to Glassberg (2001, p. 113), 'Children place identity with *their* place like they do with their mother.' During these very early years, some of the most formative experiences children encounter are connected to place. Children explore and play, mastering their environment. As they explore, children discover who they are in their place. By seeking out natural places for play, children learn to understand their world and their place in it. They continually form a relationship with their local environment as they develop. Attachment to place evolves through children's active exploration and inter- action with their surroundings. Adults often weave their life stories into favorite childhood memories of places (Glassberg, 2001).

There is an interaction that takes place when we are young which creates a sense of 'intertwining' of the individual and the landscape. The landscape becomes part of the individual and the individual becomes part of the land- scape. We develop an embodied experience with our locality through our everyday connection to the textures, sounds, smells, and sights that make it a part of us. This association is particularly consequential when we are chil- dren (Bourke, 2017; Wylie, 2007). A child's continual encounters with their outdoor environment is in a sense engaging with a 'phenomenal landscape' which is 'known and felt' (Bourke, 2017, pp. 94–95; Hart, 1979, p. 9). Even the seemingly inconsequential act of walking encapsulates an experience of place for children. The 'footprints' they make are described as 'part of the weave' that makes up the 'tapestry' of the environment they get to know that 'imprints itself on the body' (Bourke, 2017 p. 95; Ingold and Lee Vergunst, 2008, p. 8; Lee Ingold and Vergunst, 2008, p. 73). This is a very subtle process of developing a knowledge of their place without being totally aware that it is happening (Cele, 2006). The more a path is traveled, the more meaningful it becomes – bursting with the memories of the people and experiences which imprint themselves in the echoes of a child's consciousness and imagination (Lee Vergunst and Ingold, 2008). As a path becomes well-worn, 'thick lines' of memory form – creating a meaning which is described as 'unthought,' but felt and experienced (Lee Vergunst and Ingold, 2008).

Our fondest recollections are attached to these familiar spots where we spent most of our time. Intellectual development flourishes through this exploration, fostering greater independence and creativity (Derr, 2006; Vander Ark, Liebtag and McClennon, 2020). Children experience their world, and the culture associated with it, as part of the place they live. The lessons they learn are often tied to that way of life and their broader community. The everyday experiences children encounter that combine their family and neighborhood are in places they hold on to, and these places are the most likely to assimilate into their own identity. The people and places they know best make up their entire world (Derr, 2006; Filene, 2010).

Children Experiencing Sense of Place at Heritage Sites

When visiting historical sites and places of heritage interpretation, children continue to use play and exploration as a means of understanding their surroundings and the history that happened there. Children learn, examine, and evaluate through their senses. The ability of a child to use imagination at historic sites has been described as 'the latch that must be unhooked to open the door to the past' (Craig, 1989, p. 107). This 'embodied imagination' is multidimensional, and prompts multisensory engagement with a place by involving seeing, hearing, smelling, and touching as the child investigates the landscape (Chronis, 2005, p. 396). As they walk, explore, and play, children encounter a sensory experience of sounds, visions, smells, and touch. Their journey produces an experience of the senses where even the rhythm of their feet on a pathway encourages an understanding of their sense of place (Pink, 2009; Ross, 2007; Yi'En, 2014).

Younger children want to know what things *feel* like (Tilden, 2007). Young tourists comprehend by listening, doing, touching, and scrutinizing (Kerr, Shaffer and Hartman, 2014). Many children explore through movement as well. They climb on boulders at Gettysburg National Battlefield, move their hands down every rung in the length of chain that partitions the burying ground from the walking path at Arlington National Cemetery, or dip their fingers in the cool water of the large central fountain at the National World War II Memorial in Washington, D.C. (Gallman, 2019; Kerr, Price and Savine, 2017). At times this exploration and curiosity may even seem out of line with what adults may deem proper behavior at more poignant and somber sites (Kerr, Stone and Price, 2021; Price and Kerr, 2018). A child may playfully skip down a path while adults quietly and solemnly approach the National AIDS Memorial Grove in San Francisco. They may busily inspect and sit on the memorial benches at the Oklahoma

City Memorial, while their parents reservedly and respectfully sit in the grass beside them (Stevens and Franck, 2016).

The sense of touch is also utilized by older children as they explore, learn, and develop a sense of place and remembrance at heritage sites. They will use their fingertips to trace the names engraved on memorials 'as if they were making a Braille reading' (Abramson, 1996, pp. 679–709). Even when the engravings on stone memorials have been eroded by time and weather, children will trace the almost indistinguishable names with their fingers, appearing to draw understanding and comfort from the feel of the rough inelegant etchings. When their fingertips touch the names on the memorials, the haptic sense is engaged in remembrance and understanding. For many, the act of simply being there forges a connection between the children and the landscape, which also plays a part in their remembrance and understanding (Marshall, 2004).

Adults and children will care more about the past if they can tie it to their own lives and the familiar places they identify with as their own. Children make history come alive by connecting it to the world in which they are committedly engaged. They connect their sense of place with the new terrain to which they have been introduced. Children expand their sense of the new place by starting with themselves and expanding outward (Filene, 2010). Meaning is inspired by the landscape, artifacts, and mediated narrative that, taken together, creates a storyscape. During the experience of following a story while at the site, children will fill narrative gaps and redefine the historic events that took place at the site in terms of their own understanding and actively engage their imagination. Being surrounded at historic sites by embodied evidence of the past in the form of monuments, cannons, markers, museum artifacts, etc., child visitors' senses are enkindled for bodily exploration. Children feel the urge to explore the storyscape tactilely, like touching a cannon on a battlefield or a tombstone at a cemetery (Chronis, 2005). These tangible artifacts epitomize the world of the past, recreating it as a lived context (Turner, 1990).

Children's Sense of Place and Dark Tourism

While visiting difficult heritage and other dark sites, young visitors continue to experience sense of place through the same methods which they use at other historic landscapes. They lean on prior knowledge and experience to comprehend their setting and the storyscape, and they use their senses to experience the environment. Added to the normal acts of exploration and understanding, there is an emotional aspect as well. Feelings of empathy, compassion, and remembrance are also felt by children through the senses. Stories of the known and unknown who suffered and died at the very site that is being visited establish an imaginary link of empathy between the visitors and the deceased (Escalas and Stern, 2003). Understanding what younger visitors take away from places of difficult heritage, and even from commemorations they may experience while there, allows us to gain insight into what resounds with a generation that may have no experience with war or large-scale violence (Bird, 2016).

The sense of sound stands out as a significant motivator to spark empathy and remembrance for children visiting dark sites (Marshall, 2004). The soundscape that is presented at many of these sites evokes a range of memorable and powerful emotions, including thoughts of absence and loss (Bird, 2016). During moments of commemoration, the addition of music, speeches, and the footfalls of bands add to the haunting contrast between silence and sound. The footfalls of bandsmen or flag-bearers break the silence and may evoke the footfalls of soldiers marching off to war (Marshall, 2004). The playing of *Taps* and the bellowing of orders at the Tomb of the Unknowns at Arlington National Cemetery breaks a solemn silence and forms an aural boundary for the memorial and the ceremony. To literally touch a memorial and to imagine a closeness with history, war, and death is what often resonates with young visitors (Bird, 2016). The experiences young visitors gain by using their senses to experience difficult heritage sites allows them to trade the familiarity of the 'footprints' they take in the neighborhoods of their everyday lives with 'walking in the footsteps' of people they never had a chance to meet.

When children visit dark sites in larger groups (on school trips, for example), younger tourists adhere to many of the same cognitive precepts as their adult counterparts. Individual children may become influenced by, and persuaded about, the tragic events within which they are immersed. Difficult heritage sites take on a distinctiveness as communication spaces for collective memory (Glassberg, 2001). Often these spaces convince children of an interpretation of tragic events that convey a shared morality. This occurrence, which Stone (2009, pp. 63–64) identifies as 'collective effervescence,' is stimulated by assembled social groups that mobilize people's passions to form a symbolic order of society. The shared emotional experiences of the group allow individuals to connect with one another on the basis of common purpose and motivation. Collective effervescence allows individuals to lose themselves in a world that is more moral and just than that which actually exists (Stone, 2009).

Sites such as the 9/11 Memorials in New York, Virginia, and Pennsylvania require children to not only remember and commemorate the victims of the attack, but to project morality and reaffirm American morals and values. Expectations are directed at young visitors as the next generation to 'carry the torch' of remembrance (Bird, 2016, p. 50). Collective effervescence at these dark sites promotes, for example, ethical discussions about 9/11 and its ramifications (Stone, 2009).

History can be an abstract concept for children to comprehend. It is something that took place a long time ago 'somewhere else' (Filene, 2010, p. 181). Sense of place is a uniquely powerful tool for captivating young tourists in historical thinking. Children may have no understanding of historical causation or precedent, but they do comprehend the notion that there are layers of the past beneath them (Filene, 2010).

Historic places are endowed with *atmosphere* established by the events and memories that took place there (Uzzell and Ballantyne, 1998). Emotions illuminate our memories and experiences while we encounter historic sites, directing our focus to information and understanding of the place (Uzzell and Ballantyne, 1998). When students witness first-hand the places where historic events happened, they begin to engage with the past and use historical thinking as a foundation for grappling with contemporary problems. Historic sites offer the enticement of geographic authenticity. Young people can witness the actual places where history was made by real people (Filene, 2010).

Any museum, historic site, or other institution that attempts to bring the past to life will eventually succeed or fail based on how well it manages to interpret, in a clear meaningful way, to children (Tilden, 2007). Children want to learn differently at a destination site than they do in the classroom. Instead of lectures, they want to lose themselves in stories of where they are and why that place is important (Tilden, 2007). Children have the unique ability to identify themselves with the historical scene they explore (Manucy, 2009, cited in Tilden, 2007).

Preparing for a Visit: Considerations for Incorporating Sense of Place

The emotional connections to a historic site and the knowledge of place that young tourists bring to it offer a solid starting point for engagement, alternative perspectives, and new relationships (Vergeront, 2013). Preparation for visits to these locations, especially to difficult heritage sites, first involves the preparation of background knowledge and facts about the place. Many historic sites make creation of pre-visitation materials a priority, to help teachers prepare children for what they will encounter. Popular American historic education sites like Colonial Williamsburg, George Washington's Mount Vernon, the National 9/11 Pentagon Memorial, the National 9/11 Memorial and Museum, and the numerous sites run by the National Park Service have extensive materials created especially for teachers and students who visit (Tilden, 2007). Of course, these materials work best when they supplement a teacher's lesson plan, allowing teachers to foster a broader understanding of the place.

The second part of planning for the visit involves psychological preparation. As was discussed earlier, young tourists will care more about the past, and the places associated with it, if they can relate it to their own experiences and identities. Difficult heritage sites are places of 'hot interpretation,' or locations which grab us, engage our senses, and make us consider the emotional side of understanding what took place there (Uzzell and Ballantyne, 1998). Therefore, children must be emotionally prepared before their site visit takes place.

An important aspect of this psychological preparation is to condition young tourists' intersensorality. Intersensorality describes how the senses work together to give events and moments in the past meaning and texture

(Smith, 2010). It has already been established that children learn tactilely and through their senses. It is important for them to know that their connection to place can take on many forms, including experiencing such things as smell, warmth, moisture, and even isolation. Identifying themselves with a place allows them to develop a deep personal bond that attaches them to that place. It compels them to proudly and intensely share their sense of place with others (Vergeront, 2013).

Psychological preparation before the visit can be cultivated through a variety of methods. Visual representations such as photographs, multimedia presentations, primary source material, and historiographic exercises can all play a role in establishing a mindset of 'thinking emotionally' through the senses. For example, a photograph of Confederate dead awaiting burial at Gettysburg may prompt adolescent students to address the feel of the humid weather, the smell of decomposing corpses, or the sounds following a battle. This reflection requires that students engage their senses and evince what it may have felt like to have been there.

As a teacher prepares for the visit, it is important for them to understand that sense of place is linked to a geography of rhythms. Resonance is fostered by the significance of the place. Interconnecting the sensory feelings from the experience needs to begin before the trip. Students must be made ready to *feel* the space when they arrive.

The power of a space can be created by its vastness, its emptiness, or even its haunting silence. These open spaces create a bond between self and place that forms while the visit takes place. Many experiences at the site create stories and insights, or 'conversations with the landscape' (Bird, 2013, p. 170). Returning to our Gettysburg example, preparing students to have conversations with the landscape before they arrive will create a more meaningful connection when they find themselves standing at the site where the photograph of Confederate dead was taken, and where the rocks still remain as witnesses to the battle.

During the Visit: Experiencing Sense of Place

Linking the actual experience of the visit to a difficult heritage site through preparation requires that young tourists reflect on what they already know, and then tie that to what they experience as they tour the site. Sense of place is enabled by sensory elements, such as the sound of the wind rustling through the trees or the colors in the sky as the sun sets. These elements act as an intersection between time and place, creating a bond between those who died and those who came to bear witness and remember their story (Bird, 2013). Although they are prepared for their visit, children need guidance when connecting with these 'landscapes of ghosts.' Teachers and interpreters stimulate the relationship between children and dark landscapes by unlocking the stories that go with the spaces in an appropriate and methodical way.

Younger children's experiences at difficult heritage sites should incorporate learning strategies that work best for their age group and level of cognitive understanding. For example, at the Flight 93 National Memorial in Shanksville, Pennsylvania, younger tourists explore tactilely the smoothness and coarseness of the marble at the Wall of Names, which displays the names of the victims of the attack. The Junior Ranger Program booklet at the site even gives directions on how the memorial textures can inform young children as they: 'Look, Find, Read, Touch and Describe.' Museums and historic sites typically provide more hands-on learning than any classroom setting, when learning about the past (Cowan and Maitles, 2011).

The ultimate power of place comes from learning to walk the ground, think about the history-shaping significance of the space, and ponder the decisions that went into making it significant (Gallman, 2019). While visiting Washington, D.C., on an 8th-grade school trip in 2019, my students demonstrated their encounters with place with the comments shown in Table 8.1.

After the Visit: Reflecting on Sense of Place

Children need the opportunity to reflect upon and share the experiences and emotions encountered during the visit, whether it is through conversation, composition, or any other form of communication. This must be done in an opened-ended way where they not only talk about what they took from the trip, but also explain where it will take them. We benefit when we discover from the child's perspective how the visit changed them, and how they view the legacy of the event and place.

What young tourists take away from their sense of place while visiting difficult historic sites can speak to their own experiences that they brought into them. They identify with the people's stories and sacrifices. They feel a sense of loss and empathize with those who suffered. Most importantly, our children learn from those people from long ago – and through the places where they are remembered – that our choices matter and our actions can touch generations to come.

Conclusion

Every day thousands of young visitors stand at dark tourism sites that memorialize and commemorate tragedy and death. These dark spaces are often emotionally demanding and complicated, requiring deliberate attention to how children engage with them and learn their difficult histories. Sense of place binds children spiritually and cognitively to a space so that they can make meaning of dark history in a relevant, consequential manner. Young tourists will bring previous experiences and a sense of their own self-identities to every place they go. During these visits children will explore the landscape through their senses. This coalescence of knowledge and senses shapes a new understanding of the place and the experience, establishing a comprehension of both.

It is the author's hope that this chapter encourages future researchers to understand that a more holistic approach needs to be utilized when preparing children to visit difficult heritage and other dark sites. By inviting young

Table 8.1 Examples of Post-Visit Comments Reflecting Sense of Place

Site	Verbatim Comment	Note
George Washington's Mount Vernon	'It was cool being where George Washington was and going inside his and his wife's house. And getting to see what they saw and feel what they felt.'	This student made a direct connection with the lives of George and Martha Washington and with the generation in which they lived.
	'But touching the original hand rail (on the staircase leading to the second floor) was mad neat.'	This student made a connection using the sense of touch. He developed a personal connection with George Washington by touching the same object that Washington touched. This sensorial experience elicited a hauntological response in the young visitor.
Ford's Theater	'…I wish you could see the room it (the assassination) happened in.'	This student wanted to create an interconnection with the space of Abraham Lincoln's assassination by experiencing its area and room through multisensorality. He is disappointed that he could not experience more.
United States Holocaust Memorial Museum	'They had a place in the museum where you got to see the actual shoes and other things that they used. You could still smell the smoke when you walk in there.'	This student connected historical knowledge to activate his sense of place. He mistook the smell on the shoes of musty old leather for the smoke of the crematoriums. His senses were being activated by prior understanding and imagination to place a smell on an object where it did not belong.
	'When going through the museum we saw a thing of shoes and it really hit me.'	The emotional reaction this student felt, due to the deep sadness of the place, heightened his emotive connection to the place through the objects he associated with it.
	'Thank you for preserving these terrible acts against humanity. I feel as if I've been to a huge funeral and then entered Hell. For me, the heaps of shoes were quite powerful as I thought of the now dead feet that once walked the earth in them.'	This student imagined, not just the shoes, but 'the now dead feet' to whom the shoes belonged. She associated the aspects of death and loss with the shoes, and then contemplated what those losses meant for humanity.

tourists to share how they experience a sense of place in these dark spaces, we can foster more meaningful experiences with and for them. Centering young tourists in our research will not only advance tourism scholarship but also contribute to a greater understanding of how children experience place at dark sites throughout the world.

References

Abramson, D. (1996). 'Maya Lin and the 1960s: Monuments, time lines and minimalism', *Critical Inquiry*, 22, pp. 679–709.

Bird, G.R. (2013). 'Place identities in the Normandy landscape of war: Touring the Canadian sites of memory', in White, L. and Frew, E. (eds.) *Dark tourism and place identity: Managing and interpreting dark places*. New York: Routledge, pp. 167–185.

Bird, G.R. (2016). 'Landscape, soundscape and youth: Memorable moments at the 90th commemoration of the Battle of Vimy Ridge', in Reeves, K., Bird, G.R., James, L., Stichelbaut, B., and Bourgeois, J. (eds.) *Battlefield events: Landscape, commemoration and heritage*. London and New York: Routledge, pp. 48–63.

Blair, C., Dickinson, G., and Ott, B.L. (2010). 'Introduction: Rhetoric/Memory/Place', in Dickinson, G., Blair, C., and Ott, B. (eds.) *Places of public memory: The rhetoric of museums and memorials*. Tuscaloosa: The University of Alabama Press, pp. 1–54.

Bourke, J. (2017). 'Children's experiences of their everyday walks through a complex urban landscape of belonging', *Children's Geographies*, 15(1), pp. 93–106.

Bowman, M.S. (2010). 'Tracing Mary Queen of scots', in Dickinson, G., Blair, C. and Ott, B. (eds.) *Places of public memory: The rhetoric of museums and memorials*. Tuscaloosa: The University of Alabama Press, pp. 191–215.

Briggs, L.P., Stedman, R.C., and Krasny, M.E. (2014). 'Photo-elicitation methods in studies of children's sense of place', *Children, Youth and Environments*, 24(3), pp. 153–172.

Carmichael, P.S. (2019). 'An unknown grave', in Gallagher, G.W. and Gallman, J.M. (eds.) *Civil war places: Seeing the conflict through the eyes of its leading historians*. Chapel Hill: The University of North Carolina Press, pp. 42–48.

Cele, S. (2006). *Communicating place: Methods for understanding children's experience of place*. Stockholm: Stockholm University.

Chronis, A. (2005). 'Co-constructing heritage at the Gettysburg Storyscape', *Annals of Tourism Research*, 32(2), pp. 386–406.

Clinton, C. (2019). 'A room of his own', in Gallagher, G.W. and Gallman, J.M. (eds.) *Civil war places: Seeing the conflict through the eyes of its leading historians*. Chapel Hill: The University of North Carolina Press, pp. 177–182.

Cowan, P. and Maitles, H. (2011). '"We saw inhumanity up close". What is gained by school students from Scotland visiting Auschwitz?', *Journal of Curriculum Studies*, 43(2), pp. 163–184.

Craig, B. (1989). 'Interpreting the historic scene: The power of imagination in creating a sense of historic place', in Uzzell, D.L. (ed.) *Heritage interpretation*. London: Bellhaven, pp. 107–112.

Dallago, L., Perkins, D.D., Santinello, M., Boyce, W., Molcho, M., and Morgan, A. (2009). 'Adolescent place attachment, social capital, and perceived safety: A comparison of 13 countries', *American Journal of Community Psychology*, 44, pp. 148–160.

Dehoorne, O. and Jolliffe, L. (2013). 'Dark tourism and place identity in French Guiana', in White, L. and Frew, E. (eds.) *Dark tourism and place identity: Managing and interpreting dark places*. New York: Routledge, pp. 156–166.

Derr, T. (2006). '"Sometimes birds sound like fish": Perspectives on children's place experiences', in Spencer, C. and Blades, M. (eds.) *Children and their environments: Learning, using and designing spaces*. Cambridge: Cambridge University Press, pp. 108–123.

Derrida, J. (1994). *Specters of marx: The state of the debt, the work of mourning, and the new international*. trans. Peggy Kamuf. New York: Routledge, p. 6.

Escalas, J. and Stern, B. (2003). 'Sympathy and empathy: Emotional responses to advertising dramas', *Journal of Consumer Research*, 29, pp. 566–578.

Filene, B. (2010). 'Are we there yet? Children, history, and the power of place', in McRainey, D. and Russick, J. (eds.) *Connecting kids to history with museum exhibitions*. Walnut Creek: Left Coast Press, pp. 173–195.

Gallman, J.M. (2019). 'The triangular field and the Devil's Den', in Gallagher, G.W. and Gallman, J.M. (eds.) *Civil war places: Seeing the conflict through the eyes of its leading historians*. Chapel Hill: The University of North Carolina Press, pp. 23–28.

Glassberg, D. (2001). *Sense of history: The place of the past in american life*. Amherst: University of Massachusetts Press.

Hart, R. (1979). *Children's experience of place*. New York: Irvington.

Ingold, T. and Lee Vergunst, J. (2008). 'Introduction', in Ingold, T. and Lee Vergunst, J. (eds.) *Ways of walking: Ethnography and practice on foot*. London: Routledge, pp. 1–19.

Kerr, M.M., Price, R.H., Savine, D., Ifft, K., and McMullen, M.A. (2017). 'Interpreting terrorism: Learning from children's visitor comments', *Journal of Interpretation Research*, 22(1), pp. 83–100.

Kerr, M.M., Shaffer, A., Hartman, M. (2014). Interpreting the Flight 93 crash for children: A collaborative evaluation project. *Legacy: The Magazine of the National Association for Interpretation*, July/August, 2014.

Lengen, C. and Kistemann, T. (2012). 'Sense of place and place identity: Review of neuroscientific evidence', *Health & Place*, 18(5), pp. 1162–1171.

Lim, M. and Barton, A.C. (2010). 'Exploring insideness in urban children's sense of place', *Journal of Environmental Psychology*, 30(3), pp. 328–337.

Marshall, D. (2004). 'Making sense of remembrance', *Social & Cultural Geography*, 5(1), pp. 38–54.

Pink, S. (2009/2015). *Doing sensory ethnography*. London: Sage.

Price, R.H. and Kerr, M.M. (2018). Child's play at war memorials: Insights from a social media debate. *Journal of Heritage Tourism* 13(2): 167–180.

Ross, N.J. (2007). '"My journey to school…": Foregrounding the meaning of school journeys and children's engagements and interactions in their everyday localities', *Children's Geographies*, 5(4), pp. 373–391.

Smith, M. (2010). 'When seeing makes scents', *American Art*, 24(3). Chicago: University of Chicago Press, pp. 12–14.

Stevens, Q. and Franck, K.A. (2016). *Memorials as spaces of engagement: Design, use and meaning*. New York: Routledge.

Stone, P.R. (2009). 'Dark tourism: Morality and new moral spaces', in Sharpley, R. and Stone, P.R. (eds.) *The darker side of travel: The theory and practice of dark tourism*. Bristol: Channel View Publications, pp. 56–72.

Tilden, F. (2007). *Interpreting our heritage*. 4th ed. Chapel Hill: The University of North Carolina Press.

Turner, R. (1990). 'Bloodless battles: The Civil War reenacted', *The Drama Review*, 34(4), pp. 123–136.

Uzzell, D. and Ballantyne, R. (1998). 'Heritage that hurts: Interpretation in a post-modern world', in Uzzell, D. and Ballantyne, R. (eds.) *Contemporary issues in heritage and environmental interpretation*. London: The Stationary Office, pp. 152–171.

Vander Ark, T., Liebtag, E., and McClennon, N. (2020). *The power of place: Authentic learning through place-based education*. Alexandria: ASCD.

Vergeront, J. (2013). 'Place Matters', *Museum Notes*, [online]. Available at https://museumnotes.blogspot.com/2013/07/place-matters.html (Accessed: 31 January 2019).

Willard, P., Lade, C., and Frost, W. (2013). 'Dark beyond memory: The battlefields at Culloden and Little Bighorn', in White, L. and Frew, E. (eds.) *Dark tourism and place identity: Managing and interpreting dark places*. New York: Routledge, pp. 264–275.

Wineburg, S. (2001). *Historical thinking and other unnatural acts: Charting the future of teaching the past*. Philadelphia: Temple University Press.

Wylie, J. (2007). *Landscape*. London: Routledge.

Yi'En, C. (2014). 'Telling stories of the city: Walking ethnography, affective materialities, and mobile encounters', *Space and Culture*, 17(3), pp. 211–223.

Youngblood, L. (2004). 'A sense of place', *Legacy Magazine*, 15(4), pp. 8–9.

9 Difficult Heritage and the Digital Child

Challenges and Opportunities

Gregory J. Wittig

Introduction

Digital platforms have transformed the way we explore and learn about the world, and have enabled virtual tourism, which eliminates barriers such as financial expense and physical accessibility. Through multimedia technology, touristic sites can offer captivating immersive experiences and present authentic audio and video (e.g., the Black Friday video and the Phone Bank Exhibit at the Johnstown Flood Museum and Visitor Center that will be discussed later). This influence was documented by Roche and Quinn (2017), who note that digital media on heritage sites can inform pre-teens' preconceptions of dark tourism.

Today's young people are growing up in an interconnected world that provides instant access to other people, information, and entertainment. With the advent of smartphones, the digital child is often one click away from anything they want – anytime, anywhere. Research on children's understandings of death and dying has pointed out the prominence of thanatechnology in shaping understanding, including technologies such as social media and internet gaming (Sofka, 1997, 2009). Digital media is ubiquitous, and the incessant flow of information rewires the neural structure of the brain (Cavanaugh et al., 2016). The ability to take in and synthesize a rapid stream of information appears to come at the cost of deep processing and mindful knowledge acquisition (Greenfield, 2009). However, those particular skills seem to be required in order to meaningfully process dark moments in history. As Šorgo et al. (2017) ironically posits, digital natives are not necessarily information literate.

Since dark tourism sites present disturbing, difficult, and important content, it is crucial that educators understand the digital learner. This chapter will address the questions: (1) Who is the digital child? (2) How do we prepare them for dark tourism sites? (3) How should dark tourism sites respond to this new type of learner?

Digital Natives in the Digital World

It has been 20 years since Marc Prensky wrote the prescient piece *Digital Natives, Digital Immigrants* – coining the term digital native, which represents an often overused and misunderstood concept. Simply stated, the

DOI: 10.4324/9781003032199-12

students to whom he referred were native speakers of the digital language of computers, video games, and the internet. Digital natives do not know of a time without ubiquitous technological connectivity. As much as Prensky was bullish on how it would positively transform education in the next 20 years, he warned, 'Our students have changed radically. Today's students are no longer the people our educational system was designed to teach' (2001). As such, he added, 'We need to invent Digital Native methodologies for all subjects, at all levels, using our students to guide us' (6). Prensky's warning in 2001 feels more like an admonishment in 2020.

Since Prensky's challenge to educators two decades ago, the speed of technological advancement has exponentially exploded. It has typically outpaced the speed of society's ability to process and effectively utilize the new technology. That has ushered in what is now referred to as disruptive technologies: Inventions or innovations that change the way consumers, industries, and businesses operate. In doing so, they displace the existing systems. For example, the smartphone replaced landlines and made a legacy industry obsolete, while creating an entirely new market. Other disruptive technologies include smart devices, driverless cars, drones, nanotechnology, robotics, renewable energy, cognitive science, medical innovation, social media, and AI.

The change has been so dramatic that historians refer to the current era as the *Fourth Industrial Revolution*. As was the case in previous revolutions, there are both positive and negative impacts. Disruptive technologies have already changed the world on a global level by reorganizing and driving the world economy. On an individual level, they have altered how we live, socialize, and manage our day-to-day lives. Digital technologies are not passive tools. Each tool influences and shapes interests, ways of seeing, and habits of thinking. For digital natives, the terrain of their existence is mitigated through such technology, shaping the very people they become. Cavanaugh et al. (2016) state it this way: 'So, in what amounts to a vast unrehearsed experiment, digital technology has culturally embedded itself in not only our students' social development but also their educational foundation from Kindergarten forward' (2016, p. 378).

Julie M. Albright's *Left to Their Own Devices: How Digital Natives Are Reshaping the American Dream* describes the world of the digital native as *untethered* and disconnected from the institutions, customs, and expectations that have held society together for millennia (2019). 'New types of values are emerging, facilitated by our digital connectivity, associated with young, untethered adults, values that are also beginning to spread to connected individuals of other generations' (73).

In her book, Albright identifies some of these emerging values: (1) *The Desire for a Technologically Mediated World* in which young people prefer to interact with others through technology than in real life; (2) *Experiences over Acquisitions*, in which young people value experiences over the ownership of things; (3) *Transactors versus Owners*, in which sharing and trading is valued over cash currency; (4) *I Want It Now*, in which young people expect immediate gratification; (5) *Customizable World*, in which young people expect to get what they want the way they want it; and (6) *The End of Trust*, in which the deregulation of the media fragments shared values – resulting in tribalism.

How embedded into this new technology are young people? Here is one perspective: The U.S. 2019 Common-Sense Census report stated that 'On average, 8-to-12-year-olds in this country use just under five hours' worth of entertainment screen media per day (4:44), and teens use an average of just under seven and a half hours' worth (7:22)—not including time spent using screens for school or homework' (Rideout and Robb, 2019, p. 3).

The report states that 53% of children in the United States own a smartphone by the age of 11, and 84% of U.S. teenagers now have their own phones. 'More than twice as many young people watch videos every day than they did in 2015, and the average time spent watching has roughly doubled' (Rideout and Robb, 2019, p. 5).

As for social media usage, 'Among 16 to 18-year-olds who use social media, the median age of first use is 14; twenty-eight percent say they started before age 13, 43% at 13 or 14 years old, and 30% not until they were 15 or older.' In 2019, the average amount of time teens reported spending on social media each day was 70 minutes. Sixty-three percent said they used it every day (Rideout and Robb, 2019, p. 39). According to the Pew Research Center, 'YouTube, Instagram and Snapchat are the most popular online platforms among teens. Fully 95% of teens have access to a smartphone, and 45% say they are online "almost constantly"' (2019). As each year passes and each new innovation emerges, young people are increasingly plugged in and living in what Albright describes as a technologically mitigated reality.

We cannot understand the digital child unless we have an idea of the digital world in which they live. The speed of technological advancement is historically unprecedented and growing at exponential rates. On the horizon are more disruptive innovations. How do we as educators, parents, and institutions both understand the digital world and the digital child living in it? How do we help them learn to thrive and grow in such an unprecedented time? The digital child living in the digital world is also a digital learner, and research on digital learners may shed light on how we do this.

The Digital Learner

Even though much has been said about *digital natives*, very few studies carefully investigated the characteristics of this group. Furthermore, there has been no consensus concerning what exactly is a digital native. However, *Let's Talk about Digital Learners in a Digital Era* (Gallardo-Echenique, Marqués-Molías, Bullen, and Strijbos) provides a comprehensive review of literature on the term *digital native*. To make things clearer for the rest of this chapter, we will borrow from them the term *digital learner*. Using the logic of Gallardo-Echenique et al., *digital learner* is better because 'it offers a more global vision of the 21st century student in the digital age … and it is more readily suited/usable in practice…' (172).

There is no shortage of information offered by experienced educators on the characteristics of the digital learner. Although that offers beneficial insight, a great deal of it is anecdotal and has not been verified through rigorous and peer-reviewed research. However, there are consistent overlaps in observations across multiple sources. Those reveal that digital learners prefer receiving information from multiple/hyperlinked digital sources and prefer parallel processing and multi-tasking. They prefer processing pictures, sounds, color, and video before they process text. They unconsciously read text on a page or screen in a fast pattern and prefer to network and collaborate simultaneously with many others. Digital learners prefer just-in-time learning or learning that is needed in the moment. They are looking for instant solutions to immediate problems. Digital learners are trans-fluent between digital and real worlds, and prefer learning that is simultaneously relevant, active, instantly useful, and fun (Jukes and Schaaf).

The following section offers a case study involving a school trip with students who are digital learners. The study illustrates the impact of a well-designed and organized dark heritage site upon students who are prepared ahead of time to help ensure that they are more receptive to the experience.

Case Study

It takes a school bus an hour and a half to drive the 60 miles from Pittsburgh to Johnstown – pretty much a straight shot on 22-East. We went on a very warm and sunny morning in late May. The bus windows were down and the wind whipped kids' hair around. Everyone was enjoying the ride, the sunshine, and their friendships. The bus was carrying my eighth-grade U.S. History class, Dr. Mary Margaret Kerr from the University of Pittsburgh who hosted the trip, my student teacher, and myself.

Dr. Kerr had invited our class to tour the Johnstown Flood Museum as a part of her continuing work on children and dark heritage sites. This wasn't the first time I worked with Dr. Kerr. Five years earlier, one of my sixth-grade classes (as well as a colleague's fourth-grade class) piloted and gave feedback on the Junior Ranger Program booklets that Dr. Kerr and her team developed for the National Park Service. The goal of those booklets was to explain the topics surrounding Flight 93 and 9/11 to elementary school-aged children. My eighth graders were third graders when the trip to the Flight 93 memorial site occurred, so they had a sense that this trip was also important. Although it was the end of the school year of their final year at Falk School, they balanced their end-of-the-year crazies with a respect for the space they were about to experience. On the bus, it was all joyful chatter with the occasional, 'How much longer? I have to go to the bathroom?'

The Johnstown Flood occurred on May 31, 1889, after the catastrophic failure of the South Fork Dam, located on the south fork of the Little Conemaugh River, 14 miles upstream from the town of Johnstown, Pennsylvania. The plan was to go directly to the museum at the site of the dam break, then take a short detour to the South Fork Fishing and Hunting Club, which is historically relevant to the flood. We would stop for lunch at the Little Conemaugh river overlook at the bottom of the dam site, then continue to the Flood Museum in Johnstown itself.

In class, we had been discussing the impact of industrialization on the working class and the rise of the robber barons as a part of our Gilded Age unit of study. The Johnstown Flood fit perfectly into our studies because the catastrophic failure of the South Fork Dam was the result of the negligence and apathy of the industrialist owners and wealthy patrons of the South Fork Hunting and Fishing Camp. The actions and inactions of the South Fork Hunting and Fishing Club led to (and perpetuated) the devastation and disregard for the poor. Club patrons included Andrew Carnegie, Henry Clay Frick, Andrew Mellon, and Martin F. Scaife. One of our overarching questions was whether we lived in a second Gilded Age based on new technologies, greater wealth disparity, and diminishing working conditions and rights.

We arrived at the top of the dam and unloaded the buses around 10:30 that morning. As Dr. Kerr went in to meet the staff at the Museum, the students and I gathered in a grassy area to the side of the museum building, where there was a view of the top of the dam and the flood path down the valley toward Johnstown. In the opposite direction, the Hunting and Fishing Club still stands in the spot it did when the dam failed. Students quickly wondered where the dam was. When I pointed to the area where the dam collapsed and the pitch heading down the valley toward Johnstown, the scope and horror of the event became immediately apparent. The distance between the hunting lodge across the valley and where we stood next to the dam was more than a mile away. All of that space was once filled with water. The void left behind was staggering.

In my 30 years of teaching middle school, I have learned a lot about how young people occupy the space in which they are placed. It is controlled by several variables, only one of which teachers have any influence – namely, the work put into it ahead of time. First, the instructional team can immerse students in authentic ways to grasp the content, and this step is indispensable. Authenticity for a middle schooler growing up in the digital world is no easy task. Second, the thought and work put into the visit ahead of time from those managing the host site are critical. Similarly, it is important that the site itself is authentic. Was the exhibit organized and designed to authentically engage young people? Did site designers and organizers consider the developmental stages of the young visitors? Finally, the actual space itself is crucial. In relation to dark heritage sites, the physical space itself induces a deeper look, and the view we had overlooking the dam is a good example. The sun hits the spot in a certain way, the silence invites one to be quiet, and the size and scope make one feel small and insignificant.

Many variables go into how each child responds to a particular space. The teacher provides the doorway, the site hosts provide the content, and the space provides the magic. Once a space has been designed, we have zero control over it and can only help our students be open and receptive to it.

Opening to space is no easy task, especially for adolescents who have grown up with ubiquitous connectivity. They feel more comfortable with noise and movement, prefer learning on demand, and are socialized into a state of continuous partial attention.

Adults often want kids to pay attention, and typically what they mean by that is they want the kids to narrow the focus so much that everything but what is to be attended to is in view. However, that is the exact opposite of the attention needed to open oneself to a space. For that purpose, being attentive means being open to whatever happens. The focus is soft, not directed, allowing one to receive any and all ideas and impressions. It is an emptying of intent, of goals, and of self that is only achieved over time. Students go through stages of discomfort, self-consciousness, boredom, impatience, and the urge to talk or interact with a peer. Ultimately, if all goes well, those things evaporate like a morning fog and are replaced by the potential and possibility to perceive things differently.

In preparing my students for the trip to Johnstown, I introduced them to the story of Victor Heiser, the 16-year-old son of a local merchant. On the day the dam broke, Victor waded through the knee-deep water between his house and barn. It was just after 4 p.m. on May 31, 1889, and his father had instructed him to untie the horses so they wouldn't drown if the water rose higher. When Victor started back toward the house, he caught sight of his parents in a second-story window, frantically motioning for him to turn back. Victor scrambled to the barn's roof just in time to see a mountain of water and debris crush his family home.

There was something about Victor Heiser that resonated with my students. Every day, they see, hear, or read about some tragedy like this one, or even things more horrific. The cultural surround and the continuous 24-hour news cycle plays and replays spectacle after spectacle. Movies and TV shows display the most terrifying and gruesome details imaginable. Hyper-real video games, unfiltered online videos of violence, and the ability to always change the channel can result in a detached view of real events. However, the story of Victor Heiser somehow stuck with these kids, where a story of a suicide bombing or mass shooting had not. Was it because it was written and not viewed? Was it because it was a young person as the protagonist?

After we got everyone off the buses, we lined up to go inside, where we would meet our guide. The museum sits at the top of where the dam once stood. In the distance, the Hunting Camp sat like a mute witness to greed. The first thing that catches a visitor's attention upon entering the main hall is a replica of the infamous photograph of a house with a tree jabbed into the side of it. The students had seen this picture before, but the scale of the display was jarring. The museum occupies two floors and this exhibit uses both floors to make its impact. Students were instantly drawn to it. Maybe it was something about the tableaux of it, the incongruent shape, or the house decontextualized from its surroundings. The exhibit includes a bank of phones that featured the actual voice of Victor Heiser recounting his story. This proved to be one of the more impactful features of the exhibit, and the juxtaposition of the house and the sound of Victor Heiser's voice woke up some students in a profound and connective way.

My students had been in many museums before and glossed by the displays pushing ever forward to the exit as if on a conveyer belt. Field trips are coded as easy days, blow-off days, and opportunities to be with friends. Students typically look forward to sitting with friends on the bus, eating lunch in a different space, and hitting the gift shops. Trying to break through this force field is difficult. If a heritage site or museum doesn't find a way to change the energy, then armies of young people will march in and out and take away little, if any, connection.

Before we left the Lake View Visitor Center, we watched a documentary called Black Friday. As we were getting settled, a representative from the visitor center came in and set up the context of what we were about to see. They discussed that this was a very powerful, and at times frightening, film. The docent told us that if it became too much for anyone, they were free to leave and sit in the hall. For our group, this was more of a sales pitch than a warning. But this isn't always the case when dealing with graphic violence and destruction and young people.

In the digital world, there are young people who have consumed so much violent and graphic imagery that they are more desensitized to it. Others, who have been kept away from such content, lack the ability to confront the ugly details. Both types of youth experience raise questions about the impact of the digital landscape. That is why, on any trip to a space like this, it is necessary to know your students (or own child). Ask and seek answers to the compound question, 'How do I ease this young person to confront such content and, conversely, what do I have to do ahead of time to help the desensitized wake up to the content, so it is not just another screen to swipe?'

Historical empathy is difficult to achieve. The conditions need to be right, and the set-up needs to be well-thought-out and sequenced. The movie *Black Friday* was made for the museum and was designed to make the audience have a visceral experience. When the auditorium went completely dark, only the sound of raindrops – slow at first and in the distance, flooded the room. The sound of thunder and the advancing storm permeated the darkness. In a visual world where children are bombarded by images and spectacles on their screens, it is often sound that shocks them to consciousness. The video began and followed the water to Johnstown, taking my students along. Victor Heiser became more real; maybe for a second students became individual Victor Heisers in the darkness of the auditorium, as they imagined floating away from their families on a barn door.

Twenty minutes later we were on the bus for the short ride down the hill to the bottom of the dam. It was around 80 degrees and sunny – a beautiful May afternoon that was a prelude to summer vacation. The energy was big. Kids spread out in the grassy area on the hill overlooking the Little Conemaugh river and let go of anything related to the flood, the hunting lodge, or the loss of life. The most interesting thing to them was their food, their friends, and the path that led down to the river, which was only a foot deep and 20 feet wide – hardly the 70-foot wall of carnage they had learned about. When the bus wouldn't start, we were faced with a two-hour wait and the relentless temptation to play in the water.

Things began as they do with young people. They skipped stones on the water, and then one student named Jacob plunked a big rock off the bank into the small river. It didn't go too far in, but where it landed the currents quickly rushed around it. Before long, Jacob and some friends began to carry more rocks, but instead of chucking them in, they began to stack them side by side and on top of each other. Dr. Kerr and I watched in amazement as the students' play became something more than we could have expected and at the time didn't have words for. Over the next hour, they constructed a dam at the base of the dam that had broken so long ago.

There was some talk about how the water ran through this narrow channel and how terrifying it must have been, as they stood in the Little Conemaugh, quietly passing boulders down the line to build their own dam. Looking up at the corridor where 20 million tons of water careened and banked its way to Johnstown, I realized that play is more than play, and learning is deeper than facts. Field trips to dark heritage sites require prep work on both ends, but also benefit significantly from an openness to the story the space itself wants to tell.

Conclusions and Implications

The above case study provides a glimpse into how one group of digitally native middle school students interacted with a dark heritage site, and the role educators and curators played in that experience. This section will use the information from the opening of the chapter as a lens in which to examine the case study Then we will suggest areas for further research and provide practical recommendations to educators planning student trips to heritage sites.

As Jukes and Schaaf (2019) point out, digital learners are trans-fluent between digital and real worlds, and they prefer learning that is simultaneously relevant, active, instantly useful, and fun. In many ways, the Johnstown Flood field trip addressed these preferences and, as a result, students were invited to learn and understand at a deeper level. Students learned from intellectual, empathetic, physical, and emotional spaces.

Many things made this trip successful. It flowed easily between the digital and concrete worlds. For example, it started outside by surveying the area of the dam, then it moved inside the visitor center to interact with the exhibits, many of which were digital. Next, it went to the South Fork Fishing and Hunting Club lodges and housing, and finally transitioned back outside into the natural space.

The digital tools engaged the intellect as well as the senses, too, and were effectively married to the content. According to the Smithsonian Learning Lab Research Team Report (2017), 'Learning flow is more consistent and more readily supports achievement when there is a high level of correlation or alignment between content, objects/resources, visual supports or media, and tasks to aid in persistence and minimize cognitive load' (p. 21). The museum's digital exhibits authentically engaged the students. For example, for the phone bank exhibit, the curator chose to use the actual audio recording of Victor Heiser retelling his experience of surviving the flood by riding it out on a barn door. That was noted by a few of the students. The multisensory movie presentation of the flood narrative is a similar example of technology serving and supporting content. The content was appropriately chosen for the digital medium and the students engaged with it organically.

The content set-up was thorough, on both the school front and the museum's part, so that students had a deeper understanding of the tragedy. The instructors provided extensive background, engaged the students in critical debate on culpability, and helped students make connections to current events and issues. Each exhibit is an engineered and orchestrated semiotic construction *ordered and controlled* by the curators. Sometimes the narratives do not align. When they do, the learning experience is dynamically enhanced. The fact that the museum's narrative matched that of the educator cannot be overlooked. Dark tourism encounters are, in reality, not encounters with death but *remembrances* of the dead induced by symbolic representations – what Seaton calls engineered and orchestrated remembrances (2018).

Finally, one thing that made the trip fruitful was that the physical space was accessible and students could interact with it. Being able to inhabit the actual space in which the tragedy occurred transformed the museum trip into an experience. For those students, the memory of moving those rocks at the base of the dam will likely never go away. They know the feel of the water on their feet in the same path where the 70-foot wave flooded the site Students were armed with the appropriate level of historical background and engaged with the underlying issues through relevant and timely discussions and debate. Then they were placed in the actual physical space of the tragedy. That kind of complete, holistic experience can be the closest one is able to get to bearing become the closest witness to the past.

Areas for Exploration

One area for further exploration is the effect of contested historical narratives around dark heritage sites. How does the digital child navigate competing ideologies and narratives?

How does the commercialization of dark narratives for profit and political gain impact the young digital learner's experience and intellectual and emotional takeaways? In the case of the Johnstown Flood, did the heritage site curators frame the owners of the Hunting and Fishing Club as robber barons or captains of industry? Did they frame it as an act of nature or the result of human neglect by members of an elite club? Who is culpable? How will the digital learner negotiate competing frames, and what will that mean to their understanding of what constitutes the truth?

Another area worthy of exploration is the degree to which digital learners prefer interacting with others through technology, rather than face-to-face, as Albright asserts. Oftentimes, students become distracted by technology as a toy and not a tool. Students in this case study, however, connected deeply with the digital interface. It remains unclear how much of that was a result of teacher preparation and how much can be attributed to the impact of the technological medium. Since the students had already built a connection to the Heiser narrative, was it simply serendipitous that the phone bank exhibit was effective? Would it have been equally successful if the narrative was told through another medium or if a different personal testimonial was used?

To complicate Albright's assumption further, the spontaneous play at the base of the dam was, in many ways, the one thing that engaged the students most viscerally. This was direct, not technological engagement. However, Albright's assertion that young people value experiences over the ownership of things is seen more clearly in the spontaneous play example.

A third area worthy of further exploration is the role teachers, heritage site creators (curators), and parents play in how the digital child processes dark heritage sites. To what extent is a digital child's experiences at a dark heritage site a product of the efforts put in ahead of time by the educators and curators? How much of it is created by the impact of the site itself? What more can be done to prepare the digital child to encounter a dark heritage site?

Recommendations for Educators and Curators

Since current theories and research methods in dark tourism do not account for the differences between children and adults, what are some ways stakeholders can respond to the digital child? As Kerr et al. (2021) observed,

> To safeguard children while exploring fully what they think, feel, do, and remember at dark sites, we need additional research and different approaches. Fortunately, other disciplines have worked for decades to develop child-centered research methods, and we can adopt these to advance our efforts.

(p. 4)

The Smithsonian Learning Lab is the Smithsonian's major rethinking of how the digital resources from across the Smithsonian's 19 museums can be used together for learning. The Learning Lab has stated that

> Developing and sustaining an online learning community focused on inquiry and learning is crucial in helping students access both their instructors and peers. Sharing their thinking, their findings, and their learning processes, and having access to those of their peers, helps validate work approach, keeps students engaged, and provides an opportunity to blend social, cognitive, and teaching dynamics.

(2017, p. 21)

Educators need to become more actively engaged in preparing young people to encounter dark heritage sites through multiple methods and mediums. For example, the posing of powerful essential questions to guide student connection and understanding is fundamental. Essential questions make for an effective springboard for initial engagement into the complexity of dark heritage sites. Essential questions are defined in the sense that they are genuine, important, and necessarily ongoing lines of inquiry (Katifori et al. 2019). In *Approaching 'Dark Heritage through Essential Questions*, Katifori identifies essential questions as relatable to personal experiences, views of the world, and issues that feel relevant to people's lives. They are open-ended, there are no right or wrong answers, they are intellectually engaging, and are able to spark meaningful discussion among a group of people. They require high-order thinking such as analysis, inference, debate. When preparing for the Johnstown Flood trip, essential questions may have included the following: To what extent does extreme wealth blind a person to the suffering of the poor? What role should the government play in national tragedies? How do people handle, cope, and manage the tragic and horrific loss of family?

Educators could follow the lead of some organizations that have effectively engaged in digital storytelling for heritage sites. Digital storytelling at its most basic level is the use of multi-media tools to tell a story. For example, the CHESS Experience advertises their digital storytelling as 'mobile and augmented reality technologies to turn an ordinary museum visit into an extraordinarily personalized storytelling experience.' Co-funded by the European Commission, the CHESS project 'aims to integrate interdisciplinary research in personalization and adaptivity, digital storytelling, interaction methodologies, and narrative-oriented mobile and mixed reality technologies.' Emotive, an EU-funded heritage project, 'aims to use emotional storytelling to dramatically change how we experience heritage sites.' Emotive believes that 'drama based narratives containing careful references to a site's cultural content has the power to transform heritage and museum visitor experiences through a deepening knowledge transfer.' Digital storytelling does not have to be created by professionals; students can do it to make sense of their dark heritage site visits. Digital storytelling is already being practiced in schools, colleges, libraries, community centers, and businesses. Inviting students into the storytelling process can only deepen their connection.

Curators and heritage site developers need a finer understanding of the child. This would include first a detailed and robust understanding of the developmental stages of childhood, and too a process to collect the voices and opinions of digital learners as they encounter the curated space. The child can no longer be an afterthought and must be included as a stakeholder. This is especially relevant since heritage sites and other spaces of collective memory are using more and more technology to construct their narrative. The numbers of digital natives grow exponentially each year as does the desire for dark heritage sites to embed technology into their interface. By not soliciting the expertise of digital natives and inviting them into being co-creators of the heritage experience, invaluable insight is lost. Digital learners will want the experience to be authentic and not phony. The Smithsonian Learning Lab recommends 'mutual problem-solving or co-development of learning products helps young students make more meaningful connections to their learning and to one another through establishing relationships focused on learning outcomes' (2017, p. 21). Phoniness is a deal-breaker for young people.

Curators and heritage site developers should explore problem-based exhibits and technologies that allow students to use *just-in-time* research as well as the dark heritage site's content to solve *real-world problems*. Making it both peer collaborative and technologically interactive, students could use the site as a resource to resolve real-world problems in a manner that respects their agency and autonomy and is developmentally appropriate. Activities would be developed to match the developmental ages of the children: Something for early childhood, primary, middle, and high school levels. There are questions about the use of smart devices in the classroom that extend to museums, heritage sites, and memorials. Like schools, dark heritage sites will also have to address how these can be used effectively in concert with the site's content. For example, smartphones could be intentionally used for *just-in-time* learning that encourages the students to use their smart device as a tool and not a toy. This may make more sense for heritage sites to employ since most sites don't prohibit the use of smart devices like most schools do. Digital native learners already know how to use their phone as a tool for problem-solving in their individual and social lives. Schools, museums, parents, and heritage sites could tap into this already advanced skill and use it to more deeply connect and understand their worlds, both present and past.

With the ever-increasing use of digital technology and the numbers of digital natives growing by the year, it will become increasingly necessary for schools, parents, heritage sites, and museums to adjust and adapt to how they educate and engage young people. We must look at technology as a tool (and not a toy) that can integrate into a real-world experience without obscuring or replacing it.

References

Albright, Julie M. *Left to Their Own Devices: How Digital Natives Are Reshaping the American Dream.* 1st ed. Amherst, NY: Prometheus Books, 2019.

Cavanaugh, J.M., Giapponi, C.C., and Golden, T.D. Digital Technology and Student Cognitive Development: The Neuroscience of the University Classroom. *Journal of Management Education.* 2016;40(4):374–397. doi:10.1177/1052562915614051.

Frew, E. Exhibiting Death and Disaster: Museological Perspectives. In: Stone, P.R., Hartmann, R., Seaton, T., Sharpley, R., White, L. (eds) *The Palgrave Handbook of Dark Tourism Studies.* London: Palgrave Macmillan, 2018. doi:10.1057/978-1-137-47566-4_28.

Gallardo-Echenique, E.E., Marqués-Molías, L., Bullen, M., and Strijbos, J.-W. Let's Talk about Digital Learners in the Digital Era. *The International Review of Research in Open and Distributed Learning.* 2015;16(3). doi:10.19173/irrodl.v16i3.2196.

Greenfield, P.M. Technology and Informal Education: What Is Taught, What Is Learned. *Science: New Visions in Neuroscience.* 2009;323:69–71.

Jukes, Ian and Schaaf, Ryan L. *A Brief History of the Future of Education: Learning in the Age of Disruption.* Corwin: Sage Publishing Co. London, UK, 2019.

Katifori, Akrivi, Restrepo Lopez, Klaoudia Marsella, Petousi, Dimitra, Karvounis, Manos, Kourtis, Vassilis, Roussou, Maria, and Ioannidis, Yannis. Approaching "Dark Heritage" Through Essential Questions: An Interactive Digital Storytelling Museum Experience. *MW19: MW 2019.* Published January 13, 2019. Consulted June 28, 2021.

Kerr, Mary Margaret, et al. Young Tourists' Experiences at Dark Tourism Sites: Towards a Conceptual Framework. *Tourist Studies.* 2021;21(2):198–218. doi:10.1177/1468797620959048.

Prensky, M. Digital Natives, Digital Immigrants Part 1. *On the Horizon.* 2001;9(5):1–6. doi:10.1108/10748120110424816.

Rideout, Victoria and Robb, Micahel B. *The Common Sense Census: Media Use by Tweens and Teens.* San Francisco, CA: Common Sense Media, 2019.

Roche, D. and Quinn, B. Heritage Sites and Schoolchildren: Insights from the Battle of the Boyne. *Journal of Heritage Tourism.* 2017;12(1):7–20.

Seaton, T. Encountering Engineered and Orchestrated Remembrance: A Situational Model of Dark Tourism and Its History. In: Stone, P.R., Hartmann, R., Seaton, T., Sharpley, R., and White, L. (eds) *The Palgrave Handbook of Dark Tourism Studies.* London: Palgrave Macmillan, 2018. doi:10.1057/978-1-137-47566-4_1

Smithsonian Center for Learning and Digital Access and Navigation North Learning Solutions. Characteristics of Digital Learning Content, Pedagogies, and Platforms That Support Young Learners, An Analysis of Existing Literature and Research. Washington, DC, 2017.

Sofka CJ (1997) Social support "internetworks," caskets for sale, and more: Thanatology and the information superhighway. *Death Studies* 21(6): 553–574.

Sofka CJ (2009) Adolescents, technology, and the internet: Coping with loss in the digital world. In: Balk D and Corr C (eds) Adolescent Encounters with Death, Bereavement, and Coping. New York: Springer, pp.155–173.

Šorgo, A., Bartol, T., Dolničar, D., and Boh Podgornik, B. Attributes of Digital Natives as Predictors of Information Literacy in Higher Education. *British Journal of Educational Technology.* 2017;48:749–767. doi:10.1111/bjet.12451.

Part IV

Children within Dark Tourism

Contexts and Experiences

10 'Why Is It So Fun to Be Scared?'

Entertainment in Dark Tourism

Margee Kerr

Introduction

If one were to approach an educator or parent with the suggestion that they take their children to 'sites associated with death, suffering, and the seemingly macabre' (Stone 2006: 146), many would likely respond with a sentiment along the lines of 'you must be joking!' They may further suggest that children are not capable of understanding such content or, worse still, it could cause lasting trauma. While children may not have the cognitive capabilities to engage in deep contemplation of mortality and fatality, starting at about the age of 4 years, they do begin to grasp that death is irreversible or permanent. Between the ages of 5 and 7 years, children begin to start to understand the concept of *nonfunctionality* – whereby a thing that is dead can't think, feel, or dream. By around the age of 10 years, children begin to comprehend the universality of death, that is – everything living eventually dies. Prior to this, it is not uncommon for children to think that only some people or things die: For example, they may only think family pets die, or the elderly (Feldman, 2013). In other words, while parents and teachers may not like to think about their children or students engaging with morose or macabre ideas, they likely already are. However, children's and adolescents' dark tourism experiences, especially at visitor sites of recent or distant trauma, as well as memorials to the significant dead, are given little attention in the dark tourism literature (Kerr and Price, 2018). Therefore, the purpose of this chapter is to explore how and why youth engage in dark tourism. Specifically, this chapter explores how youth use recreation as a means of engaging with overwhelming, often frightening realities including mortality and fatality. Through a frame of recreation, youth can learn about these difficult inevitabilities, practice how to manage them, and cultivate resilience.

Dark Tourism and Children: Toward a Fearful Delight

Concerns regarding traumatization even through vicarious exposure to suffering are warranted. Indeed, studies of children who watched the events of 9/11 unfold live on television found some children develop symptoms of

DOI: 10.4324/9781003032199-14

post-traumatic stress disorder (Holmes, Creswell, and O'Connor, 2007). However, exposure to the macabre does not happen in a vacuum but, rather, in a social context which works to frame how children construct meaning of the experience. Presumed 'negative' content (i.e., mortality and fatality) does not inherently elicit 'negative' emotional experiences (for instance, fear, sadness, anxiety, etc.) Rather, our emotions are constructed in the moment from previous experiences and learning, sensory inputs, as well as social context (Gendron, Crivelli and Barrett, 2018; Hoemann, Xu and Barrett, 2019). For example, emotional reactions in the moment of a child watching a terrorist attack live on television, along with how they feel recalling the experience in the future will be different from reactions and meaning-making of a child visiting a memorial of the same attack during a school field trip. Indeed, in their study of artifacts left by children visiting the 9/11 Flight 93 memorial near Shanksville, Pennsylvania, Kerr and Price (2018: 572) found that '...when children visited this dark site, they left cheerful toys, simple words, and brightly colored drawings in striking contrast to somber wreathes, mourners' messages, religious medals, and photographs of the passengers and crew'. Consequently, children displaying cheerful sentiments are a reminder that engaging with mortality and fatality can result in positive emotional experiences.

However, much of dark tourism research to date has focused on sites with a history of tragedy and identifying and investigating the emotions that result from such experiences, which are often categorized as negative emotions (e.g., suffering, sadness, fear) (Seaton, 2018). This is in part due to the tendency to give more attention and notoriety to death that occurs under tragic conditions, but it is also in part due to early conceptualizations of dark tourism that centered definitions on encounters with death (Seaton, 2018). This is problematic, Seaton (2018: 13) argues, because 'death is unknown and permanently unknowable as no one returns to report back on the experience'. Therefore, dark tourism is better understood as the remembrance of death through symbolic representations (Seaton, 2018). Specifically, Seaton (2018: 14) defines dark tourism as 'encounters through travel with the engineered and orchestrated remembrance of mortality and fatality'. This definition offers a critical shift in focus from previous definitions. Firstly, understanding dark tourism as a *remembrance* of death shifts the focus of inquiry to the living, to the people who create such remembrances, and to those who engage with them. Secondly, describing dark tourism as *engineering* and *orchestration* highlights that it is an ongoing production, one carried out by the living who bring with it their own motivations and goals, and who are influenced by the larger socio-cultural factors of time and place. This definition then allows for layered investigations into how a site or experience may change over time under new producers and the role of agency in both production and engagement. Attention to human agency in dark tourism research is important because, as discussed further below, a sense of agency is critical in confronting and managing overwhelming ideas and emotions like mortality and fatality (Maier, 2015).

Finally, the definition distinguishes between mortality, which refers to the inevitability of death, typically from natural causes, and fatality, which refers to unexpected death, typically resulting from violent or extraordinary conditions (Seaton, 2018). Specifying *mortality* and *fatality* acknowledges that while the conditions of death significantly influence the form and content of remembrance, both fall under the umbrella of dark tourism. In other words, dark tourism is not exclusive to the remembrance of only tragic, unjust, or untimely death. Rather, this definition allows for *all* engineered and orchestrated remembrances of death, from war memorials to haunted attractions, thereby also making space for understanding and investigating a wide range of intentions, motivations, and emotional experiences within dark tourism.

Numerous studies examine the diverse motivations for adults choosing to visit dark tourism sites, where the engineering and orchestration of remembrance (EOR) of mortality and fatality is designed to encourage quiet and meaningful reflection (Stone et al., 2018). This often results in patrons experiencing a sense of empathy, sadness, and mourning for victims, or pride and gratitude for fallen heroes (Walter, 2004, 2009). Hence, choosing to engage with content expected to produce 'negative' emotional states – for example, watching 'tearjerker' films – can be thought of as opportunities where we are 'glad to be sad' (Rozin et al., 2013). That said, however, too often left out of research are EOR of mortality and fatality are not meant to inspire *fearful remorse* but *fearful delight*, where the tone of the environment is not somber but spirited and animated. Of course, social contexts in which this is considered appropriate are strictly defined. Indeed, according to the norms and values of Western society, the subject of death, dying, suffering and brutality are not to be taken lightly; to do so is considered highly disrespectful, crass, and vulgar. However, within the context of fiction and fantasy, even the most sensational depictions of the macabre and engagements with mortality and fatality are suitable for consumption, and consume we do. With this in mind, we now turn to studies investigating the landscape of recreational horror.

Recreational Horror

Recreational horror here is defined broadly as dark tourism encounters which inspire or intend to inspire both positive and negative effects (e.g., fear and joy, terror and thrill, anxiety or excitement). This includes horror-themed entertainment and content designed with the intent of delivering a fearful and fun experience. While not everyone will experience this content as both fun and scary, horror-themed entertainment has significantly grown in popularity over recent decades. Moreover, adolescents are increasingly a target demographic for such content including films, video games, escape rooms, and haunted attractions (Hollywood Reporter, 2006; Gershon, 2019). Indeed, horror films held 12.04% market share in 2020, up from 1.96% in 1996 (The Numbers, 2020), and according to the National Retail Federation, 68% of

Americans planned on celebrating Halloween in 2019, with an average of 20% of the population planning to attend a haunted attraction each year from 2015 to 2019 (National Retail Federation, 2019). After Christmas, Halloween is the largest consumer spending holiday in the US, reaching a record high of $9.1 billion in 2017 (National Retail Federation, 2019). Consumer and market research tracks these changing trends and identifies the common motivations for engaging with horror entertainment which are similar to those of other genres of entertainment, namely having fun with friends, interest in advances in video and gaming technology, wider availability of affordable special effects decorations (e.g., inflatables, animatronics, fog machines, lighting effects), and of course effective marketing campaigns (National Retail Federation, 2019). However, the horror entertainment industry is not the only venue where youth engage in a recreational manner with mortality and fatality. Recreational horror also includes instances where an engineered and orchestrated remembrance is experienced in a rewarding or 'fun' manner, even if that was not the explicit intent; for example, playing hide-seek inside a cemetery. Turning attention to these instances reveals how recreational horror can aid in the learning process.

Learning through Play

Children actively attempt to investigate their curiosities, to learn about themselves, ideas, and the world through imagination and play. Within the context of recreation, ideas and experiences, especially those that are complex or considered dark, can be introduced, explored, and even practiced from a safe distance (Clasen, 2012). Similar to observations in animal models that rough and tumble play is a means of exploring aggression (Panksepp, 1981; Panksepp and Yovell, 2014), recreational horror can be a safe means of learning about fear, danger, disgust, risk, or even mortality (Andersen et al., 2020). For example, studies show children are known to play at dark tourism sites like battlefields and war memorials (Bowman and Pezzullo, 2009; Kerr and Price, 2016). My own observations consulting with museums and attractions support this: I have frequently witnessed and made note of the variety of forms of recreation at dark tourism sites. For instance, in my work as a research consultant for Eastern State Penitentiary Historic Site in 2016 and 2017, I have regularly observed children (under the age of 12) engaging with site elements (e.g., prison cells, shackles, iron gates) through play. Frequent forms of play include one child taking the role of prisoner, while another takes the role of warden and then role-playing scenarios of attempting escape, fleeing capture, of demanding or being denied food, clothing, or water, etc. In adopting these roles, it appears that they are attempting to make sense of the larger context – in this case notions of who and what is 'good' and 'bad' – of what a penitentiary is, what it means to be confined, and who has the ability and authority to confine others. In one representative exchange, a child (a 6- or 7-year-old white male) role-playing the 'prisoner' locked inside one of the prison cells attempted to exit the cell and was quickly stopped by his role-playing partner (also a white male, age 7 or 8) who had adopted the role of the 'warden'. However, before censoring the 'prisoner', the child playing the 'warden' appeared to break character and say to the 'prisoner' in a softer voice a phrase along the lines of 'you're not supposed to do that, you're the prisoner, I'm supposed to tell you when to leave'. Nodding his head, the 'prisoner' returned to his cell and their role-play resumed.

I have also observed instances where similar kinds of role-play are met with a very real censorship from a chaperone or scolding from a parent who deems the behavior inappropriate and disrespectful, thus teaching dominant contextual social norms regarding good and bad behavior. Regardless of whether they were scolded, this kind of physical role-play offers critical learning experiences, especially for children who are still developing abstract thinking and higher-level cognitive processing skills. To comprehend what it means to be a prisoner, they physically confine themselves, to understand what it means to be imprisoned for theft, they pretend to be a thief. In these instances, children can maintain a sense of agency in how meaning is made of new and frightening environments and contexts. Indeed, through recreation, a sense of control can be brought to overwhelming and frightening realities, especially the realities of mortality and fatality.

Recreational Horror as a Means of Managing Fear

In addition to learning about threats and frightening realities, engaging with recreational horror content can be a means of increasing a sense of competency when confronting new or unknown threats (Kerr, Siegle and Orsini, 2019; Scrivner et al., 2021). A study conducted during the COVID-19 pandemic found that self-identified horror fans scored higher on measures of resilience during the pandemic compared to non-horror fans, with fans of apocalyptic themes specifically exhibiting greater resilience and preparedness (Scrivner et al., 2021). The study also assessed participants on trait morbid curiosity, defined as the extent to which an individual is motivated to learn about the dangerous aspects of life (Scrivner, 2020). Results suggest that trait morbid curiosity was also related to resilience (Scrivner et al., 2021). Further, a study investigating how attending a haunted attraction impacts emotions found that overall mood ratings improved significantly from before to after experiencing the attraction, and reports of feeling tired and anxious declined (Kerr et al., 2019). Critical for the discussion here, this study also found that those with high mood ratings were more likely to report that they had challenged their fears, learned something about themselves, and rate the experience as highly intense, thrilling, and scary (Kerr et al., 2019). While not exclusive to children, these studies offer insight into the positive psychological gains that can result from engaging with fears and challenges through recreational horror.

Taking on difficult, but not too difficult challenges in a supportive environment is fundamental to childhood development (Feldman, 2013). Recreational horror can be one means of accomplishing this. My experience working with eighth- and ninth-grade students offers an example of what this process might look like for adolescents, specifically how young minds use fiction to engage with mortality and fatality. My popular-science book, *'Scream: Chilling Adventures in the Science of Fear'* (Kerr, 2015), gained popularity among middle and high school students and, as a result, teachers have frequently requested my advice on teaching fear-related concepts. For instance, a ninth-grade humanities teacher from a technical high school in California asked if I would offer guidance to her students who she had assigned the task of designing and building their own mini-haunted attractions. I met virtually with her classes in 2018 and 2019, advising and brainstorming with groups of four to six students at a time on how to create their mini-haunted attractions. Adopting Seaton's (2018) conceptualization of EOR, a great deal of insight can be gained in examining this student assignment as a case study of adolescence engagement with mortality and fatality.

The mini-haunted house assignment illustrates how EOR is an active process; in this case, one originating with the teachers' own motivations and goals. Her primary motivation, as she explained to me, was to create an engaging and fun opportunity to practice their creative and critical thinking skills, along with skills in engineering and construction. Meanwhile, her goal was to encourage students to think more deeply about fear, about what they are afraid of and why. Before any building could take place, the mini-haunted attraction assignment required students to complete a workbook that included five design sections. Firstly, *Focus Fears*, in which students had to identify four specific fears their attraction would feature (e.g., claustrophobia, fear of the dark, etc.). Secondly, *Background Setting* whereby students were required to decide on and describe the physical location of their story, and thirdly, *Background Characters* in which they had to create and describe the story's characters. The fourth section, *Suspense Arc*, required students to list the specific 'scare points' or startles that would work to build suspense throughout the attraction and that would build to a climactic ending. The final section called *Floor Plan*, entailed students mapping out the physical footprint of the attraction.

My first observation was that students overwhelmingly appeared to spend more time on the *Background* sections compared to the sections *Suspension Arc* and *Floor Plan*. In the *Background* sections, students wrote rich and detailed descriptions of their story, setting, and characters. I also observed that a majority of groups choose fictional locations that fit the description of Erving Goffman's total institutions: Places of confinement, set apart from society where life is formally governed by authority figures and where individuals have limited control of their daily activities (Goffman, 1961). Specifically, students set their attractions in abandoned historic prisons, psychiatric facilities, and boarding schools. Students offered vivid descriptions of dilapidated institutions, making liberal use of dramatic and theatrical adjectives and onomatopoeias. Similarly, in addition to developing the overall plot, students wrote rich character sketches, going into detail about the backgrounds of the antagonists which often featured a tragic death or loss.

The following is a representative example of content and theming for the *Background* sections observed among the groups (presented exactly as written by students):

Background: Setting

A dark room surrounded by bars of metal. rough to the touch yet loud to the ears when hit by an object. Large metal cages built to confined people like animals. Paint on the walls chipping away, and slipping away the bars start to fade away little by little the bars start to weaken. The sounds of wind was angry on this fateful day. The smell of body odor, vomit, and many other undetectable, putrid smells withdrew from each prison cell. Cries, screams, wails, whispers fill the air. This place is a living hell to those who were assigned to this place. The IAEI insane asylum was established in 1966 to treat the most criminally insane person after 10 years of experimenting with unethical techniques the prisoners revolted during a tour that was given to other psychologists wanting to learn the techniques.

Background: Character

The vivid memory of those deafening screams, the sensation of pure fear and awful perception of grief have now been filling up her once empty void. At just the age of fifteen, the unexpected happened. She lost the only person she loved more than life itself; her mother. It has been two years since, and she has never smiled again. No silly joke or comedic action has been able to brighten up her dark, dull eyes. She felt more alone every moment that passed her by. Being weak, she then turned to harsh chemicals, in assumption and full belief that they would help her feel content once more. Instead, they only drove her mad and she reacted in ways no one ever would.

Because of her horrid behaviour, she was imprisoned in a psychotic nightmare. An insane asylum. During her time there, they were finally able to make her laugh. The disturbing thing is, she didn't smile and she held a malevolent expression. At times, she would argue with the walls. Randomly cry so abruptly. Yell in the middle of the night. Soon enough, she grew to be sick of such claustrophobic space. Her only motive now was to escape. A monster trying to escape. Who knows what she would do if she ever had the chance to escape...

Here we have a compelling example of EOR from the adolescent perspective. In this one narrative, students confront the death of loved ones, the emotional toll of that loss, and the dangers of turning to substance abuse as a coping mechanism. Adolescence is a time of discovering and building an independent identity and imagining what it will mean to be responsible for one's own well-being (Feldman, 2013). Imagining life without the support and love of a parent is often a daunting and even frightening realization and may seem entirely impossible. This EOR functions as a means of not only confronting mortality and fatality, but also as a cautionary tale and a reminder that in the wake of loss, one must be strong.

Despite the gruesome scenarios and tragic histories, students shared their ideas with me through excited smiles, and read aloud their background narratives with animated voices, often improvising more details than what they had written on their worksheets. Their enthusiasm, however, often waned when getting to the *Suspension Arc* and *Floor Plan*. They outlined their ideas for the 'scare points', and their floor plan in a more neutral, matter-of-fact tone. Lack of attention or development of the actual environmental elements can be explained, in part at least, by lack of expertise in the technical aspects of haunted attraction design. However, it may also be the case that the *engineering* part of EOR was perceived as less intriguing, with fewer opportunities for their own creative exploration. Seaton (2018: 13) writes that 'engineering is the choice of form and medium (headstone, memorial tablet, epitaph, etc.), and orchestration is their content, layout, and style (gravestone design, memorial speech, mausoleum features, etc.)'. The form and medium had already been determined by their teacher: A haunted house. How hallways twisted and what kind of startle effects to include was perhaps less interesting as it offered little opportunity to explore the frightening realities of mortality and, more frequently in this case, fatality. The opportunity to orchestrate the content and write the story was an invitation to explore and engage with their fears and anxieties on their own terms, within the context of recreation, and in a supportive environment.

In these narratives, students are confronting and exploring the emotionally weighted and difficult ideas of loss, loneliness, confinement, and social exclusion. As authors, they were in control of the content and narrative and had agency over the entirety of the project. Research shows that control and a sense of agency are critical in influencing our experience of fear in the moment and overcoming fears in the future (Hancock and Bryant, 2020; Wanke and Schwabe, 2020; Limbachia et al., 2021). Agency is intimately related to control, but they are not the same thing: Agency is the *perception* of how much control we feel over actions and consequences (Caspar, Cleeremans and Haggard, 2018). Both are important in influencing how we manage stress. For example, a study manipulating real and perceived stressor controllability, in this case, when or if a participant would receive an electric shock, revealed that those who had control of the stressor or who believed to have control, extinguished conditioned fears more quickly than those who do not (Hartley et al., 2014). Moreover, those who did not have control over a stressor demonstrated increased fear responding to conditioned fears (Hartley et al., 2014). Studies of risky play, which is a play that involves a degree of fear and where the chance of getting hurt is high, also point toward benefits in physical and emotional health in the present, and better stress management skills in the future (Brussoni et al., 2015). Collectively, this research suggests that engaging with *controllable* stressors may help us deal with future *real* stressors (Hartley et al., 2014; Maier, 2015). Recreational horror offers this opportunity, along with opportunities for social rewards.

Building Friendships through Fear

In addition to learning about and confronting difficult and complex concepts like death, playful engagement with horror-related content often offers a positive social experience. Youths typically attend horror movies and haunted attractions with friends and family, serving as opportunities to form stronger social bonds and feel a sense of collective belonging (Clasen et al., 2020; Andersen et al., 2020). Recreational horror may be particularly well suited to forming social bonds as they are designed to elicit strong emotions which research shows contributes to forming strong memories (Johansen et al., 2011). Indeed, highly emotional events become more dominant memories compared to the everyday, routine, and neutral life experiences (Johansen et al., 2011; Lawson, Gauer and Hurst, 2012). This is likely to be an evolutionary adaptation. In short, it is necessary to remember the contexts and events which cause the body great stress so as to avoid them, and survive, in the future (LeDoux, 2013). Additional studies show that individuals form stronger social bonds with others, including strangers, when emotions are heightened – for example, during ritual or watching a scary movie (Xygalatas et al., 2011; Bastian, Jetten and Ferris, 2014). While not contagious in the strict sense, studies suggest that emotions are often shared – for instance, many feel compelled to scream or laugh while witnessing their friend do the same (Haj-Mohamadi, Fles and Shteynberg, 2018). This is in part due to how we make sense of what others are feeling, namely by recreating in ourselves what we think the other is experiencing. Neuroimaging research confirms overlapping patterns of brain activity when engaging in self-evaluation and evaluation of others (Lee and Siegle, 2012). This is why we may cringe when watching a friend chip a tooth on a hard candy, because we are creating, or imagining what that must feel like for ourselves. In the context of recreational horror that inspires both fear and fun, it may be especially beneficial to have a group of friends who are also experiencing a mix of emotions, thus buffering against negativity dominating the experience. For example, turning to a friend who is laughing following a particularly frightful scare can lighten what might have otherwise been experienced as an unpleasant startle.

Conclusion

Prevailing wisdom offers that youths should be protected from harsh realities and that topics approaching mortality and fatality should be kept at a distance when possible or approached with careful consideration. And indeed, early exposure to violence and death can be traumatizing and lead to lifelong challenges. Research into how and why children engage with dark tourism, then, has not received the same attention compared to adult engagement. Yet, children and adolescents do engage with this material often through recreation. Engaging in recreational horror may be motivated by a desire for entertainment and fun with friends, but evidence also points toward recreation aiding in the process of learning about difficult or frightening ideas. In the context of 'play', the negative emotions often inspired by EOR of mortality and fatality are offset by the rewarding, positive emotions that associate playful engagement. Rather than avoiding these realities, which are ultimately unavoidable, youth can approach and learn about them on their own terms. Confronting frightening material from a safe distance and in the context of recreation allows youths to 'practice' stress management and cultivate resilience. It further offers opportunities for youth to gain a sense of control and agency over difficult inevitabilities such as the death of parents and loved ones.

Opportunities to investigate children's engagement with dark tourism are vast, including analysis of the content designed to inspire both fun and fear, those who do the engineering and orchestrating, and how it changes based on time and place. Likewise, further research into how children experience such content and how it shapes their worldview would be exceedingly useful. Also needed is research into how, when, and why children introduce elements of play or recreation in the context of dark tourism sites which are not designed for such engagement, such as war memorials. These avenues of inquiry will lead to a greater understanding of childhood development and changing cultural attitudes surrounding education, entertainment, and parenting. There is another benefit to this research, one relevant beyond the field of dark tourism literature, and that is in the reminder that finding or creating rewarding elements in remembrances of mortality and fatality is not necessarily a sign of disrespect or 'bad'. Following in the footsteps of children, we can harness the positive to help manage the overwhelming, frightening, and inevitable realities of life and death.

References

Andersen, M. M. et al. (2020) 'Playing With Fear: A Field Study in Recreational Horror', *Psychological Science*, 31(12), pp. 1497–1510.

Brussoni, Mariana, et al. (2015) 'What Is the Relationship between Risky Outdoor Play and Health in Children? A Systematic Review', *International Journal of Environmental Research and Public Health*, 12(6), pp. 6423–6454.

Bastian, B., Jetten, J., and Ferris, L. J. (2014) 'Pain as Social Glue: Shared Pain Increases Cooperation', *Psychological Science*, 25(11), pp. 2079–2085.

Bowman, M. S. and Pezzullo, P. C. (2009) 'What's so "Dark" about "Dark Tourism"?: Death, Tours, and Performance', *Tourist Studies*, pp. 187–202. doi: 10.1177/1468797610382699.

The Numbers 'Box Office Performance History for Horror Movies' (2020). Available at: https://www.the-numbers.com/market/genre/Horror (Accessed: July 29 2020).

Caspar, E. A., Cleeremans, A., and Haggard, P. (2018) 'Only Giving Orders? An Experimental Study of the Sense of Agency When Giving or Receiving Commands', *PLoS One*, 13(9), p. e0204027.

Clasen, M. (2012) 'Monsters Evolve: A Biocultural Approach to Horror Stories', *Review of General Psychology*, pp. 222–229. doi: 10.1037/a0027918.

Clasen, M., Kjeldgaard-Christiansen, J., and Johnson, J. A. (2020) 'Horror, Personality, and Threat Simulation: A Survey on the Psychology of Scary Media', doi: 10.1017/s0954579413000849.

Feldman, R. S. (2013) *Development Across the Lifespan*. New York, NY: Pearson College Division.

Gershon, Livia. (2019) 'Selling Slashers to Teen Girls'. *JSTOR Daily*. Available at: https://daily.jstor.org/selling-slashers-to-teen-girls/ (Accessed: February 1 2021).

Gendron, M., Crivelli, C., and Barrett, L. F. (2018) 'Universality Reconsidered: Diversity in Making Meaning of Facial Expressions', *Current Directions in Psychological Science*, 27(4), pp. 211–219.

Goffman, Eriving. (1961) *Asylums: Essays on the Social Situation of Mental Patients and Other Inmates*. New York: Anchor Books.

Haj-Mohamadi, P., Fles, E. H., and Shteynberg, G. (2018) 'When Can Shared Attention Increase Affiliation? On the Bonding Effects of Co-experienced Belief Affirmation', *Journal of Experimental Social Psychology*, pp. 103–106. doi: 10.1016/j.jesp.2017.11.007.

Hancock, L. and Bryant, R. A. (2020) 'Posttraumatic Stress, Stressor Controllability, and Avoidance', *Behaviour Research and Therapy*, 128, p. 103591.

Hartley, C. A. et al. (2014) 'Stressor Controllability Modulates Fear Extinction in Humans', *Neurobiology of Learning and Memory*, 113, pp. 149–156.

Hoemann, K., Xu, F., and Barrett, L. F. (2019) 'Emotion Words, Emotion Concepts, and Emotional Development in Children: A Constructionist Hypothesis', *Developmental Psychology*, 55(9), pp. 1830–1849.

Holmes, E. A., Creswell, C., and O'Connor, T. G. (2007) 'Posttraumatic Stress Symptoms in London School Children Following September 11, 2001: An Exploratory Investigation of Peri-traumatic Reactions and Intrusive Imagery', *Journal of Behavior Therapy and Experimental Psychiatry*, 38(4), pp. 474–490.

Johansen, J. P. et al. (2011) 'Molecular Mechanisms of Fear Learning and Memory', *Cell*, pp. 509–524. doi: 10.1016/j.cell.2011.10.009.

Kerr, M. (2015) *Scream: Chilling Adventures in the Science of Fear*. New York, NY: PublicAffairs.

Kerr, M. M. and Price, R. H. (2016) 'Overlooked Encounters: Young Tourists' Experiences at Dark Sites', *Journal of Heritage Tourism*, pp. 177–185. doi: 10.1080/1743873x.2015.1075543.

Kerr, M. M. and Price, R. H. (2018) '"I Know the Plane Crashed": Children's Perspectives in Dark Tourism', in *The Palgrave Handbook of Dark Tourism Studies*, pp. 553–583. doi: 10.1057/978-1-137-47566-4_23.

Kerr, M., Siegle, G. J., and Orsini, J. (2019) 'Voluntary Arousing Negative Experiences (VANE): Why We Like to be Scared', *Emotion*, 19(4), pp. 682–698.

Lawson, A. L., Gauer, S., and Hurst, R. (2012) 'Sensation Seeking, Recognition Memory, and Autonomic Arousal', *Journal of Research in Personality*, pp. 19–25. doi: 10.1016/j.jrp.2011.10.005.

LeDoux, J. E. (2013) 'The Slippery Slope of Fear', *Trends in Cognitive Sciences*, 17(4), pp. 155–156.

Lee, K. H. and Siegle, G. J. (2012) 'Common and Distinct Brain Networks Underlying Explicit Emotional Evaluation: A Meta-analytic Study', *Social Cognitive and Affective Neuroscience*, 7(5), pp. 521–534.

Limbachia, C. et al. (2021) 'Controllability Over stressor Decreases Responses in key Threat-related Brain Areas', *Communications Biology*, 4(1), p. 42.

Maier, S. F. (2015) 'Behavioral Control Blunts Reactions to Contemporaneous and Future Adverse Events: Medial Prefrontal Cortex Plasticity and a Corticostriatal Network', *Neurobiology of Stress*, pp. 12–22. doi: 10.1016/j.ynstr.2014.09.003.

Panksepp, J. (1981) 'The Ontogeny of Play in Rats', *Developmental Psychobiology*, 14(4), pp. 327–332.

Panksepp, J. and Yovell, Y. (2014) 'Preclinical Modeling of Primal Emotional Affects (SEEKING, PANIC and PLAY): Gateways to the Development of New Treatments for Depression', *Psychopathology*, pp. 383–393. doi: 10.1159/000366208.

Rozin, Paul, et al. (2013) 'Glad to be Sad, and Other Examples of Benign Masochism', *Judgment and Decision Making*, 8(4), pp. 439–447.

Scrivner, C. (2020). 'The Psychology of Morbid Curiosity' *PsyArXiv*. doi: 10.31234/osf.io/xug34.

Scrivner, C. et al. (2021) 'Pandemic Practice: Horror Fans and Morbidly Curious Individuals are More Psychologically Resilient During the COVID-19 Pandemic', *Personality and Individual Differences*, 168, p. 110397.

Seaton, T. (2018) 'Encountering Engineered and Orchestrated Remembrance: A Situational Model of Dark Tourism and Its History', in Stone, P. R., Hartmann, R., Seaton, T., Sharpley, R., and White, L. (eds.) *The Palgrave Handbook of Dark Tourism Studies*. London: Macmillan, pp. 9–31. doi: 10.1057/978-1-137-47566-4_1.

National Retail Federation *Social Media Influencing Near-Record Halloween Spending* (2019). Available at: https://nrf.com/media-center/press-releases/social-media-influencing-near-record-halloween-spending (Accessed: July 31 2020).

Stone, P. R. et al. (2018) *The Palgrave Handbook of Dark Tourism Studies*. London: Palgrave Macmillian.

Hollywood Reporter 'Teen Movie Marketing' (2006), 18 July. Available at: https://www.hollywoodreporter.com/news/teen-movie-marketing-138425 (Accessed: February 1 2021).

Walter T. (2004) 'Body Worlds: Clinical Detachment and Anatomical Awe', *Sociology of Health and Illness*, 26(4), pp. 464–488.

Walter, T. (2009) 'Dark Tourism: Mediating Between the Dead and the Living', in Sharpley, R. and Stone, P. R. (eds.) *The Darker Side of Travel*. Bristol: Channel View, pp. 39–55. doi: 10.21832/9781845411169-004.

Wanke, N. and Schwabe, L. (2020) 'Dissociable Neural Signatures of Passive Extinction and Instrumental Control over Threatening Events', *Social Cognitive and Affective Neuroscience*, 15(6), pp. 625–634.

Xygalatas, D. et al. (2011) 'Quantifying Collective Effervescence: Heart-rate Dynamics at a Fire-walking Ritual', *Communicative & Integrative Biology*, pp. 735–738. doi: 10.4161/cib.17609.

11 'Edutainment' in Dark Tourism

Toward a Child's Perspective

Daniel W.M. Wright

Introduction

Arguably, Western societies are restricting, curbing, or perhaps even distancing themselves from actual encounters with death – at least, in comparison to the exposure to death experienced by our ancestors. The question, therefore, is do we want our children to be exposed to scary stories of recent or distant past tragic or death events? Significantly, if we continue to shield children from scary or dangerous situations and narratives, we are potentially limiting them from a normal human emotion – that is, fear. The potential to be frightened is important because fear is a critical human emotion. Exposing children to fearful situations, during their youth, can better equip them to understand and manage fear as they grow into adults (Floyd, 2016). Important lessons for adult life are often gained in childhood, and these are often taught through effective stories and narratives. If children are not engaging with *scary* and *dark* stories, then important life lessons are not being taught from an early age. Thus, children may develop into more anxious adults, because they have limited experiences of managing fear, because fear is not seen as something normal, or to be experienced.

Society is seeing an increasing tendency to intellectualize holidays and a growing desire to offer study and learning opportunities during leisure time is also increasing within the current touristic landscape (Light, 2000). Consequently, tourism in general and dark tourism in particular provides visitors with experiences that increase their own cultural and educational capital about 'heritage that hurts' (Roberts, 2018). Importantly, dark tourism also utilizes entertainment in addition to educative techniques, in order to interpret difficult heritage. Indeed, Herbert Marcuse (1964: 66–67) remarked that 'entertainment and learning are not opposites; entertainment may be the most effective mode of learning'. As social beings, we are accustomed to telling and listening to stories. Stories (traditionally) have been communicated via (more) passive methods, such as narration, reading, or watching visual content. However, we are now on the frontier of new technologies and these are increasingly becoming popular in tourism attractions. Such technologies allow visitors to experience virtual and augmented reality environments. Significantly these technologies provide new opportunities for immersive storytelling. The utilization of virtual and augmented reality (VR/ AR) technologies could

DOI: 10.4324/9781003032199-15

be a powerful tool in which to place users (children) into dark environments, worlds, and events from different viewpoints and perspectives.

However, the messages being communicated via digital content can significantly impact a child's experience and understanding. Research is increasingly showing that a balanced approach between education and entertainment is an effective method of child learning (Puckett and Diffily, 2004; Chang, Hwang, Chen and Müller, 2011). Arguably, this may be even more effective when the experience is immersive and offers engaging narratives. Dark tourism sites and attractions are 'safe' places where children can learn about difficult heritage and, importantly, experience emotions related to fear. With new technologies come new opportunities for teaching and learning. By balancing the educational and entertainment-centric approaches and with gaming technology, managers of dark sites can further enhance a child's understanding of the difficult heritage.

Hence, the purpose of this chapter is to explore critical intersections between 'entertainment' and 'education' in dark tourism (re)presentations and children's experiences thereof. In particular, the chapter offers original ideas surrounding the potential use of immersive technologies at dark tourism sites for children. Ultimately, the study suggests that key stakeholders, including managers and owners, must consider the most effective methods of introducing immersive technologies, the content and stories, and consider gaming techniques to further engage children. In doing so, *edutainment* – that is, education and entertainment – experiences offered could provide valuable lifelong lessons that not only benefit children but also society more generally. Firstly, the chapter briefly discusses children's exposure to deaths, particularly through literature and popular culture, as underpinning discourse for dark tourism interpretation and edutainment.

Exposing children to death through literature and popular culture

As Floyd (2016: online) notes, 'growing up has always been a scary proposition [and] kids have to face the unknown'. Indeed, society should not shield children from fear and its difficult heritage. Instead it should continue to consider how exploring such difficult pasts and events can be beneficial to children. Fear is an emotion that is real and fearful situations are something all adults at some stage will encounter and attempt to overcome. Society can only do so much to mitigate the risks of life and to protect children but, arguably, this could lead to more harm. For example, scary tales of dangerous circumstances can trigger an emotional response in the *amygdala* which releases adrenaline, a chemical that stimulates people to fight or flee. Importantly,

> the amygdala is a collection of cells near the base of the brain. There are two, one in each hemisphere or side of the brain. This is where emotions are given meaning, remembered, and attached to associations and responses to them – 'emotional memories'.
>
> (Healthline, 2005–2021: online)

Through storytelling in safe environments, children can be exposed to fear and darkness; narratives that expose them to the anxiety that is stimulated by dangerous, life-threatening circumstances. Significantly, such exposure can teach children important valuable lessons of the past through the emotion of fear.

Therefore, children who are exposed to scary stories are likely to be better equipped when confronted with fears in a world that exposes them to uncertainty. Put simply, children reading scary narratives should become more resilient and better prepared for difficult challenges faced in adulthood. For instance, an internet search on Google or Amazon presents a plethora of dark content within books, videos, and games for children. The stories and their narratives range from historical non-fictional accounts to fictional events. Of course, telling children scary stories is nothing new. Our ancestors told stories of courage and danger around campfires. Today, parents read stories from the *Horrible Histories* book collection, a series recommended for children aged 12 years and above, that retells difficult and macabre aspects of history. *The Mystery of the Haunted House* (Baxter and Goulding, 2012), a 3D popup book (for ages 6–11 years) dares children to peek through the windows of a scary old mansion with a ghost-themed narrative. Unsurprising, therefore, from a tourism attraction perspective, children can engage and visit dark sites and even here, the narratives are wide-ranging.

Dark tourism attractions: Toward an education and entertainment orientation

The coverage of dark tourism attractions has now been well documented in the academic literature. The scope of dark attractions from a supply perspective was outlined in Philip Stone's (2006) taxonomical model – 'The Dark Tourism Spectrum: towards a typology of death and macabre related tourist sites, attractions and exhibitions'. Within the spectrum typology, Stone identifies key product features of dark sites. The features of the spectrum are central to visitor attractions being perceived as either 'darkest' (sites of death and suffering) or 'lightest' (sites associated with death and suffering). One of the central features of the spectrum focuses on the importance of being either educationally orientated or entertainment orientated (that is, a visitor attraction more centered on the latter would be perceived as a lighter). Sites that are on the 'darkest end' of the spectrum model include places such as the Cambodian Killing Fields, Ground Zero, and Auschwitz-Birkenau – all of which are places of death and suffering. Sites at the 'lightest end' of the spectrum may include locations such as the Dungeon visitor attractions, offered by Merlin Entertainments PLC in London, York, Hamburg, and elsewhere. These types of tourist attractions provide exaggerated theatrical performances and exuberant narratives that portray heritage of death and suffering in an entertaining manner.

Another feature of Stone's spectrum model, which is seen as central to sites moving from the darkest end to the lightest end, is the spatial timeframe. Here, the focus is on when the event took place – or, how much time has passed since the event took place. The suggestion is that as time progresses, an attraction and its narratives can be perceived as 'less dark'. Of course, the complexities of such chronological distance are much greater than this, as noted by Stone (2006), and each dark event/attraction must be considered independently. However, an interesting consideration by Wright (2018) who adopted a futurology perspective, explored the potential of so-called future Terror Parks. Wright (2018) explores how dark events are (or could) become sensationalized more frequently, that is, closer to the time of the actual event, as different genres of media content are created for audiences to engage with. He provocatively outlines how events like 9/11 'might' become commercially valuable for the movie industry in the future. Significantly, narratives of the 9/11 event within television and film today range from non-fictional and documentary-focused (educational-centric) to fictional narratives (entertainment-centric). Consequently, Wright's (2018) Terror Park discussions ask the question, as dark events become global and available to a wider audience – how will they be presented to audiences? He also suggests that recent dark events could be used to generate entertainment-based content at visitor attractions. Put another way, our present-day tragedies could become our future entertainment.

Importantly, however, Wright (2018) stresses that more needs to be done in order to consider and manage how recent difficult events are engaged with. Of course, as adults, we have the right to view different forms of content, but from a child's perspective, there should be greater consideration on what content they are exposed to and why (Kerr and Price, 2016). As already suggested, exposing children to difficult heritage is important, but how this is achieved is much more challenging. The stories at dark tourism attractions (both in the darkest and lightest sense) are often difficult in nature, due to the suffering of victims or the nature and context of death. However, even at the lightest of dark tourism attractions, where entertainment is emphasized as central to the guest experience, there are still educative lessons to be learned about topics of difficult heritage.

Increasing engagement through the 'edutainment' concept

Sharpley and Baldwin (2009) suggest visits to dark visitor attractions present different motives (and meanings) to that of general leisure attractions, as they have a focus on education and learning. They go on to argue that children's visits to battlefields provide an opportunity to bring history to life, as children can begin to understand what it would have been like to be a soldier. Similarly, Stone (2009, 2011) suggests that the Dungeon visitor attractions in the UK, Holland, and Germany, or Gunther Von Hagen's Body Worlds exhibition, offer entertaining 'death mediators' whereby visitors are educated and entertained by encountering real and imagined corpses, as well as experiencing simulated atrocities of the past (Walter, 2009; Stone, 2012). Consequently, Israfilova and Khoo-Lattimore (2018) note that children who visit dark tourism sites can gain knowledge through the interpretation of history, artifacts, and objects. Therefore, visual interpretations for children require a kinaesthetic learning approach to understanding subject matters and difficult events. In other words, visual and interpretive demonstrations play a significant role in learning in comparison to more traditional text-based learning. While dark tourism offers educational and entertainment-centric approaches to interpretation, arguably a more balanced approach could support children's understanding and engagement with difficult heritage.

Edutainment is the combination of two words – *education* and *entertainment*. The notion of edutainment focuses on attaining better education by incorporating broader facets of entertainment. However, the idea of learning through play is not new. John Comenius (birth name: Jan Amos Komensky, 1592–1670) became famous for his pedagogical work exploring the educational development of children. In his book, *School of Infancy* (1631), he discusses the importance of early education and stressed the most valuable service parents can give their children is to encourage them to play (Puckett and Diffily, 2004). He is often associated with the term 'school as play'. More recently, the focus has moved to the integration of media content within education such as video games, television, and other multi-media material which, subsequently, intends to enhance educational experiences through enjoyable entertainment practices. As noted by Chang et al. (2011), edutainment has been widely employed and explored and experts continue to promote edutainment by integrating education and entertainment. These include, but are not limited to, the technological contribution of mobile devices, computers, software, and games, as well as augmented applications and virtual reality headsets and content which is ensuring the edutainment concept is being widely accepted as an effective means of learning (Chang et al., 2011). Therefore, new technologies are playing a significant role in edutainment and 'storytelling' and will continue to evolve and impact in the future.

'Edutainment' and technology: Toward the art of storytelling

As Boris (2017: online) notes, 'telling stories is one of the most powerful means that leaders have to influence, teach, and inspire'. According to Lindgren and Bandhold (2009), humans create stories to help us understand life, to generate meaning, to establish order and spaces to live collectively, and to consider our past, present, and future. Thus, storytelling plays a significant role in humanities education and, moreover, through stories people can better understand certain issues because stories increase involvement and motivation (Šisler and Brom, 2008). Consequently, Azuma (2015: 260) suggests that 'storytelling is fundamentally important, and any advancements in media technology that enable people to tell stories in new and potentially more compelling ways can have profound impact'. Azuma (2015: 260) goes on to highlight the power of storytelling by suggesting that 'telling a story is an important method of education and instruction. Stories can contain lessons, codified bits of wisdom that are passed on in a memorable and enjoyable form'.

However, many stories from the past often focus on 'horrible (hi)stories'. As noted by Floyd (2016: online):

> ...scary stories have a timeless quality, and the appeal of their emotional power has survived countless generations. Before cable television and ubiquitous internet connectivity, scary stories were used as cautionary tales to help children become aware of dangerous people, situations to avoid, or places to bypass. Scary stories were often passed along through the oral tradition to reinforce a moral lesson, so like folktales they served a purpose beyond pure entertainment.

In some form or another, dark tourism sites and attractions are essentially exposing educational and/or entertaining stories through difficult heritage narratives. Some of the stories may be uncomfortable or scary, especially from a child's perspective. As society progresses, technology will provide new platforms of engagement and, consequently, new story content. Thus, according to Azuma (2015: 260), 'technological developments that make the story clearer and more memorable can aid retention and understanding'. Moreover, Bimber, Encarnação and Schmalstieg (2003) discuss how interactive digital story techniques when applied with new media forms can allow for a shift into the third dimension – virtually and in the real/physical world. Therefore, a significant advantage of this transition is the ability to communicate information more effectively via digital means. As Bimber et al., (2003: 87) go on to state, 'the user experience is thus transformed from relating different pieces of information to one another to *living through* [original emphasis] the narrative'. While focusing on mixed reality (MR) and augmented reality (AR), Azuma (2015: 259) argues that future technology will allow innovative methods of storytelling that will allow 'virtual content to be connected in meaningful ways to particular locations, whether those are places, people or objects'. Azuma (2015) also notes the likelihood of AR and MR being used as a new form of location-based media that enables new storytelling experiences. Significantly, with new technologies come new forms of storytelling, and these have the potential to be used at dark tourism sites to engage, educate as well as entertain children.

Immersive technologies: Virtual, augmented, and mixed reality

Society is witnessing significant strides in the development of virtual and augmented reality technologies. While there is growing use of other terms, such as mixed reality, the focus in this chapter is on the *spectrum of reality* (Figure 11.1). The emphasis is more on the environment and how society while living in the real world can now, through technological means, experience virtual worlds, and even blended worlds where the real and virtual environments exist together. The three commonly used terms are:

- *Virtual reality (VR)* – Immerses users in fully artificial digital environment computer-simulated environments, which can simulate places in the

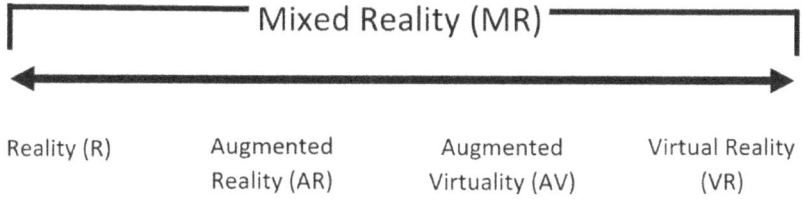

Figure 11.1 The Mixed Reality Continuum.

Source: Adapted from Milgram and Kishino, 1994.

real world as well as fantasy and imaginary worlds, offering users the ability to immerse themselves into artificial environments that stimulate sensory experiences.

- *Augmented reality (AR)* – Overlays virtual objects on the real-world environment that allows its users to interact with the real world, while (augmented) digital content is added to the real world around us. Here, users remain present in the real world but can add objects of a real or fantasy nature to real-world surroundings (of which only the user can see).
- *Mixed reality (MR)* – Overlays and anchors virtual objects to the real world where the real-world and virtual-world objects are presented together in a single display. Mixed reality is arguably a more recent development in immersive technologies.

The 'Mixed Reality Continuum' model in Figure 11.1 presents an overview of how these technologies allow us to move from the real world to the virtual world, and the types of augmented environments in between. To the left of the spectrum is the natural world (the real one experienced every day). To the right of the spectrum is VR whereby users are fully immersed into computer-generated environments. The rest of the spectrum is made up of mixed reality experiences; ones where the virtual and real world are merged together. Hence, the focus in this chapter is on the ability to move from the real world to more augmented and virtual reality environments. The technology is important because it has the capability to transport users into new worlds and new experiences. The more advanced the technology, the more immersive the experience and, ultimately, the greater the sensation of presence in different worlds.

Immersive technologies, human presence, and education

Slater, Usoh, and Steed (1995a: 6) suggest that the 'presence in a real or virtual environment is a psychological state of consciousness. It is a sense of *being there* [original emphasis] in the environment which in the case of the virtual environment is displayed by a computer through appropriately connected hardware channels'. Therefore, *immersion* is a description of the technology and *presence* is an emergent property of immersion (Slater, Usoh, and Steed, 1995b). Significantly, immersion in an environment can lead to the user having presence in that environment. This is where the technical quality of the technology is significant, because the better the technology the greater potential for a person to be immersed and, consequently, feel present in the environment.

As technological landscapes evolve, humans are exposed to ever-more immersive experiences. Indeed, VR and AR are moving users away from the traditional flat-screen observational linear narratives. Drue Kataoka (an artist and technologist) notes that AR and VR offer entirely innovative creative mediums, where content creators can build worlds, pixel by pixel (cited in Hall and Takahashi, 2017). These new technologies are likely to replace

> rectilinear devices with technologies that depict worlds in ever-expanding concentric circles, providing a level of immersion and experience that has never been seen before. This could be game-changing: users will no longer view content but will be placed inside ever-expanding virtual worlds and find themselves at the center, hence the 'immersive' nature of the technology.
>
> (Hall and Takahashi, 2017: online)

In less-immersive media such as traditional two-dimensional (2D) television screens, children experience content as real, more so than their older counterparts and apparently, this can affect how children behave and, perhaps importantly, what they learn (Richert, Robb, and Smith, 2011). This is also important when considering children's experiences within virtual environments. However, there is limited research into immersive virtual reality and children's experiences, despite their frequent media use and willingness to adopt new technologies (Lauricella, Cingel, Blackwell, Wartella, and Conway, 2014). Nonetheless, children experience a virtual environment differently to adults. As noted by Reeves (1989), the body responds to digital media technology as if it were real because the mind has not evolved to respond to it any differently from the physical world. However, when using virtual reality as adults, we are aware that we are safely located in a room using a headset, even though we are likely to feel our hearts race or palms sweat when looking over a virtual cliff (Blascovich and Bailenson, 2011). Younger children, however, are likely to respond differently, both cognitively and behaviorally to sensory salient and immersive media than to adults (Bailey and Bailenson, 2017). Sharar et al. (2007) found that children (between 6 and 18 years of age) reported higher levels of presence and 'realness' in virtual environments compared with adults (aged 19–65 years). Significantly, this suggests young children are more likely to be influenced by the immersive content, both positively (such as educational motives) and negatively (increased materialism). Previous research has highlighted the importance of presence in virtual environments (VE). It is suggested that a highly present individual is more likely to behave in the VE. Consequently, the individual is more likely to behave in a manner like their everyday reality. Thus, the more immersive the technology the greater sense of presence and ultimately, higher levels of user engagement (Grigorovici, 2003; Baños, et al., 2004; Aymerich-Franch, 2010; Cummings and Bailenson, 2016).

Virtual reality has long been recognized as a valuable teaching tool, due to its fundamentally different mode of communication between computer and person, and by offering highly interactive and dynamic forms of simulation (Hoffman and Vu, 1997; Çiflikli, PIsler, and Güdükbay, 2010; McGloin, Farrar, and Krcmar, 2011; Wright, 2020). Indeed, the extremely visual nature of VR and the user's ability to interact in *other worlds* in ways not possible in the real world make VR a compelling and motivating learning environment (Hoffman and Vu, 1997). VR can be applied when generating environments that facilitate knowledge-building experiences (Regian, Shebilske, and Monk, 1992). Augmented Reality has also proven to be an effective means of educating and entertaining young people of all ages and grade levels, and the technology is gaining tract as a valuable learning and immersive tool (Golosovskaya, 2020). Traditional systems focus on the transfer of knowledge from teacher to pupil; however, AR (and VR) has the ability to turn the learning process into a multidimensional virtual experience (AugBrite, 2020).

Furthermore, it is suggested that AR can create an effect of presence which clearly reflects the connection between the real and virtual world. The technology is said to psychologically attract the user and activate their attention and susceptibility to the information being transferred (AugBrite, 2020). Guttentag (2009) notes how VR offers (dark) tourism many useful applications and as the technology continues to evolve, the number and significance of such applications undoubtedly will increase. In particular, the area of tourism in which VR may prove valuable is in planning and management, marketing, accessibility, heritage preservation, and, significantly, entertainment and education. With this latter point in mind, Kysela and Štorková (2015) suggest that augmented reality can be used as an immersive storytelling (gaming) medium for teaching history in tourism locations.

Gaming and immersive storytelling

Games tell and take users through stories, and the manner in which they do impact users' experience. In contrast to other media – books, television, and film – computer games project interactivity onto the player, where they can take the role of either the protagonist or antagonist. As discussed by Egenfeldt-Nielsen (2010), a very simple principle of playing computer games is to engage the user with an unknown universe and allow them to slowly learn more about that universe. Unlike other media, such as movies and books, where the reader can move forward within the narrative without necessarily learning more, computer games demand a higher level of interaction and structural complexity. Computer games have a prerequisite for learning, and the variety of in-game experiences and modalities will allow for greater levels of interaction.

At their most effective, computer games will draw users into the experience as they demand more focus and energy, as users lose a sense of time and place. Thus, users are more invested in the learning experience. Indeed, users learn about the new universes, which are driven by the actions within the game (Egenfeldt-Nielsen, 2010). Games are based on a system of rules, participations, and goals. Game-based stories, particularly in the narratives of many video games, take place within interconnected systems – often referred to as *systems thinking*. In other words, it is a way of viewing the world as a series of interlocking causal feedback loops. For example, in a novel, if the protagonist takes an action, it will always have consequential effects (Farber, 2014). However, the actions of the user in a game have consequences and the narrative of stories evolves as a direct result of the user's actions, alongside the potential of the in-game playability and stories modes. It is these characteristics that make researchers and educators believe in the power of learning through gaming over other forms of more passive media content (Egenfeldt-Nielsen, 2010).

The gamification concept is one that has seen growth within the tourism industry. Deterding *et al.*, (2011) argue that gamification is the contextualizing of game design outside its original domain. Significantly, gamification of tourism is said to contribute to more rewarding interactions and a higher level of satisfaction for users (Xu, Buhalis and Weber, 2017). Consequently, 'gamification involves applying elements of gamefulness, gameful interaction, and gameful design with a specific intention in mind' (Deterding *et al.*, 2011: 10). Meanwhile, Zichermann and Cunningham (2011) note that gamification can be applied when trying to engage users and influence their behavior by using game mechanics in areas other than traditional gaming contexts. It is also suggested that gamification can add fantasy, immersion, and fun to a user's experience (Xu, Buhalis and Weber, 2017). In short, through gamification techniques, tourism site managers can further deliver multi-dimensional and multifaceted experiences to visitors, something that is arguably significant within dark tourism attractions and children's experiences thereof. Thus, the application of game technology and the concept of gamification can further support and enhance dark tourism in becoming *edutainment centric* when delivering children experiences and inherent dark or difficult (his)tories.

Of course, stories are often formulated around historical or fictional narratives, and narratives are a way of presenting connected events in order to tell a good story. Narratives have been around since the beginning of storytelling and the following are four common categories and approaches to structure narratives when presenting stories (MasterClass, 2019):

- Firstly, *linear narratives* which present the events of the story in the order in which they happened. This can include first/second/third-person narration. The linear narrative aims to immerse the reader into the daily life of a protagonist (or antagonist) and the reader observes the events of the character's life unfold in chronological order.

- Secondly, *non-linear narratives* present events of a story out of order. The use of flashbacks and other literary devices are used to shift the chronological order often to emphasize emotional mindsets based on specific events, such as past trauma.
- Thirdly, *quest narratives* focus on the reader observing a character working toward a goal and having to overcome insurmountable obstacles.
- Finally, a *'viewpoint narrative'* is applied when expressing the points of view or subjective personal experience of a main character or of other relevant characters in the story. This narrative style often takes the form of first-person narration or third-person omniscient narration. Points of view can switch between private thoughts of different characters, but can often be subjective as a narrator will personally reflect on their experiences and the behavior of others.

To enhance the sensation of fun in a game, the gaming industry develops methodologies to manage the game execution. Significantly, this refers to the storytelling within the game. Two areas of storytelling and narrative interaction arise when playing games. *'Emergent narratives'* allow players to be free, to do what they want. It is here that the quality of the game depends on the player's (cap)ability in navigating the game, and more recently, interacting with the technology. *'Story-based games'* are more restrictive, as players have a limited influence on the course of the game (Champagnat, Delmas, Augeraud, 2008). It is important to identify these narratives as they give the reader a deeper understanding as to how users engage with stories and, perhaps more importantly, how stories (and those within games) allow users to interact with the 'other' universe. Within dark tourism attractions, and interpreting difficult heritage for children, this is crucial because the story being told can be scary and upsetting. Ultimately, the perspective and narrative in which it is being told can impact children's overall engagement and understanding.

Future immersive edutainment in dark tourism

Playing digital games is not only about fun, the learning and educational potential is greater. In dark attractions, the focus should be on more 'serious games' and using new technologies as a persuasive technique, because computer/video games can use a set of cognitive design properties that focus on changing a user behavior and transferring knowledge, instead of the mere focus on entertainment commonly applied with traditional gaming (Ghanbari, Simila, and Markkula, 2015). According to expert game designer Jane McGonigal, games have the potential to change the world and importantly game developers have a responsibility to make it happen. McGonigal says reality is broken and can only be fixed if we make the real-world work like massive, multiplayer games and consequently, game developers have a responsibility to steer gamers (users) toward improving the real world. Games can inspire large groups of people to pool their knowledge and skills together,

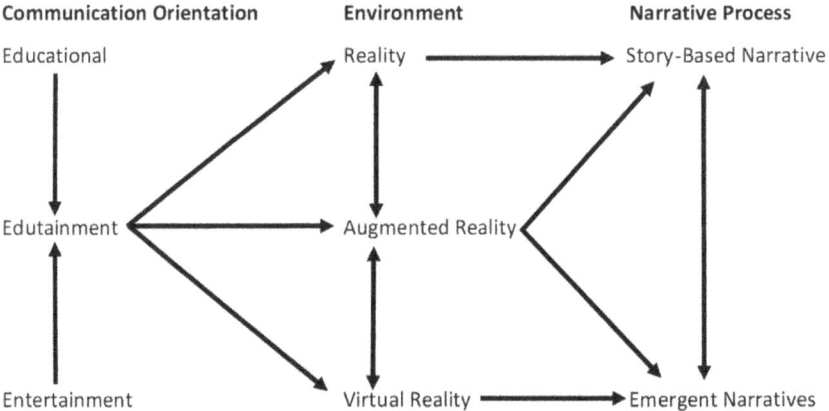

Figure 11.2 Edutainment Narratives in Real, Augmented, and Virtual Environments.

in so doing, overcome difficult obstacles. According to McGonigal, this approach to gaming is what humanity needs to tackle global social issues, such as poverty, hunger, disease, and climate change (Zetter, 2018). Figure 11.2 presents a model in which the discussions in this chapter are brought together. It emphasizes the importance of communication, environment, and narrative. Through gaming and immersive technologies, narratives can and should be created that establish a balance between education and entertainment; narratives for children that focus on 'edutainment'.

Dark attractions have the potential authority to influence a child's knowledge, understanding, and appreciation of dark events. Therefore, key stakeholders (such as managers and owners) hold a significant responsibility in the experiences available to children at dark sites and the messages they are communicating. Beyond the individual potential of exposing and preparing children for difficulty in their adult lives, visiting dark heritage attractions and memorials plays a significant role in transporting the heritage on, from one generation to the next. Therefore, it is important that children are engaged and learning about the past. The dark narratives and tragic events displayed at such sites are ones that humanity would do well to avoid repetition of in future, and through immersive storytelling and new technologies, attractions can begin to create more edutainment-based content that engages and educates children. Kerr and Price (2016) called for more research that examines and explores children's experiences at sites that commemorate tragic and sometimes gruesome heritage. As noted, the significant value of this is for stakeholders who manage exhibits, memorial sites, and dark attractions, as young visitors' experiences are highly valuable. This is not only as children as repeat (adult) guests in the future (Cullingford, 1995; Docket et al., 2011; Kerr and Price, 2016), but also because managers should value the messages, experiences, and ultimately lessons gained by children.

Conclusion

> Reality is broken. Game designers can fix it.
>
> Jane McGonigal (cited in Zetter, 2018)

This chapter suggests the importance of exposing children to dark and difficult experiences might be a means to becoming more resilient and understanding of cultural trauma and acts of atrocity. Dark tourism, while still largely aimed at adults, is identified as a potentially safe and socially sanctioned space in which children can learn about 'horrible history'. The chapter also recognized the growing importance of edutainment (gaming) based experiences, using novel technologies (AR and VR) in order to tell engaging and immersive stories. Importantly, the study identified dark tourism as places where children engage, for fun or educational purposes, with heritage that hurts.

As society progresses further into the 21st century, (current) sites of difficult heritage and new attractions will emerge, and they are likely to be deep-rooted with complex, dark realities. However, they will continue to be a platform, and a place of education, where visitors can begin to understand the difficult narratives that surround their existence. From a child's perspective, the knowledge value can be enhanced through increased communication. This is based on a balance between educational and entertainment – that is, *edutainment*. Enhanced immersive technologies can provide users with a greater sense of presence in (mixed) augmented environments and virtual worlds. Therefore, by creating effective digital media content, (dark) visitor attractions can offer more immersive experiences and ultimately, not only engage children in the dark histories through entertainment practices but also enhance their understanding of the difficult narratives. Thus, this ideally provides a more holistic edutainment approach through the application of immersive technologies, which are centered on complementary levels of entertainment and education.

Sadly, the world is littered with dark tourism sites, and will undoubtedly continue to be so. Moreover, uncertainties of life and mortality events and subsequent anxieties are potentially going to be more varied for children across the world. Growth in new technologies, robotics, and artificial intelligence could result in wider mass unemployment, global warming could destroy habitats and the environment which sustains our existence, depleting natural resources we need to live. Society could be exposed to more natural and manmade disasters, floods, droughts, lack of food and water, poor oxygen levels, social unrest and terrorism, disease and viruses, or biochemical wars. Hence, if these continue to be part of humanity's challenges, then society must continue to expose children to difficult and scary narratives surrounding these and other potential cultural trauma. Indeed, future generations – our children – might need to be one of the most resilient societies in order to overcome future uncertainties.

That said, it is arguably an exciting time as visitor attractions are likely to see increased inclusion of immersive technologies and digital content as a method to further engage and enhance the guest experience. However, from a child's perspective, the digital content being created should be one of edutainment. Ultimately, this chapter has stressed how (dark) visitor attractions are places where our children can learn from the (tragic) past to better prepare them for the (hopeful) future. Managers (of sites) and content creators should now consider the creation and application of effective edutainment experiences via immersive technologies (VR and AR) at their sites as a means to connect children with difficult heritage more informatively.

References

AugBrite (2020) *Augmented Reality as a Tool for Children Education: School of the Future*. Available at: https://augbrite.com/blog/augmented-reality-as-a-tool-for-children-education-school-of-the-future/ [Accessed on: 24.07.2020].

Aymerich-Franch, L. (2010) Presence and Emotions in Playing a Group Game in a Virtual Environment: The Influence of Body Participation. *Cyberpsychology, Behavior, and Social Networking*, *13*, 649–654.

Azuma, Ronald. (2015) Location-based Mixed and Augmented Reality Storytelling. In W. Barfield (Eds.) *Fundamentals of Wearable Computers and Augmented Reality*, (2nd Edition; pp. 259–276). London: CRC Press.

Bailey, J. O. and Bailenson, J. N. (2017) Immersive Virtual Reality and the Developing Child. In F. C. Blumberg and P. J. Brooks (Eds.) *Cognitive Development in Digital Context by Blumberg* (pp. 1026–1035). London: Academic Press.

Baños, R. M., Botella, C., Alcañiz, M., Liaño, V., Guerrero, B., and Rey, B. (2004) Immersion and Emotion: Their Impact on the Sense of Presence. *CyberPsychology & Behavior*, *7*, 734–741.

Baxter, N. and Goulding, J. (2012) *The Mystery of the Haunted House: Dare You Peek Through the 3D Windows?* London: Anness Publishing.

Bimber, Oliver, Encarnação, Miguel L., and Schmalstieg, Dieter (2003) The Virtual Showcase as a new Platform for Augmented Reality Digital Storytelling. *EGVE '03 Proceedings of the Workshop on Virtual Environments*, pp. 87–95, Bauhaus University, Weimar, Germany.

Blascovich, J. and Bailenson, J. (2011) *Infinite Reality: Avatars, Eternal Life, New Worlds, and the Dawn of the Virtual Revolution*. New York: William Morrow and Co.

Boris, V. (2017) *What Makes Storytelling So Effective For Learning?* Available at: https://www.harvardbusiness.org/what-makes-storytelling-so-effective-for-learning/ [Accessed on: 15.07.2020].

Champagnat, R., Delmas, G., and Augeraud, M. (2008) *A Storytelling Model for Educational Games*. Available at: http://ceur-ws.org/Vol-386/p01.pdf [Accessed on: 06.07.2020].

Chang, M., Hwang, W.-Y., Chen, M.-P., and Müller, W. (2011) Edutainment Technologies: Educational Games and Virtual Reality / Augmented Reality Application. *6th International Conference on E-learning and Games, Edutainment 2011*, Taipei, Taiwan, September 7–9, Springer, New York.

Çiflikli, B., PIsler, V., and Güdükbay, U. (2010) Increasing the Sense of Presence in a Simulation Environment using Image Generators Based on Visual Attention. *Presence: Teleoperators and Virtual Environments*, *19*, 557–568.

Cullingford, C. (1995) Children's Attitudes to Holidays Overseas. *Tourism Management*, *16*(2), 121–127.

Cummings, J. J. and Bailenson, J. N. (2016) How Immersive Is Enough? A Meta-Analysis of the Effect of Immersive Technology on User Presence. *Media Psychology*, *19*, 272–309.

Deterding, S., Dixon, D., Khaled, R., and Nacke, L. E. (2011) From Game Design Elements to Gamefulness: Defining "Gamification". *Mind Trek 2011 Conference Proceedings*. ACM Press, Tampere.

Dockett, S., Main, S., and Kelly, L. (2011) Consulting young children: Experiences from a museum. *Visitor Studies*, *14*(1), 13–33.

Egenfeldt-Nielsen, S. (2010) *Beyond Edutainment: Exploring the Educational Potential of Computer Games*. PhD Thesis, Available at: https://books.google.co.uk/books?hl=en&lr=&id=snupBAAAQBAJ&oi=fnd&pg=PA9&dq=edutainment&ots=0rWyQGcZrv&sig=fHnh6yWq_BtDEWiPsEjC-QoB4l4&redir_esc=y#v=onepage&q=edutainment&f=false [Accessed on: 06.07.2020].

Farber, M. (2014) *Game-based Storytelling*. Available at: https://www.edutopia.org/blog/game-based-storytelling-matthew-farber [Accessed on: 06.07.2020].

Floyd, S. (2016) *Fear is Normal: Childhood Life Lessons in Scary Stories*. Available at: https://www.augusthouse.com/single-post/Fear-is-Normal-Childhood-Life-Lessons-in-Scary-Stories [Accessed on: 15.02.2020].

Ghanbari, H., Simila, J., and Markkula, J. (2015) Utilizing Online Serious Games to Facilitate Distributed Requirements Elicitation. *Journal of Systems and Software*, *109*, 32–49.

Golosovskaya, A. (2020) *Augmented Reality for Kids: Concepts, Technical Side, and Use Cases*. Available at: https://invisible.toys/augmented-reality-for-kids/# [Accessed on: 24.07.2020].

Grigorovici, D. (2003) Persuasive Effects of Presence in Immersive Virtual Environments. In G. Riva, F. Davide, and W. IJsselsteijn (Eds.) *Being There: Concepts, Effects and Measurement of Presence in Synthetic Environments* (pp. 192–205). Amsterdam, The Netherlands: Ios Press.

Guttentag, D. A. (2009) Virtual Reality: Application and Implications for Tourism. *Tourism Management*, *31*(5), 637–651.

Hall, S. and Takahashi, R. (2017) September. Augmented and virtual reality: the promise and peril of immersive technologies. *World Economic Forum* (Vol. 2).

Healthline (2005–2021) *Amygdala Hijack: When Emotion Takes Over*. Available at: https://www.healthline.com/health/stress/amygdala-hijack [Accessed on: 03.02.2021].

Hoffman, H. and Vu, D. (1997) Virtual Reality: Teaching Tool of the Twenty-first Century? *Academic Medicine*, *72*(12), 1076–1081.

Israfilova, F. and Khoo-Lattimore, C. (2018) Sad and Violent But I Enjoy It: Children's Engagement with Dark Tourism as an Educational Tool. *Tourism and Hospitality Research*, *9*(4), 478–487.

Kerr, M. M. and Price, H. R. (2016) Overlooked encounters: young tourists' experiences at dark sites. *Journal of Heritage Tourism*, *11*(2), 177–185.

Kysela, J. and Štorková, P. (2015) Using Augmented Reality as a Medium for Teaching History and Tourism. *Social and Behavioral Sciences*, *174*, 926–931.

Lauricella, A. R., Cingel, D. P., Blackwell, C., Wartella, E., and Conway, A. (2014) The Mobile Generation: Youth and Adolescent Ownership and Use of New Media. *Communication Research Reports*, *31*(4), 357–364.

Light, D. (2000) Gazing on Communism: Heritage Tourism and Post-communist Identifies in Germany, Hungary and Romania. *Tourism Geographies*, *2*(2), 157–176.

Lindgren, M. and Bandhold, H. (2009) Scenario Planning: The link between future and strategy. London: Palgrave Macmillan.

Marcuse, H. (1964) *One Dimensional Man: Studies in the Ideology of Advanced Industrial Society*. Boston: Beacon Press.

MasterClass (2019) *4 Types of Narrative Writing*. Available at: https://www.masterclass.com/articles/types-of-narrative-writing#what-is-narrative [Accessed on: 16.07.2020].

McGloin, R., Farrar, K. M., and Krcmar, M. (2011) The Impact of Controller Naturalness on Spatial Presence, Gamer Enjoyment, and Perceived Realism in a Tennis Simulation Video Game. *Presence: Teleoperators and Virtual Environments, 20*, 309–324.

Milgram, P. and Kishino, F. (1994) A Taxonomy of Mixed Reality Visual Displays. *IEICE Transactions on Information and Systems, E77-D*, 1321–1329.

Puckett, M. B. and Diffily, D. (2004) *Teaching Young Children: An Introduction to the Early Childhood Profession* (2nd Edition). Canada: Thomson.

Reeves, B. (1989) Theories about News and Theories about Cognition: Arguments for a More Radical Separation. *The American Behavioral Scientist, 33*(2), 191–198.

Regian, J. W., Shebilske, W. L., and Monk, J. M. (1992) Virtual Reality: An Instructional Medium for Visual-spatial Tasks. *Journal of Communication, 42*(4), 136–149.

Richert, R. A., Robb, M. B., and Smith, E. I. (2011) Media as Social Partners: The Social Nature of Young Children's Learning from Screen Media. *Child Development, 82*(1), 82–95.

Roberts, C. (2018) Educating the (Dark) Masses: Dark Tourism and Sensemaking. In P. R. Stone, R. Hartmann, T. Seaton, R. Sharpley, and L. White (Eds.) *The Palgrave Handbook of Dark Tourism Studies* (pp. 603–637). London: Palgrave Mcmillan.

Sharar, S. R., Carrougher, G. J., Nakamura, D., Hoffman, H. G., Blough, D. K., and Patterson, D. R. (2007) Factors Influencing the Efficacy of Virtual Reality Distraction Analgesia during Postburn Physical Therapy: Preliminary Results from 3 Ongoing Studies. *Archives of Physical Medicine and Rehabilitation, 88*(12 Suppl. 2), S43–S49.

Sharpley, R. and Baldwin, F. (2009) Battlefield Tourism: Bringing Organized Violence Back to Life. In R. Sharpley and P. R. Stone (Eds.) *The Darker Side of Travel: The Theory and Practice of Dark Tourism* (pp. 186–206). Aspect of Tourism Series. Bristol: Channel View Publications.

Šisler, V. and Brom, C. (2008) Designing an Educational Game: Case Study of 'Europe 2045. In Z. Pan, A. D. Cheol, and W. Müller (Eds.) *Transactions on Edutainment I* (pp. 1–16). New York: Springer.

Slater, M., Usoh, M. and Steed, A. (1995a) Taking steps: the influence of a walking technique on presence in virtual reality. *ACM Transactions on Computer-Human Interaction (TOCHI), 2*(3), pp.201–219

Slater, M., Usoh, M. and Steed, A. (1995b) The virtual treadmill: A naturalistic metaphor for navigation in immersive virtual environments. In *Virtual environments' 95* (pp. 135–148). Vienna: Springer.

Stone, P. R. (2006) A Dark Tourism Spectrum: Towards a Typology of Death and Macabre Related Tourist Sites, Attractions and Exhibitions. *An Interdisciplinary International Journal, 54*(2), 145–160.

Stone, P. R. (2009) 'It's a Bloody Guide': Fun, Fear and a Lighter Side of Dark Tourism at The Dungeon Visitor Attractions, UK. In R. Sharpley and P. R. Stone (Eds.) *The Darker Side of Travel: The Theory and Practice of Dark Tourism* (pp. 167–185). Aspect of Tourism Series. Bristol: Channel View Publications.

Stone, P. R. (2011) Dark Tourism and the Cadaveric Carnival: Mediating Life and Death Narratives at Gunther von Hagens' Body Worlds. *Current Issues in Tourism*, *14*(7), 685–701.

Stone, P. R. (2012) Dark Tourism and Significant Other Death: Towards a Model of Mortality Mediation. *Annals of Tourism Research*, *39*(3), 1565–1587.

Walter, T. (2009) Dark Tourism: Mediating between the Dead and the Living. In R. Sharpley and P. R. Stone (Eds.) *The Darker Side of Travel: The Theory and Practice of Dark Tourism* (pp. 39–88). Aspect of Tourism Series. Bristol: Channel View Publications.

Wright, D. W. M. (2018) Terror Park: A Future Theme Park in 2100. *Futures*, *96*, 1–22.

Wright, D. W. M. (2020) Immersive Dark Tourism Experiences - Storytelling at Dark Tourism Attractions in the Age of 'The Immersive Death'. In M. H. Jacobsen (Eds.) *The Age of Spectacular Death* (pp. 89–109). London: Routledge.

Xu, F., Buhalis, D., and Weber, J. (2017) Serious Games and the Gamification of Tourism. *Tourism Management*, *60*, 244–256.

Zetter, K. (2018) *TED 2010: Reality Is Broken. Game Designers Must Fix It*. Available at: https://www.wired.com/2010/02/jane-mcgonigal/ [Accessed on: 21.07.2020].

Zichermann, G. and Cunningham, C. (2011) *Gamification by Design: Implementing Game Mechanics in Web and Mobile Apps*. Canada: O'Reilly Media.

12 'Deconstructing Dark History and Difficult Heritage'

Engaging High School Students in the Use of Historiographical Analysis Techniques

Michael Lovorn

Introduction

As a high school teacher and social studies methods instructor (in the USA), I have regularly implemented discussions and lessons on dark history and difficult heritage in my classes. For me, dark and difficult elements of our collective historical journey have always held tremendous student engagement potential. As a result, throughout my nearly 30 years in the profession, I have devoted many semester hours to engaging students and teacher candidates in human tragedy examinations and subsequent narratives, counternarratives, and the multiple perspectives that surround them.

This chapter, therefore, is a reflection on a dark history/difficult heritage lesson unit I taught in a public high school just outside Pittsburgh, Pennsylvania. As a Guest Lecturer I taught a six-day unit on the 1963 assassination of President John F. Kennedy. Each hour-long lesson introduced 11th grade students (aged 16–17) to pictographic displays and discussions of various circumstances leading up to the event, the event itself, the space where the murder occurred, individuals and groups known and believed to have been involved, as well as national reaction to the assassination. This chapter culminates with a review of successes and shortcomings of the lesson(s) and summarizes student reactions to the dark history/difficult heritage unit. The chapter begins with a brief orientation of how I have utilized dark history and difficult heritage in my teaching.

Teaching dark history and difficult heritage

As a social studies teacher, and realizing the regularity with which history textbooks omit particularly shocking or gruesome details about carnage or death, I have occasionally felt compelled to include age- and developmentally appropriate elaborations to intrigue my students and provide clearer pictures of human tragedy. Of course, I realize such imagery is not easy to present or digest, but I feel strongly that students must be exposed and trained to grapple with difficult concepts and events and, thus, have integrated targeted techniques both inside and outside the classroom. Over the years, I have led students on numerous virtual tours of dark history sites. For example, I have

DOI: 10.4324/9781003032199-16

taught about the Sultana steamboat explosion in 1865 which is the worst maritime disaster in US history, the 1890 massacre of Lakota Indians by US troops at Wounded Knee, and the fatal maiden voyage of the Titanic in 1912: All complete with stark imagery and spatial orientation, emphasizing the tremendous human tragedy in each event.

I have also taken young people to many historical sites commemorating dark events in history. Domestic school excursions have included the Lorraine Motel in Memphis, Tennessee (assassination site of Martin Luther King Jr in 1968), Centennial Park in Atlanta, Georgia (site of a domestic terrorist attack in 1996), the controversial "Battle of Liberty Place" memorial in New Orleans, Louisiana (commemorating the attempted insurrection by White League nationalists in 1874), and the Branch Davidian compound outside Waco, Texas (site of the Waco siege and massacre in 1993). I have also led tours of international dark history sites including at Bogside in [London] derry, Northern Ireland (site of the 1972 "Bloody Sunday" killings by British forces against civilian protesters), the Sharpeville Massacre site in Soweto, South Africa (where 69 apartheid protesters were killed by police in 1960), and Nazi German death camps across Germany and occupied Poland. Regardless of the topic, I always preempt my lessons with a declaration to all students that I do not revel in death and destruction, nor do I want them to do so. However, there are times when educated people should study and discuss these ideations in more than abstract terms. Recently, I had the good fortune of inviting world-renowned forensic pathologist Cyril Wecht to my graduate social studies education class to discuss his work on the 1963 assassination of President John F. Kennedy. Dr. Wecht's research continues to fuel discussion and invite conclusions that some consider remarkably conspiratorial in both tone and content. His visit was an exciting departure from typical social studies methods protocol, and students thoroughly enjoyed his detailed simulation of Kennedy's limousine as it passed through Dealey Plaza, culminating in an argument on the implausibility of the so-called "magic bullet" that purportedly caused seven entry/exit wounds on its occupants.

Moreover, my pre-service teachers were familiar with Kennedy's assassination and varying conspiratorial theories surrounding it. However, few of them had heard a first-person account connecting dots in this manner. They were rapt with Dr. Wecht's counter-narrative and their reflections included sentiments such as "…I had no idea… and wonder how many shooters there really were", "I think for the first time I understand what a coup d'état is…" and "He taught us like it was a mystery. I want to teach like this!" Research reinforces this theory that students enjoy learning about historical events, particularly dark ones, that receive relatively little coverage in traditional textbooks (Israfilova and Khoo-Lattimore, 2019; Kerr and Price, 2016), and center on critical analysis of historical events deemed difficult, unsolved, or even "conspiratorial" in nature (Lovorn, 2012; Morrison, 2019). Importantly, these teaching experiences in secondary and post-secondary settings have informed my practice and grounded my approach to dark and difficult history in three interwoven foundational understandings. Firstly, like most of us, students are intrigued by topics that delve into the morose or macabre, particularly those related to political chicanery, human tragedy, betrayal, murder, or other depravities. Secondly, our examinations of dark history are made richer and more sustainable when teachers implement a structure for students' critical analysis. Finally, connections are made more powerful when students are enabled to visit sites of significant dark history or difficult heritage, either in person or via virtual tour.

In the case of the JFK assassination, as outlined earlier, my study examined the application of these understandings by qualitatively analyzing a historiographical analysis engaging high school students over six days in a deep examination of the assassination of President John F. Kennedy as he rode in a motorcade in Dallas, Texas on November 22, 1963. After a brief history lesson on the times and context, including the contentious presidential election of 1960, the disastrous 1961 Bay of Pigs Invasion, and the subsequent Cuban Missile Crisis in 1962, students undertook a virtual tour of the Dallas Dealey Plaza, where the assassination took place. Students were then invited to reflect upon and share thoughts on the act of assassination and discuss similarities and differences with other high-profile murders. Next, I prepped students for the unit by explaining we would focus on JFK's visit to Dallas, theories surrounding his assassination, the chaos that ensued at the scene, and identify various suspects. For the remainder of this chapter, what follows is a day-by-day account of our case study lessons, and my analysis of the aforementioned understandings, as driven by two research questions:

1   What are high school students' perceptions of a virtual tour and lesson unit centered on a dark history site?
2   To what extent does a structured historiographical analysis protocol advance students' understandings as related to this dark history site?

Trending toward dark history/tourism

As the Editors of this volume have demonstrated both with their respective bodies of work prior to this volume and the chapters collected herein, the past two decades have seen a dramatic increase in general interest in examining dark tourism and the ways in which we process difficult heritage (Seaton, 2018). Undoubtedly, much of this has been the direct result of an ever-growing repository of video clips and commentary on the internet. Indeed, unfettered 24/7 internet access coupled with an ability to explore atrocities and tragedy in relative anonymity also seems to spur our interest in dark events and related tourism sites.

We also agree there is tremendous educational potential in both in-classroom (Lovorn, 2012) and outside-the-classroom settings (Olwell, Murphy and Rice, 2016). As tourist visits to sites of historical morose or macabre have increased over this time (Becker, 2019; Cui *et al.*, 2019), many have evolved into – first organic and eventually structured – teaching and learning opportunities (Price and Kerr, 2018). My interest in dark tourism sites emerged from personal fascination and my own analysis of historical events and, naturally, it seems, to have found its way into my lectures and school-related touring. I remain intrigued by sites of human tragedy, and this interest has only deepened since the advent of the internet, as I am now able to locate even entirely un-commemorated sites. For instance, in recent years, and with the assistance of GPS coordinates posted on the internet, I have visited the then-unmarked field just northwest of Clear Lake, Iowa, where the plane carrying famed rock-and-rollers Buddy Holly, Ritchie Valens, and J.P. "Big Bopper" Richardson crashed in February 1959. I have also visited the modestly commemorated site in Paris where a guillotine of the 1789–1799 French Revolution once stood, and the heavily commemorated (even "touristy") Choung Ek Genocidal Centre – often referred to as the "Killing Fields" – a mass grave of 1970s' Khmer Rouge victims in Phnom Penh, Cambodia.

Interestingly, I have also visited other sites of tragic events that have developed over time into significant memorials. I can think of no less than three distinct visits I have made to the site of the 9/11 World Trade Center attack in New York City. The first visit was in late 2002 when "Ground Zero" was still undergoing a massive clean-up and victim recovery effort. At this time, and for years to come, the entire area remained fenced off and was completely inaccessible to the public. However, the worksite fence and area establishments were pixelated with touching and very personal memorials and mementos left by family and friends of the victims. I visited again in 2011 when the official 9/11 memorial opened to the public. At this time, the plaza was dominated with two solemn pools constructed in the footprints of WTC1 and WTC2, each hallowing names of the victims. I made a third visit to the site in 2017, during which I was able to tour the World Trade Center museum memorializing the tragedy. From the organic and impromptu, to the formal and interpretive, each of these visits elicited a unique response to the events of that terrible day in 2001. It is worth noting that on the latter of these visits, I was accompanied by my father on his first and only trip to New York. It was the memorial and commemoration of this national tragedy that prompted his visit. Indeed, it has always been these sorts of awe-inspiring memorials to victims of atrocious acts that have spurred my interests in historiography and caused me to incorporate them into my teaching and lessons.

A structure for students' critical analysis: My historiographical analysis technique

Each year, I begin my social studies methods courses with a simple line I appropriated from renowned history education scholar James Loewen (2009): "History is not what happened. History is what we say happened." I then engage students in an ice-breaker discussion on what this statement means: not only for history teachers, but for the students they teach. This purposefully provocative conversation is intended to introduce future social studies teachers to the tremendous potential of historiographical analysis in the classroom. Historiography is the study of how history is presented and how historical events or people are memorialized or commemorated. There are countless applications of historiography in classroom settings, but I wish to explore the means by which teachers can implement these techniques in engaging students in the examination of dark history.

Historiographical analysis techniques of this design are not new to studies in history. History professors have engaged their college students in deep investigation of the connections between historical commemoration and the act of narrative-building for decades (Lovorn, 2012). However, the concept is still relatively new for middle and high school teachers. My contribution has been to structure the technique so as to take students of these developmental levels outside the classroom and into their local communities to perform historiography skills-based evaluations of historical commemorations.

Commemorations in this context refer to public displays including landmarks, monuments, spaces, historical markers, or similar. Considering ongoing movements to remove statues due to controversial depictions of individuals or ideals (Block, 2020), perhaps the topic could not be timelier. Certainly, many of the statues being toppled indicate a dramatic and real-time broadening of our collective conceptions of difficult heritage. To this end, I have devised a multiphase technique focusing on examining how dark history is presented in the public arena. The technique is designed for implementation at the high school level, and the phases intend to systematize students' training in collecting and critiquing dark history commemorations. In doing so, students hone skills to recognize perspective and agency, evaluate sources and evidence, explain causal relationships, and formulate strong arguments and counterarguments. Each of these phases is described briefly in Table 12.1.

Of course, depending on desired results or students' developmental levels, the teacher may make adaptations to these steps or the depth with which students perform their investigations. For instance, once students demonstrate command of the delineated skills, they could be enabled to duplicate these steps by historiographically analyzing a passage of text, primary/secondary sources, images, music, arts, or other sources of historical information. A historiographical analysis of a work of art, folk lyrics, or even an eyewitness account of a historical tragedy, for example, could be quite revealing and engaging for middle and high school students.

The unit: A historiographical analysis of the assassination of President John F. Kennedy

The suburban Pittsburgh school where this lesson unit was taught has a total enrollment of about 800 students, most of whom are from lower-middle to middle working-class homes. The class I visited was titled "Contemporary History and Culture" and was made up of 24 students: 14 of whom were male and 10 of whom were female. From my visual observation, I deduced 19 students were White and five students were Black. The regular classroom

Table 12.1 Phases of 'Difficult Heritage' Historiographical Analysis Technique

Phase	Duration	Activities
1	1–2 class periods (1–3 hours)	The teacher presents students with a virtual example of a dark history space or commemoration. After a brief orientation to the space/commemoration, students are encouraged to verbalize any evaluative observations about its size, shape, inscription, and other commemorative elements. The teacher is prepared to direct students' attention to particular elements of interest, with the general objectives of prompting them to perform informal evaluations, introducing techniques to maximize information gatherings, and inviting them to formulate opinions and provide initial feedback. The teacher introduces the process for making bolder, evidence-based evaluative judgments about the commemoration.
2	1–2 class periods (1–3 hours)	Students are trained in making six distinct observations when evaluating historical spaces or monuments: (1) Design, size, and shape; (2) location or spatial orientation; (3) date and origin of commemoration; (4) recognized perspectives and agendas; (5) historical accuracy and completeness; and (6) overall "take-home" message for students or anyone else who might encounter the commemoration. Students are presented a data collection sheet to be used for the duration of the project and practice completing it with the data they gathered from the examples in Phase One. The sheet is made up of seven categories of data to be collected; the six listed above, plus one more about historical advocacy.
3	1–2 class periods (1–3 hours)	Under the teacher's guidance and supervision, students visit a local commemoration – in person or virtually – and, in pairs or small groups, perform at least one part of a thorough evaluation of the site using their newly acquired historiographical analysis skills. During raw data collection activities, the teacher serves as a resource for historical context, inquiry pointers, and advice on how to find obscure information on the commemoration.
4	1–2 class periods (1–3 hours)	In their pairs or small groups, students collaborate on iterations to (1) share their perceptions and reactions to the difficult heritage site; (2) select and research similar sites of historical significance in other parts of the country or world; 3) perform an independent historiographical analysis of a particularly relevant site using the newly acquired skills.
5	1–2 class periods (1–3 hours)	Students prepare to share their comparative findings in a formal setting by crafting presentations in the form of PowerPoints or trifold posters. The trifold poster option has proven useful in a variety of settings because they may be displayed anywhere in the school, and students can serve as presenters in a face-to-face setting. The "historiography fair" as I call it, can be held in a central location of the school, perhaps during lunch, and students, other teachers, and administrators can attend.

teacher was wrapping up a two-week unit on the "Tumultuous 1960s" when I was invited to join the class. For full and generalized accounts of the JFK assassination, readers are directed to peruse pertinent webpages on the websites of the Sixth Floor Museum at Dealey Plaza (2021), and the John F. Kennedy Presidential Library and Museum (2021). Readers may dive deeper into the historical events, controversy, and conspiracy theories surrounding the assassination by examining Jim Marrs' *Crossfire: The Plot That Killed Kennedy* (2013) and James Douglass' *JFK and the Unspeakable: Why He Died and Why It Matters* (2008).

Phase One: Tragedy in Dallas

As the unit was presented near the end of the school year, there was a degree of schedule flexibility; however, I still had to adapt (abbreviate) the aforementioned phases a bit. Phase One (and the unit itself) began on a Friday with me building upon lessons the classroom teacher had already presented on Kennedy's administration, the times, and his legacy. I added pertinent details about Kennedy's rise to power, the election of 1960, the then-popular comparison of the Kennedy family's to "Camelot", his particular popularity among young people, and by contrast, various people and political entities (including Cuban and Russian Communists, the FBI and CIA, and the American mafia) who harbored disdain for him. Next, we briefly surveyed a timeline of events in Fort Worth and Dallas on what would become a fateful Friday (November 22, 1963). Day one ended with the playing of Walter Cronkite live, frazzled broadcast announcing to the nation that "...three shots rang out in Dallas..."

Phase Two: Inconsistencies in initial reports

Day two (Monday) began with an introduction to Abraham Zapruder, Lee Harvey Oswald, and several others who would end up in or around Dealey Plaza on the day of the assassination. Next, I presented students with a spatial orientation to Dealey Plaza itself, including photos from various angles, including Houston and Elm Streets, the Texas School Book Depository, the triple-underpass, as well as the infamous "Grassy Knoll". Students were invited to verbalize their evaluative observations about the space, which several did.

When asked for initial assessments of the area, one student opined, "It looks like a normal place". Another added, "Yeah, but normal places are where bad things happen". As a group, we discussed these comments for a few minutes during which time several students made pertinent connections to recent dark events and unexpected tragedies including mass shootings in Las Vegas in 2017 and the 2011 earthquake/tsunami disaster in Japan. We also spoke briefly on similarities and differences in these examples and our central focus – that is, the JFK assassination. We returned to our informal evaluations of Dealey Plaza, and as I shared more period and contemporary photos and diagrams, I invited students to observe particular features relating to the size, location, and spatial orientation of this area. Students observed that the plaza was large, surrounded by buildings, allowed the motorcade to make "a bunch of strange turns", and would "make it hard to watch everybody there". To reiterate, clarify and focus all students' attention on each of these observations, I repeated key points and summarized them on the board.

Up to this point, students had been taught about the end result of the assassination (the murder of an American President), and that there had been significant public opinion deviations from the official report. I reminded students that, while they have merely undertaken a couple of days investigating this topic, the American public at the time did so for months and, for some, even years. We revisited the timeline of events. I ended day two by introducing the Warren Commission and reading about five telling excerpts from its 1964 summary report. Students were presented with a predesigned data collection sheet made up of a graphic organizer that enabled them to catalog these excerpts (as well as other observations) and weigh their veracity. I explained that when they returned the next day, we would dig deeper into the cries of conspiracy.

Phase Three: The power of (unexpected) visual imagery

On the third day (Wednesday), I opened Phase Three with a lesson with two disclaimers. The first was a reminder that, while there are limitations to virtual tours, there are also often some advantages. In our case, we were enabled to utilize the internet not only to make several particularly relevant observations about the time and space in which the assassination occurred, but we could also verify or cross-reference personal accounts, "official" details, and conspiratorial commentary in real-time. The second disclaimer was that I was about to show the class the short-but-graphic video clip that would become a centerpiece in the 50-year debate challenging the official series of events as articulated in the Warren Report. The video clip, shot on an 8-mm camera by Dallas shoe salesman Abraham Zapruder, would later become known as the "Zapruder Film". Students were warned about the particularly graphic images in the silent 26-second clip and were invited to step out of the room if they preferred not to see it. None did so.

We accessed the uncut, unedited Zapruder Film on YouTube, and watched it three times in relatively quick succession. The first viewing was uninterrupted from start to finish. Students gasped and groaned as the fatal shot exploded into Kennedy's head and chaos ensued. I asked for a quick show of hands of those who had seen the clip before. None had, and for a moment, I relished in the setting of being able to discuss these images with a group of young people who had really no preconceptions about the assassination. Subsequently, I invited reactions. Several students spoke. One stated, "Oh my god, why was he in a convertible if he knew people hated him?!" Another student said, "Well, it was a murder from being shot for sure". Another added, "He died instantly, didn't he?"

I contextualized and elaborated on each of these observations, summarizing them on the board. I reminded students that, unlike today when so much of our lived experiences are immediately accessible upon being posted on social media, it would be several years before the public would see the uncut, unedited Zapruder Film. As a result, the public's summation of these events depended heavily upon press releases and newspaper accounts and was heavily influenced by their trust in government officials and agencies. When I showed the Zapruder Film for a second time, I paused the video at various points to narrate, contextualize and call students' attention to a variety of focal elements. Next, I sketched a top view of Dealey Plaza on the board, keying on various possible angles of fire, particularly those from the sixth-floor window of the book depository, the Grassy Knoll, and the triple underpass. Again, I invited students to comment on their observations of the video. Interestingly but perhaps unsurprising, there was a universal agreement among students that the "kill shot" came from a position in front of the limousine. One student asked, "How did anybody believe he was shot from behind?" and quickly added, "His head goes back, not forward!" Another student suggested, "...maybe Lee Harvey was in a different place!" This comment humored several of us and we all enjoyed a quick laugh, which in retrospect, considering the dark topic, I find very interesting.

The third viewing included another pause-and-speak approach. During this showing, I identified and provided a 30-second bio of each of the individuals in or around the limousine (complete with photos, in which the students took great interest). This group included John and Jackie Kennedy, the first couple of Texas at the time John and Nellie Connally, and Secret Service agents Clint Hill, Roy Kellerman, and Bill Greer (the driver). Next, I pointed out several intriguing individuals in the frame margins, including eyewitnesses Gordon and Mary Arnold, Ann Atterberry, Ernest Brandt, Officer Bobby Hargis, Officer H.B. McLain, Bill and Gayle, Newman, and Malcolm Summers. I briefly described each of them while sharing still images of them in Dealey Plaza, and adding them one-by-one to my overhead diagram. I did so while describing their unique vantage points and how several of them would share their memories of that day for many years.

Finally, to deepen their interest, I also introduced several provocative or mysterious bystanders who may not have appeared in the video, but were known to have been in Dealey Plaza that day, including the "three tramps", the "Umbrella Man" (who actually did make a brief appearance in the Zapruder Film), and reports of a curious rifle-wielding policeman standing behind the fence atop the Grassy Knoll. Grainy images of each whipped the students into a fever pitch for conspiratorial theorization. Subsequently, students entered into a very interesting and relatively elaborate side discussion on how much we were focusing upon and relying so much on the grainy and relatively two-dimensional visuals of the Zapruder Film to solve this mystery. As a group, they seemed to understand this difficulty and described it as a challenge they have not had to confront in their lifetimes. Several agreed that, by contrast, most of their lives have been meticulously documented in both still and video imagery, and posted online for all to see. One student added that there are security cameras just about everywhere these days and sometimes they capture criminal activity in progress and are used in solving crimes.

Phase Four: Breakout investigations

Day four (Wednesday) began with a brief student-led timeline of events and a delineated review of all we had discussed up to this point. As they recalled events in order, I conspicuously added them to a master timeline and summary on the board. I divided students into eight pairs or small groups and asked each group to move to one of eight research stations. Then I called on one representative from each group to select a number from 1 to 8, and assigned a corresponding area of focus for each group to research.

I strategically selected eight areas of focus to provide a comprehensive examination of Kennedy's visit to Dallas, the events and circumstances surrounding his death, and the investigations by various interest groups that followed. Thes eight areas of focus included (1) Kennedy's reason for visiting Texas and movements on the morning of November 22; (2) the parade, motorcade route, shooting, and the race to Parkland Hospital; (3) Lee Harvey Oswald's arrest, interrogation, and murder; (4) Kennedy's troublesome interactions with the Cuban government; (5) organized crime groups and reasons they might want Kennedy dead; (6) Kennedy's unstable relationships with the CIA and FBI; (7) the Warren Commission, findings and report; and (8) Jim Garrison and the trial of Clay Shaw.

Groups were given 30 minutes to perform online data collection and specific graphic organizers that facilitated their documentation of perceptions and reactions in these focused contexts, and to prepare them for a class-wide share-out activity. Each graphic organizer included topic-specific tips on relevant individuals, groups, events, and websites to research to complete these data collection activities. I moved from group to group, ensuring key elements had been appropriately examined. I also provided more relevant anecdotal information to keep each group invested in their topic. Twice during the data collection activity, I called the class back together to gauge their research pacing and remind all students they would be contributing to a larger narrative. I took the last 20 minutes of the period to model how I would like groups to aggregate and summarize collected data. To accomplish this, I presented a ninth topic – Kennedy's relationship with Vice President (and one-time rival) Lyndon Johnson – and shared a pre-completed graphic organizer. I invited students to opine as to why I had included various elements and explained them as such. This activity also served to reinforce my expectations for the share-out.

These research protocols were structured to foster a relatively seamless narrative of the assassination and ensuing conspiracy. This phase also incorporated the unit's only outside-the-classroom work. After describing my presentation expectations to the class, I asked students to work together to refine their graphic organizers and to be ready to share with their classmates.

Phase Five: Weaving together a narrative of difficult heritage

On days five and six (Thursday and Friday), student groups shared their findings. Groups were informed each presentation should last a maximum of 15
minutes; 10–12 minutes to share findings, and the balance for questions from
classmates. Groups were invited to use PowerPoint and/or to include visual
and audio clips as they deemed appropriate. They were also told that the only
"requirement" was for all group members to participate in some aspect of the
presentation.

Group 1, a pair of girls, provided a considered summary of Kennedy's reasons for visiting Dallas, including his interest in mending political and regional
fences. They shared a succinct-but-detailed timeline of his movements, including his speech to the Fort Worth Chamber of Commerce, interactions with
the public at the Hotel Texas, and his short flight from Carswell Air Force
Base to Love Field in Dallas. Their presentation was enriched with photos
from the day, and a brief video clip during which the Fort Worth Chamber of
Commerce presented JFK with a Stetson cowboy hat. While unintentional,
this opener was received by their peers as an ominous foreshadowing of events
to come. Comments and questions from the class included, "Do they know if
people were watching him when he gave the speech?" And "Maybe they
wanted to kill him then". Several students seemed interested in this thought.

Group 2, three boys, presented a PowerPoint on the motorcade route from
Love Field through downtown Dallas, the assassination, and the race to
Parkland Hospital. Their presentation was anchored in an impressive mile-
by-mile diagram on the internet. We discussed the route, estimating arrival
times and different curious events that took place. Again, students asked several questions about the layout of Dealey Plaza and its proximity to the book
depository.

Group 3, two boys and a girl, continued this examination by re-introducing
Lee Harvey Oswald as the prime suspect. They showed photographs of
Oswald's military service, his time in the Soviet Union, and of course, the
famous image that appears to show him holding the murder weapon. I walked
everyone through a closer examination of this photo, pointing out how many
scholars believe it to be a fake. Students asked several questions about Oswald
and his whereabouts on the morning before and during the assassination. At
this point, virtually all students were highly engaged and interested in a
deeper conversation on Mr. Oswald. We took time to discuss him a bit more,
and watched the video clip of Oswald's subsequent murder on 24 November
1963 by Jack Ruby at the Dallas Police Station. Students were astonished to
learn this was the first-ever televised-live murder, and were all really interested in the fact that I have been to his grave in Fort Worth!

Groups 4 and 5, both groups of three girls each, were able to present on
Thursday, and Group 6, four boys, started our Friday class. Each of these
groups continued the discussion by sharing Kennedy's interactions with their
assigned groups (Castro's agents, the mafia, and the CIA and FBI, respectively),
and why each might want the President out of the way. Each presentation was

rich in speculation, which is something I was hoping for, and brought our collective interest to a high. During these presentations, each of which was delivered via PowerPoint, students' excitement increased at the possibility that they had stumbled upon previously unknown, often backdoor, connections between Kennedy and these agents. Interestingly, multiple students questioned Kennedy's squeaky-clean image.

Group 7, two boys and a girl, informed us all on the scope and purview of the Warren Commission, including their findings and official report. One student actually found the abridged report in the library and brought it to class. Interestingly, like Group 3, this group also called our attention to the photo of Oswald holding the mail-ordered rifle. They scrutinized the photo even more than we had done previously, and seemed to doubt the veracity of the entire Warren Report based on their interpretation of this one piece of evidence. A deep class discussion ensued about the possibility of evidence manipulation. Several students in the class added to the conversation. One stated, "We know sometimes the police plant evidence. They do it on Dateline" [referring to a weekly American newsmagazine/reality legal television show broadcast by NBC]. I laughed at this remark, but then asked the class to explain the significance of manipulated evidence. Several students' opinions were summed up by one comment: "If they can change the way a jury looks at a trial, they can basically change the whole outcome!"

Group 8 wrapped up our examination of the Kennedy assassination by introducing the name Jim Garrison, the New Orleans District Attorney who, in 1967, charged local businessman Clay Shaw with lying under oath and conspiring to murder the President. Up until this point, I had purposefully left the indictment of Clay Shaw out of our discussions on conspiracies. It was my hope this would be a nice culminating revelation, sparking students' continued interest in the topic. Group 8 students introduced these two men by summarizing the 1969 trial which, as was dramatized in Oliver Stone's 1991 blockbuster film *JFK*, resulted in a dismissal of charges against Shaw. As anticipated, student comments confirmed a class-wide fascination. One student remarked, "I believe the CIA did it. This guy was in the CIA. He lied and they know it". Another student asked, "Why was he always acting so shady? He had to do it". A third student declared "I'm going to watch that [JFK] movie…"

Summary: Toward a historiographical analysis protocol

Throughout this reflective activity, I sought high school students' perceptions of a virtual tour and lesson centered on a site of dark history/tourism and difficult heritage, namely the assassination of JFK. Moreover, I wanted to ascertain the extent to which a structured historiographical analysis protocol of this design advances students' understandings of historical perspective as related to dark history. To answer these questions, I captured students' comments throughout the week and followed up with an open-ended discussion during the last ten minutes of the final class meeting. While data collection was largely anecdotal, the general sentiment was overwhelmingly positive. I share a few student responses and reactions here.

Overall, students overwhelmingly expressed their enjoyment in the lessons/unit. Several elaborated on their interest in this example of dark history and described how, early on, their intrigue sparked motivation to perform independent research to further investigate information that might have been incomplete, inaccurately presented, or outright suppressed by historians or the government. When I asked them specifically to discuss their interests in dark history, I received several interesting answers. One male student said, "I think all history is dark history! People have been getting killed for a long time". Another male student elaborated with the comment "I like it because these are the things we don't know about that change history. Like you said, there's a lot the history books don't tell us".

Next, I asked students to share their thoughts about the violence and/or gruesome details that so often define or accompany dark history events. A female student replied, "It doesn't bother me. We see it all the time…" Another female added, "I don't like to hear about all the murders, but we do need to learn about it. Especially when it is a President". A male student summed up the opinion apparently shared by many: "Halo [the video game] is more gruesome than this! I don't even consider that violence". Others reminded him that a person actually died in the Zapruder Film.

When I opened the discussion up to general comments, several students said they enjoyed taking on the roles of "history detectives". One female student said, "I wish we could do this with every lesson. I feel there are things we don't know about everything in American history. This would make it more interesting to me". A male student semi-humorously added, "What if they wrote a book just on the bad guys? I want to read that one". Still, others stated how much this lesson had made them want to visit Dealey Plaza (in terms of dark tourism). I found this an interesting sentiment, in that not only were students intrigued by the tragedy and mystery of it all, but they seemed to realize there is value in visiting a place of such historical significance. When I asked them to explain why, one male student replied, "…because being there we could see how close the shooter was". Another male student stated, "I'd like to see the view from the book [depository] window…"

Next, I asked students to elaborate on their thoughts of the historiographical analysis protocol I had used as a means of deconstructing and understanding dark history. "Well", one male student replied, "it helped us study several things at once". Another student added, "…and it made us take lots of things into consideration". I took this to mean the structure had provided a framework for them to perform a comprehensive investigation of Kennedy's assassination. "If you hadn't done it this way", a female student added, "we wouldn't have known there was a conspiracy". To this, one male student disagreed, "That's not true. The [text] book tells us there might have been a conspiracy. And I had heard of it anyway". When I asked if this style of lesson would work for other projects in the semester, the class was divided. As stated earlier, some students seemed to believe the historiographical analysis protocol would be "fun" or "more interesting" if we used it all year long. Others, however, seemed equally opposed to the idea, summed up by a student suggesting "I don't know if I would want to do this all year. There's a lot to remember and it'd get old after a while". Despite this derision, there was a great deal of uniformity in their reactions to the unit. Of all influential materials, it seemed students were most intrigued by photographs and video clips. Regarding photos, the group was quick to voice observations about individuals of interest, suspects, and the Dealey Plaza. Regarding videos, numerous students requested we watch them multiple times.

Conclusion

This project was designed to help develop students' thresholds for counter-narrative by examining an example of dark history. Overall, I found that these high school students' perceptions of a virtual tour and lesson centered on a site of dark history/tourism or difficult heritage, and determined the extent to which a structured historiographical analysis protocol of this design advances students' understandings of historical perspective as related to dark history. Based on their participation in lesson/unit activities, presentations, and contributions in our follow-up discussion, I deduced the strategy was successful in getting these high school students excited and engaged about "doing history".

Importantly, knowledge and skills related to dark history should include not only understandings of facts, but also understandings of how facts, historical narratives, and arguments are "constructed" and weaved together. Due to the sensitivity of issues at hand, this development is especially important when (de)constructing or investigating dark history. Inquiry-based and investigative strategies encourage students' historical thinking, to gather a variety of historical sources, and then use those sources to develop and defend answers to historical questions. Within the specific context of the Kennedy assassination and other dark history events, these skills take on new meaning as students grapple with challenging imagery, and even greater challenges to grand narratives.

References

Becker, A. (2019) Dark tourism: The 'heritagization' of sites of suffering, with an emphasis on memorials of the genocide perpetrated against the Tutsi of Rwanda. *International Review of the Red Cross* 101(910): 317–331.

Block, M. (2020) *Push to remove confederate monuments opens debate on other honored historical figures.* Retrieved from: https://www.npr.org/2020/07/02/886503966/push-to-remove-confederate-monuments-opens-debate-on-other-honored-historical-fi [Accessed: 2 July 2021].

Cui, R., Cheng, M., Xin, S., Hua, C., & Yao, Y. (2019) International tourists' dark tourism experiences in China: The case of the memorial of the victims of the Nanjing Massacre. *Current Issues in Tourism* 23(12): 1–19.

Douglass, J. (2008) *JFK and the unspeakable: Why he died and why it matters*. Maryknoll, NY: Orbis Books.

Israfilova, F. & Khoo-Lattimore, C. (2019) Sad and violent but I enjoy it: Children's engagement with dark tourism as an educational tool. *Tourism and Hospitality Research* 19(4): 478–487.

John F. Kennedy Presidential Library and Museum (2021) *November 22, 1963: Death of the President*. Retrieved from: https://www.jfklibrary.org/learn/about-jfk/jfk-in-history/november-22-1963-death-of-the-president [Accessed: 22 July 2021].

Kerr, M. M. & Price, R. H. (2016) Overlooked encounters: Young tourists' experiences at dark sites. *Journal of Heritage Tourism* 11(2): 177–185.

Loewen, J. (2009) *Teaching what really happened: How to avoid the tyranny of text-books and get students excited about doing history*. New York: Teachers College Press.

Lovorn, M. (2012) Historiography in the methods course: Training preservice history teachers to evaluate local historical commemorations. In T. Keirn & D. Martin (Eds.) Historical thinking and pre-service teacher preparation. *Special ed of History Teacher*. 45(4): 569–579.

Marrs, J. (2013) *Crossfire: The plot that killed Kennedy*. New York: Basic Books.

Morrison, K. (2019) *Study abroad pedagogy, dark tourism, and historical reenactment: In the footsteps of Jack the Ripper and his victims*. Chapel Hill: Palgrave Pivot.

Olwell, R., Murphy, M., & Rice, P. (2016) It changes your life: The value of field trips in African American history for high school students in an out-of-school-time program. *Teaching History: A Journal of Methods* 41(1): 40.

Price, R. H. & Kerr, M. M. (2018) Child's play at war memorials: Insights from a social media debate. *Journal of Heritage Tourism: Heritage and Identity: Technology, Values and Experiences* 13(2): 167–180.

Seaton, T. (2018) Dark tourism history. In P.R. Stone, et al. (Eds.) *The Palgrave hand-book of dark tourism studies* (pp. 1–8). New York: Palgrave Macmillian.

Sixth Floor Museum at Dealey Plaza (2021) *The assassination*. Retrieved from: https://www.jfk.org/?gclid=CjwKCAjwruSHBhAtEiwA_qCppu_PKzg9SR0F92ftJ5VzaWgNLNXEdFR3KwVhkbtoG7gEkOhP_u0eyhoCiIYQAvD_BwE [Accessed: 22 July 2021].

13 School Trips

A Unique Form of Student Learning for Dark Tourism Studies

Laura M. Burns and Daniel E. Keller

Introduction

Although the school tourism market is a significant one (Cooper, 1999; Dale & Keating, 2016), researchers have historically overlooked this context Falk and Dierking, 2012; Ritchie, 2003). More recently, researchers have begun to turn to this context with the recognition that school trips play an important role for children and young people. For instance, Dale and Ritchie (2020) uncovered both motivations and constraints influencing educators who planned travel in their study of overnight school excursions. Among the constraints, the researchers identified staff time, willingness, and staff shortages. Motivations included giving students a new and rewarding learning experience (Dale & Ritchie, 2020). The latter finding may explain why the tradition of school trips has continued around the world.

While academic benefits appear to be the primary reason for school trips, it is also thought that students benefit interpersonally or socially from school trip participation (Larsen & Jenssen, 2004). These trips provide students with an experience that cannot be replicated in the classroom, giving them more diverse learning opportunities. Indeed, over 93 percent of Australian teachers surveyed by Munday (2008: 152) reached that conclusion. Similarly, while studying Australian students through pre- and post-trip surveys, Ballantyne and Packer (2002) found that the exposure to these new experiences positively changed how the students felt about their learning experience. Moreover, helping students learn about their heritage through out-of-classroom experiences motivates educators to include dark tourism sites in their school trip itineraries (Burns, 2018; Flennegård & Mattsson, 2021; Israfilova & Khoo-Lattimore, 2018; Roche & Quinn, 2016). Therefore, the purpose of this chapter is to outline key parameters of school trips as a unique form of learning. Ultimately, this chapter offers dark tourism researchers a defined framework for studying the specific context of school trips and the implications and challenges thereof.

DOI: 10.4324/9781003032199-17

A Framework for Studying School Trips: The Contextual Model of Learning

After studying museum experiences for several years, Falk and Dierking (2000) put forth the *Contextual Model of Learning* (hereafter CML). This model stipulates that learning occurs as people process, connect with, and make meaning of their real-world experiences. The CML is composed of three overlapping contexts – the personal, the sociocultural, and the physical. Over time, our experiences weave together as we apply our understanding of what was experienced in the past to make meaning of the present. While Falk and Dierking (2018) updated their model in recent revisions to include other contexts, this chapter will primarily focus on the personal, sociocultural, and physical contexts alongside a brief mention of the context of time.

The *personal context* draws connections between a person's own motivations, interests, past experiences, and prior knowledge (Falk & Dierking, 2000). Motivation considers a person's emotions and reasoning as they make meaning of their experiences. We choose to process and make connections between those experiences that we believe help us gain a greater understanding of ourselves. Such choices, whether conscious or unconscious, constitute our motivations for learning. Personal interests refer to a person's likes and dislikes, as well as the concepts to which that person feels compelled to assign attention, persistence, and curiosity (Falk & Dierking, 2000). As we go through life, we continue to draw upon our past experiences and knowledge to make meaning of new experiences.

The *sociocultural context* addresses learning as a cultural construct, as well as the development of Self (an overlap with the personal context) and the identification of the person's place within society. Cultures use learning as a means to shape us into functioning and contributing members of society. As Falk and Dierking (2000: 39) note, 'the world has meaning for us because of the shared experiences, beliefs, customs, and values of the groups that inhabit it with us'. Many of a person's first learning experiences take the form of modeling those around them: parents, siblings, and others. The information that cannot be shared through modeling is quite often shared through stories and narratives. As people share experiences, they talk about what they know, making connections between past experiences and sharing interpretations of events. By doing so, people create what Falk and Dierking (2000) refer to as a community of learners.

Whereas the personal context addresses a person's sense of self and the sociocultural context addresses a person's place in society, the *physical context* helps a person understand the physical world in which they live: the nonself. When referencing the physical context, Falk and Dierking (2000) referred to all elements of the setting in which an event occurs: the natural and manmade world along with the emotions that setting evokes. The importance of understanding how to navigate one's physical environment dates to the origins of human existence, when it was literarily a life and death matter (Falk & Dierking, 2000). Over time, humans began to develop what psychologists

Barker and Wright refer to as 'behavior settings' (Falk & Dierking, 2000). The term refers to the notion that human behavior can be predicted more accurately from the physical context in which it occurs, rather than merely predicting behaviors based on the person's personality traits. People typically learn the expected behaviors for various settings and generally comply with those expectations. As such, physical contexts become associated with specific behaviors like learning or having fun (Falk & Dierking, 2000). With such an emphasis on the purpose of certain physical spaces, Falk and Dierking (2000) report that learning often becomes context-specific, with the learner striving to transfer new information from one environment to another.

The three overlapping contexts of the CML demonstrate that learning is an intricate process involving a person's sense of self, place in society, and understanding of the physical world. Indeed, 'in a sense, the sociocultural context... serves as a bridge between the individual's sense of self, the personal context, and the non-self, or physical context, the individual must live within' (Falk & Dierking, 2000: 56). People use what they know of their culture and past experiences with others to understand how to interact with the physical world. However, what these three contexts do not fully address is that learning is a continuous, lifelong process. Falk and Dierking (2000) suggest that time should also be considered when applying their CML. The addition of time to the model conceptualizes 'learning as being constructed over time as the individual moves through his sociocultural and physical world; over time, meaning is built up, layer upon layer' (Falk & Dierking, 2000: 11). As people move through life, they continually interact with others and with the physical environment, gaining a greater understanding of self, their place in society, and their relationship with their world.

The process of learning is a personal endeavor not experienced by any two people in the same way (Falk et al., 1978). When applied to the literature on school trips, Falk and Dierking's CML helps us better understand that learning process and experience for students. What follows are examples from the literature that illustrate each of the three contexts (personal, sociocultural, and physical).

The Personal Context

The personal context of learning takes into consideration a person's motivation, interests, prior knowledge and experiences, and the learning environment. Table 13.1 highlights studies that relate to the personal context.

The literature shows that students who participate in a school trip experience personal as well as academic growth (Cowan & Maitles, 2011). School trips provide students with the opportunity to learn through activities that they find enjoyable, while making connections between the school trip and previous experiences. The application of the personal context of the CML helps us better understand the benefits students derive from school trip participation. Moreover, school trip activities motivate student learning through real-world connections and the presentation of materials in a way that

Table 13.1 Examples of Literature Pertaining to the Personal Context

Source	Summary of Findings
Anderson et al. (2006)	The authors teamed up to compare their independent studies on teacher perspectives about field trips in Canada, Germany, and the United States. The comparison found that teachers in all nations face many of the same constraints when planning field trips.
Bamberger and Tal (2007)	The authors studied more than twenty classes of students at four museums in Israel to understand how the structure of the visit impacted student learning. They found that school trip activities that offered students limited choice allowed teachers to appropriately scaffold the activities while still providing students with the ability to self-direct their learning and increase engagement.
Behrendt and Franklin (2014)	Through this literature review, the authors argued the importance of experiential learning through school trips. This review also uses the literature to suggest a research-based methodology for planning and conducting meaningful school trips.
Burns (2018)	"The findings of the study directly connected to the Contextual Model for Learning's sociocultural and physical contexts. The study concluded that over the course of the trip, students created a community of learners, sharing their learning and working together to make meaning of the experience. While chaperones and staff members of the sites explicitly shared some of the behavioral expectations, students also learned to read the space they occupied, and identified and complied with implicit behavioral norms. However, evidence that students contemplated their place in the world as a result of this trip exists (the personal context)."
Cambourne and Faulks (2003)	This case study examines school tourism in Canberra, Australia. It provides many statistics about the Australian school tourism market in 2001.
Cooper (1999)	Cooper writes for those that provide the sites for school trips. After establishing that school trips are a segment of the tourism market that destinations should take seriously, Cooper sets about describing how tourist sites can attract school visits to their facility. Cooper sites specific data about school tourism in Europe, however that data is over 15 years old.
Cowan and Maitles (2011)	The article details Scottish student experiences in the *Learning from Auschwitz* program. The researchers use student responses to explain the effect of the experience. While this trip was educational, it was sponsored by the national government rather than a school system, and as such students often traveled independently and not with their classmates.
Dale and Ritchie (2020)	Online survey results from 1314 Australian schools examined socio-demographics, motivations, and constraints as influences on educators' planning and taking overnight school excursions.

(Continued)

Table 13.1 (Continued)

Source	Summary of Findings
DeWitt and Storksdieck (2008)	This review of literature concludes that while school trips may not provide students with the best learning environment, the opportunities student must explore and participate in first-hand experiences proves beneficial to the students' overall educational experience.
Falk et al. (2012)	The authors outline the history of tourism research as it is linked to learning while looking at learning through the lens of tourism. After doing so, they reach the conclusion that the field is under-researched and propose using Aristotle's concept of *phronesis* as a framework for the field.
Flennegård and Mattsson (2021)	This study observes how Swedish trips to the Auschwitz-Birkenau State Museum relate to their mission of instilling democratic values in their students through the perspectives of their teachers. The authors state that the pedagogical practice of focusing on a victim's suffering creates a sense of empathy but does not necessarily relate to students directly the importance of democratic values. They also identify a concern that, with a lack of primary source Holocaust materials, students are at risk of contemporary political influence from their teacher rather than an analysis of historical record.
Gottfried (1980)	The author observed student groups at a BioLab over the course of six months. Follow-up interviews, as well as the observations, demonstrated that students, when given the opportunity to explore on their own, engaged in experiments that resulted in students learning about animal biology and associated skills.
Israfilova and Khoo-Lattimore (2018)	The authors work to discover children's experiences on a school trip to the Guba Genocide Memorial Complex located in Azerbaijan. Here, they identified several key observations for children visiting dark tourist sites, including increased patriotic identifications, emotional distress at the site of tragedy, and increased intrigue due to an interest in war media.
Krepel and DuVall (1981)	This how-to manual begins by examining the history of school trips as well as school policies and other constraints that regulate the use of field trips as an instructional tool. After outlining the recommended procedure for planning a school trip, the authors conclude with a list of sample trips across school disciplines. They also cited other people's field trip experiences in showing examples of successful trips.
Lai (1999)	The article summarizes a case study of a geography field trip of Year 10 students at an affluent boy's school in Hong Kong. During the trip, students took part in two distinct activities; one was teacher-led while the other allowed students to guide their own activity in achieving a learning goal. Lai concluded that when given the opportunity to control their experience, students took more responsibility for their learning.
Lankford (1992)	This book serves as a guide to carefully planning field trips. Lankford presents a researched-based process for maximizing student learning on school trips.

(*Continued*)

Table 13.1 (Continued)

Source	Summary of Findings
Millan (1995)	Millan draws upon years of taking students on field trips as well as the literature to create a process for the planning of successful school trips.
Pace and Tesi (2004)	The study focuses on the interviews of eight adults between the ages of 25 and 31 to see what they remembered learning on field trips taken during their K-12 education. All participants remembered some aspect of a field trip experience from their youth.
Roche and Quinn (2016)	Roche and Quinn sought to understand how 11–12 year-olds explored a national heritage site, the Battle of the Boyne Visitor Center. Prior media exposure affected students' 'preconceptions' of the commemorated events. 'Media, technology, and visuals' at the site influence student enjoyment of the visit. Students explored the site through play, such as climbing. In addition, students identified opportunities to socialize during the visit as memorable.
Wolins et al. (1992)	This is an ethnographic study of student memories of museum trips. Students in the study participated in many museum trips annually, the number different each year of the study. They found that the classroom teacher's involvement in the trips impacted the quality of the students' experiences. Children remembered trips where they were more involved and had meaningful social interactions.

students often find fun, entertaining, and enjoyable (Anderson et al., 2006; Cooper, 1999). An early study found that because of participating in a school trip, students began to associate learning activities with 'fun and playful activity' (Gottfried, 1980: 173), as opposed to the formality and routine of the classroom.

To optimize this motivation and create a learning environment in which students are eager to participate in learning activities, adults planning school trips must pay particular attention to the structure of the activities (DeWitt & Storksdieck, 2008). The amount of choice and control provided to students during school trips impacts their learning. Students enjoyed the school trip better and were more fully engaged in learning activities when they were given the opportunity to make choices during the experience (Bamberger & Tal, 2007; Behrendt & Franklin, 2014; DeWitt & Storksdieck, 2008). The relevance of this finding to dark tourism field visits and school trips is that some students may need to opt-out of particularly distressing exhibits while others may be drawn to those very same exhibits.

It is no surprise that the literature suggests that students make greater meaning of the school trip experience when they are highly involved in its activities (Millan, 1995; Wolins et al., 1992). These findings suggest that teachers should plan a school trip with a specific learning goal in mind, but also allow students to identify the most motivating, interesting, and enjoyable ways to achieve that learning goal. By making selections on how they will learn, students also gain a better understanding of how they learn. To illustrate this point, students were highly involved in activities in several recent school trips to dark tourism sites (e.g., see Flennegård & Mattsson, 2021; Israfilova & Khoo-Lattimore, 2018; or Roche & Quinn, 2016).

The studies cited above lead us to conclude that school trips help students learn by connecting the classroom and the real world (Anderson et al., 2006; Behrendt & Franklin, 2014; Cambourne & Faulks, 2003; Krepel & DuVall, 1981; Lai, 1999; Lankford, 1992; Millan, 1995; Pace and Tesi, 2004). As Behrendt & Franklin (2014: 238) argue, 'field trips offer a unique opportunity for students to create connections, which will help them gain understanding and develop an enjoyment of learning'. For instance, by studying student school trips to Auschwitz concentration camp, Cowan and Maitles (2011: 166) conclude that such visits 'can provide young visitors with a greater understanding of the scale of the inhumanity and tragedy that occurred there'. Students may read and sit through classroom lessons about a topic like the Holocaust. However, when they stand where the victims stood, their ability to empathize and understand the magnitude of the historical events is significantly enhanced. Cowan and Maitles, (2011) also found that students' emotional responses intensified as they had time to reflect on their experiences. Importantly, 'time' – which is the fourth 'unofficial' component of the CML – allows students to reflect on school trip activities. They can potentially make connections with both past and future knowledge and experiences to learn from the school trip experience and make it more meaningful.

In summary, the personal context tells us that learning occurs when students can make connections between experiences, which may occur both inside and outside of the classroom and during or following a school trip experience. Students arrive at meaning in their own individual ways, by connecting previous experiences to the new experiences of their school trip (Falk et al., 2012). Complete learning may not take place until experiences after the school trip compel the student to recall the school trip experience and make those important connections.

The Sociocultural Context

While less prevalent in the literature, some research on school trips examines learning through the sociocultural context. This context addresses learning through modeling, social narratives, and the formation of communities of learners, all of which help students better understand their place in the world. Table 13.2 describes literature focused on the sociocultural context of learning derived from Falk and Dierking's CML.

Numerous studies suggest that the social interactions inherent in school trips often help students remember the experience for a longer period of time (Behrendt & Franklin, 2014). Falk and Dierking (1997) found that 98.4% of field trip participants could recall one specific memory from their trip, while 80.5% could recall three or more specific events from the trip. Even years

Table 13.2 Examples of Literature Pertaining to the Sociocultural Context

Source	Summary of Findings
Behrendt and Franklin (2014)	Through this literature review, the authors argued the importance of experiential learning through school trips. This review also uses the literature to suggest a research-based methodology for planning and conducting meaningful school trips.
Burns (2018)	The findings of the study directly connected to the Contextual Model for Learning's sociocultural and physical contexts. The study concluded that over the course of the trip, students created a community of learners, sharing their learning and working together to make meaning of the experience. While chaperones and staff members of the sites explicitly shared some of the behavioral expectations, students also learned to read the space they occupied, and identified and complied with implicit behavioral norms. However, evidence that students contemplated their place in the world as a result of this trip exists (the personal context).
Falk et al. (2012)	The authors outline the history of tourism research as it relates to learning while looking at learning through the lens of tourism. After doing so, they reach the conclusion that the field is under-researched and propose using Aristotle's concept of *phronesis* as a framework for the field.
Falk and Dierking (1997)	This article studied over 120 field trip participants to assess their long-term memories of their field trip experiences. The authors found that, even years later, the participants remembered where they traveled, who they were with, and specific things they did while on the field trip.
Flennegård and Mattsson (2021)	This study observes how Swedish trips to the Auschwitz-Birkenau State Museum relate to their mission of instilling democratic values in their students through the perspectives of their teachers. The authors state that the pedagogical practice of focusing on a victim's suffering creates a sense of empathy but does not necessarily relate to students directly the importance of demo-cratic values. They also identify a concern that, with a lack of primary source Holocaust materials, students are at risk of contemporary political influence from their teacher rather than an analysis of historical record.
Israfilova and Khoo-Lattimore (2018)	The authors work to discover children's experiences on a school trip to the Guba Genocide Memorial Complex located in Azerbaijan. Here, they identified several key observations for children visiting dark tourist sites, including increased patriotic identifications, emotional distress at the site of tragedy, and increased intrigue due to an interest in war media.
Larsen and Jenssen (2004)	Utilizing a case study of a 10th-year class in Norway, and the research of Fodness (1994) as their literary basis, the authors examine student motivation for participation in school trips. They conclude that the students' reasons for traveling with their class were overwhelmingly social.
Lavie Alon and Tal (2015)	The authors looked for student feedback as to the elements of the field trip that had the largest impact on their learning. The authors found that students believed that the field trip helped them connect the school curriculum to everyday life. In addition, they found that the stories and abilities of the guides had a significant impact on student engagement and learning.

(Continued)

Table 13.2 (Continued)

Source	Summary of Findings
Roche and Quinn (2016)	Roche and Quinn sought to understand how 11–12-year-olds explored a national heritage site, the Battle of the Boyne Visitor Center. Prior media exposure affected students' 'preconceptions' of the commemorated events. 'Media, technology, and visuals' at the site influence student enjoyment of the visit. Students explored the site through play, such as climbing. In addition, students identified opportunities to socialize during the visit as memorable.
Wolins et al. (1992)	This is an ethnographic study of student memories of museum trips. Students in the study participated in many museum trips annually, the number different each year of the study. They found that the classroom teacher's involvement in the trips impacted the quality of the students' experiences. Children remembered trips where they had meaningful social interactions.

after participating in a school trip, research has found that students 'recall what they did, what they saw and with whom and where the visit took place' (Wolins et al., 1992: 18). Further analysis of these memories found that while most of them were content-related, many also associated feelings and referred to social interactions experienced during the trip (Falk & Dierking, 1997). When participating in school trips, learning is inherently social. School trips can help students to better understand the world around them as well as their place within it by introducing students to new cultures while also inspiring personal growth and a greater understanding of self (Falk et al., 2012).

The opportunity to interact with peers and teachers in different ways that do not or cannot occur in the classroom also presents itself during school trips. Students interact with their peers and work together to overcome the anxiety of new environments. Collaborative and social learning techniques provide students with new or different learning strategies. These techniques promote the creation of communities of learners, as they work together to make meaning of school trip experiences and to better understand the world beyond their home communities (Lavie Alon & Tal, 2015).

The Physical Context

The physical context helps students understand the world around them – that is, the non-Self. Through the physical context, students learn that there is more to the world than what they experience daily. Table 13.3 features literature that pertains to this physical context in studies of school trips.

Potentially, school trips allow students to experience unfamiliar places and cultures they might never otherwise visit. Students learn more when they are exposed to new environments and cultures firsthand (Millan, 1995; Werry, 2008). Often, school trips provide students with new experiences that would be impossible to simulate in the classroom (Balling & Falk, 1980, Scarce, 1997).

Table 13.3 Examples of Literature Pertaining to the Physical Context

Source	Summary of Findings
Balling and Falk (1980)	Balling and Falk reflect on four different studies that aim to assess the effect that an environment's novelty has on student learning. They conclude that no matter a student's background, significant learning occurs on field trips as students are often more engaged in the learning process.
Batz, Wittler, and Wilde (2010)	Using Falk and Dierking's Contextual Model of Learning as the underlying framework for this study, the authors seek to prove the impact of motivation on learning in an out-of-school environment. They conclude that while girls show a little more motivation, all students showed a significant knowledge gain when participating in learning activities on school trips.
Brugar (2012)	Brugar observed student groups in three Detroit cultural institutions to see how teachers used museums as a resource to teach social studies. The author considers the museums' educational content alignment to state standards. Brugar concludes with recommendations for resources to use in planning school trips.
Burns (2018)	The findings of the study directly connected to the Contextual Model for Learning's sociocultural and physical contexts. The study concluded that over the course of the trip, students created a community of learners, sharing their learning and working together to make meaning of the experience. While chaperones and staff members of the sites explicitly shared some of the behavioral expectations, students also learned to read the space they occupied, and identified and complied with implicit behavioral norms. However, evidence that students contemplated their place in the world as a result of this trip exists (the personal context).
Falk and Balling (1982)	The authors detail research that aims to measure the effect of the same activity done both as part of a school trip and as part of the regular school day, outside of the elementary school, but on school grounds. They conclude that the novelty of the school trip sight distracted younger students while helping to motivate older students.
Falk et al. (1978)	The study focuses on the effect of setting novelty upon the child's behavior and cognition. The researchers consider both student IQ as well as income level. The authors suggest that the novelty of the field trip environment provides students with too much stimulation that then interferes with their learning.
McManus (1993)	This study collected data as to students' memories of museum experiences. The researcher mailed surveys to museum visitors during a school trip. They found that most memories of field trips were of objects and things.
Millan (1995)	Millan draws upon years of taking students on school trips as well as the literature to create a process for the planning of successful school trips.
Nespor (2000)	Nespor argues that as suburbs developed, and leisure spaces and travel became more private, the definition of public spaces, and students' access to those spaces has changed. She suggests that school trips allow students to access those spaces in a unique way.

(Continued)

Table 13.3 (Continued)

Source	Summary of Findings
Roche and Quinn (2016)	Roche and Quinn sought to understand how 11–12 year-olds explored a national heritage site, the Battle of the Boyne Visitor Center. Prior media exposure affected students' 'preconceptions' of the commemorated events. 'Media, technology, and visuals' at the site influence student enjoyment of the visit. Students explored the site through play, such as climbing. In addition, students identified opportunities to socialize during the visit as memorable.
Scarce (1997)	Scarce looks at the power of using school trips in teaching college sociology courses. While the students in this study are not K-12, it does offer the typical planning process as well as information on student learning and the social impact of school excursions.
Werry (2008)	Werry reflects on her experience teaching a tourism course that required students to participate in a school trip to Hawaii and New Zealand. She attempts to argue that shame in our lack of understanding of other cultures plays a role in how we interpret our experiences in those other cultures. She also touches on the fact that tourists often feel that tourism itself should be a leisure, not academic, activity.

Indeed, school trips take students away from the security of the known to 'expose them to the larger world' (Nespor, 2000: 39). It is one thing for students to read about unfamiliar cultures and places, however, it is completely different when students experience those places and cultures in person. Participation in school trips provides students with a valuable and incomparable means to do so.

Trends in School Trips

The final category found within the literature on school trips analyzes the logistical trends among (USA) schools that provide their students with school trip experiences. When attending field trips, groups travel an average of 40 miles by bus and stay for an average of less than two hours at the location (Cooper & Latham, 1989). The most popular time to take school trips is during the late spring and summer terms (Cooper, 2003). School groups tend to repeat trips on an annual basis, and once they find a school trip destination that addresses their goals, the teachers typically return year after year with their students (Cooper & Latham, 1988; Ritchie et al., 2009).

When students participate in school trips, they spend as much time, if not more, traveling to the site as they do experiencing it. In addition, the fact that trips tend to happen later in the school year, as the curriculum is ending, suggests that trips are not always scheduled where they would best supplement the curriculum.

Overnight School Trips

Overnight school trips represent a context different from day trips. A trip that expands beyond the confines of an average school day provides the opportunity to gain a more in-depth understanding of the various components of the CML. As noted earlier, Dale and Ritchie (2020) conducted an extensive study of educators' motives and constraints for planning and taking overnight school excursions. In this section, we highlight two studies that specifically focused on students living together on a multi-day trip: Gee (2015) and Burns (2018).

Gee (2015) studied 36 A-Level Year-12 students in England who participated, along with three of their teachers, in a weeklong school trip to a residential study center. Throughout the week-long excursion, Gee (2015) functioned as a participant-observer, ensuring the ability to study the teachers and students in a variety of activities and settings. Meanwhile, Burns (2018) as a member of a research team accompanied 48 adolescents on a 4-day school trip to Washington DC, USA.

As these two studies revealed, overnight school trips provide us with the ability to look at the personal context in greater detail because students are away from their families and outside the comfort zone of their daily routines. Suddenly, they must tackle decisions that their parents often make for them: what to have for dinner, what to buy with pocket money, and with whom to spend their time. In addition, extended time on a school trip location often provides students with the ability to explore the site beyond the activities planned by the teachers. This provides students with a choice about how and what they will learn during the trip. When students can make decisions on their own, they typically gain a better understanding of their own personal interests and values.

The sociocultural component of the CML emerges in a unique way during overnight school trips. Students often select the group with which they will socialize during meals, as well as those they room with during the overnight experience. The increased exposure to classmates may strengthen or cause tensions among friendships. The shared experiences of a school trip can also facilitate the development of new relationships that would not otherwise form within the regular school environment. Students have more freedom to make choices about how to spend their time and how much sleep they will get, with lessened adult supervision at night. Roommates often make this decision together. These social interactions give school-aged students a small taste of the independence they will gain with adulthood, while they also navigate the social structure of their school group.

One of the most prominent components of the CML for overnight experiences is the physical context. They may have the opportunity to travel to different states, and even different countries. Students from rural areas may experience an urban setting for the first time during overnight school trips, or vice versa. For students from lower socioeconomic backgrounds, school trips may be their first opportunity to experience such excursions because without school financing, such trips would be impossible for their families

(Pace and Tesi, 2004). The exposure to new cultures provides students the opportunity to witness different ways of living and may even change student perspectives on their future.

For some students, school trips may be the first experience of travel away from home, thereby exposing young travelers to new forms of transportation and lodging. Depending on the structure of the trip, students may live in small quarters that are at times less (or more) accommodating than home. For example, students may for the first time share a room/bathroom and must learn how to navigate the sharing of space with others. On the other hand, the lodging during the school trip may provide students with comforts to which they are not accustomed (e.g., air conditioning). Navigating these new environments teaches students not only about different lifestyles, but may also help them to better understand the quality of life they value and hope to achieve in their futures.

Despite these reasons to conduct research about overnight school trips, very few researchers appear to focus on the residential aspects of an overnight school trip, instead concentrating only on the destinations. That said, however, Gee (2015) does explore the issue of overnight school trips. Based upon observations, informal discussions, and semi-formal individual interviews with students and teachers, Gee (2015) conclude that students participating in the school excursion saw social aspects of the trip, including the blurring of lines between teacher and student, as the most memorable and beneficial aspect of the school trip. While many students developed relationships with students that they did not know prior to the trip, others spent the time strengthening existing relationships. Gee (2015) also observed that groups of students were quick to claim public spaces as their own, allowing these students to focus on their schoolwork in a more relaxing, social, and motivating environment. Students retreated to these areas whenever teachers assigned work. Applying the physical context to this ritual helps us to understand that by claiming a space as personal, students were starting to organize the destination. Each group had their place to work in a comfortable and motivating environment. Different areas evoked feelings of motivation and comfort for different students (Gee, 2015).

Similarly, Burns (2018) focused primarily on the sociocultural and physical contexts' impact on the studied trip. As a member of a research team, Burns (2018) accompanied 48 Midwestern eighth-grade students during a 4-day trip, stopping at the Flight 93 National Memorial in Shanksville, PA, USA, and going on to visit various memorials in Washington DC. With many students having little overnight travel experience before their trip, Burns (2018) had a unique opportunity to observe how this group experienced the trip through the lens of Falk and Dierking's CML model. Like Gee (2015), Burns (2018) found that students formed a learning community with their teachers and one other. To illustrate, students broadened their sociocultural understanding by engaging with the students and faculty when they had questions about new information or experiences gained from their visit. Students clearly recognized the community they were building, through engagement in group photos and looking after their peers to ensure everyone's safety. The students also broke down traditional relationships with their teachers, engaging with them in a more relaxed fashion. This change in the relationship between students and teachers directly mirrors Gee (2015).

Burns also investigated the way students interacted with the physical context of their visit, via the ever-changing implicit norms of the sites that they visited. While their adult guides provided explicit expectations for the students to follow, many eighth graders quickly learned and exhibited the unspoken norms of the location. This included behaviors such as being quiet at solemn sites and following appropriate pedestrian traffic patterns (Burns, 2018). The students showed their adaptability when changing behavioral settings, a concept integral to Falk and Dierking's physical context of learning. In explaining the sociocultural context of their CML, Falk and Dierking (2000) discuss the idea of a community of learners. All students participating in the school trip created such a community. By simply traveling together and sharing the experience of the trip, the students created a community that shares an experience unlike any other group of students, which can never be replicated. While forming a new community of learners, students from the Gee (2015) and Burns (2018) studies created connections to a group of people in the larger societal context, giving students one more reference point to understand their place in the world.

Conclusion

School trips provide a unique opportunity for students to engage in different methods of learning. Although rarely studied, overnight school trips disrupt each aspect of a student's typical learning, giving students a chance to explore new environments and question their own preconceptions on a variety of concepts. As a result, researchers can use this opportunity to deeply explore the ways student learning changes in these new settings.

The application of the sociocultural components of the CML helps us to understand the breaking down of barriers between students and teachers as presented by Gee (2015) and Burns (2018). The sociocultural context alludes to the fact that there is a hierarchy in society: teachers rank higher and are the supervisors of students. Supporting this hierarchy are the cultural expectations around the formality of a school. As such, the students and teachers create what Falk and Dierking (2000) referred to as behavior settings (behavioral expectations specific to each environment). However, the students and teachers who participate in a school trip create a different behavior setting. Students learn that outside the walls of the classroom, the formality of the teacher-student relationship relaxes, giving students the opportunity to view their teachers as human beings as well as providers of knowledge.

However, there is still much to learn in a broader sense about how students on overnight school trips interact in accordance with Falk and Dierking's model (2000). While the personal and to a lesser extent sociocultural context receive significant consideration by researchers when studying school trips, there is less regard for the physical context of learning. This is surprising, as one of the most fundamental aspects and definitive features of a school trip is the direct change of physical location. For example, some studies have explored how younger students were more engaged with a trip that had opportunities for physical activity or play. Therefore, a more complex consideration of this context is necessary to better understand what it is about a specific school trip destination that makes it more successful or meaningful to students.

Most studies to date have relied on participant observations or teacher interviews. As we see later in this book, newer methods invite students to share their perspectives directly, through written comments, interviews, drawings, and photographs. Because dark tourism destinations are known to evoke different emotional, cognitive, and exploratory experiences for young tourists, this expansion of methods is critical. Facilitating students to conduct their own studies not only elicits their views more directly, but also overcomes obstacles for researchers unable to travel with students, especially on overnight trips.

References

Anderson, D., Kisiel, J., & Storksdieck, M. (2006). Understanding teachers' perspectives on field trips: Discovering common ground in three countries. *Curator: The Museum Journal, 49*(3), 365–386. doi:10.1111/j.2151-6952.2006.tb00229.x.

Ballantyne, R., & Packer, J. (2002). Nature-based excursions: School students' perceptions of learning in natural environments. *International Research in Geographical and Environmental Education, 11*(3), 218–236. doi:10.1080/10382040208667488.

Balling, J. D., & Falk, J. H. (1980). A perspective on field trips: Environmental effects on learning. *Curator: The Museum Journal, 23*(4), 229–240. doi:10.1111/j.2151-6952. 1980.tb01672.x.

Bamberger, Y., & Tal, T. (2007). Learning in a personal context: Levels of choice in a free choice learning environment in science and natural history museums. *Science Education, 91*(1), 75–95. doi:10.1002/sce.20174.

Batz, K., Wittler, S., & Wilde, M. (2010). Differences between boys and girls in extracurricular learning settings. *International Journal of Environmental and Science Education, 5*(1), 51–64.

Behrendt, M., & Franklin, T. (2014). A review of research on school field trips and their Value in education. *International Journal of Environmental and Science Education, 9*(3), 235–245. doi:10.12973/ijese.2014.213a.

Brugar, K. A. (2012). Thinking beyond field trips: An analysis of museums and social studies learners. *Social Studies Research & Practice (Board of Trustees of the University of Alabama), 7*(2), 30–40.

Burns, L. (2018). Overnight school trips: An overlooked phenomenon. PhD Thesis, University of Pittsburgh.

Cambourne, B., & Faulks, P. (2003). Making the most of your assets: School excursions and tourism to Canberra, Australia's National Capital. In B. Ritchie (Author), *Managing Educational Tourism* (Vol. 10, pp. 135–141). Bristol: Channel View Publications.

Cooper, C. (1999). The European school travel market. *Travel & Tourism Analyst*, *5*, 89–106.

Cooper, C. (2003). Schools' educational tourism in Europe. In B. Ritchie (Author), *Managing Educational Tourism* (Vol. 10, pp. 168–174). Bristol: Channel View Publications.

Cooper, C. P., & Latham, J. (1988). The pattern of educational visits in England. *Leisure Studies*, *7*(3), 255–266. doi:10.1080/02614368800390231.

Cooper, C., & Latham, J. (1989). School trips. An uncertain future? *Leisure Management*, *9*(8), 73–75.

Cowan, P., & Maitles, H. (2011). 'We saw inhumanity close up'. What is gained by school students from Scotland visiting Auschwitz?. *Journal of Curriculum Studies*, *43*(2), 163–184. doi:10.1080/00220272.2010.542831.

Dale, N. F., & Keating, B. W. (2016). Size and effect of school excursions to the national capital. Centre for Tourism Research, University of Canberra.

Dale, N. F., & Ritchie, B. W. (2020). Understanding travel behavior: A study of school excursion motivations, constraints and behavior. *Journal of Hospitality and Tourism Management*, *43*(2020), 11–22. doi:10.1016/j.jhtm.2020.01.008.

DeWitt, J., & Storksdieck, M. (2008). A short review of school field trips: Key findings from the past and implications for the future. *Visitor Studies*, *11*(2), 181–197. doi:10.1080/10645570802355562.

Falk, J. H., Ballantyne, R., Packer, J., & Benckendorff, P. (2012). Travel and learning: A neglected tourism research area. *Annals of Tourism Research*, *39*(2), 908–927. doi:10.1016/j.annals.2011.11.016.

Falk, J. H., & Balling, J. D. (1982). The field trip milieu: Learning and behavior as a function of contextual events. *The Journal of Educational Research*, *76*(1), 22–28. doi:10.1080/00220671.1982.10885418.

Falk, J. H., & Dierking, L. D. (1997). School field trips: Assessing their long-term impact. *Curator: The Museum Journal*, *40*(3), 211–218. doi:10.1111/j.2151-6952.1997. tb01304.x.

Falk, J. H., & Dierking, L. D. (2000). *Learning from Museums: Visitor Experiences and the Making of Meaning*. Lanham, MD: Altamira Press.

Falk, J. H., & Dierking, L. D. (2012). *Museum Experience Revisited*. Walnut Creek, CA: Left Coast Press.

Falk, J. H., & Dierking, L. D. (2018). *Learning from Museums: Visitor Experiences and the Making of Meaning*. (2nd ed.). Lanham, MD: Rowman & Littlefield.

Falk, J. H., Martin, W. W., & Balling, J. D. (1978). The novel field-trip phenomenon: Adjustment to novel settings interferes with task learning. *Journal of Research in Science Teaching*, *15*(2), 127–134.

Flennegård, O., & Mattsson, C. (2021). Democratic pilgrimage: Swedish students' understanding of study trips to Holocaust memorial sites. *Educational Review*. doi: 10.1080/00131911.2021.1931040.

Fodness, D. (1994). Measuring Tourist Motivation. Annals of Tourism Research, 21, 555–581.

Gee, N. (2015). Creating a temporary community? An ethnographic study of a residential fieldtrip. *Journal of Adventure Education & Outdoor Learning*, *15*(2), 95–109. doi:10.1080/14729679.2013.849609.

Gottfried, J. (1980). Do children learn on school field trips? *Curator: The Museum Journal*, *23*(3), 165–174. doi:10.1111/j.2151-6952.1980.tb00561.x.

Israfilova, F., & Khoo-Lattimore, C. (2018). Sad and violent but I enjoy it: Children's engagement with dark tourism as an educational tool. *Tourism and Hospitality Research*, 1–10. doi:10.1177/1467358418782736.

Krepel, W. J., & DuVall, C. R. (1981). *Field trips: A guide for planning and conducting educational experiences. Analysis and action series*. NEA Distribution Center, The Academic Building, West Haven (Stock No. 1683-1-00, $5.25).

Lai, K. C. (1999). Freedom to learn: A study of the experiences of secondary school teachers and students in a geography field trip. *International Research in Geographical and Environmental Education*, 8(3), 239–255. doi:10.1080/10382049908667614.

Lankford, M. D. (1992). *Successful Field Trips*. Santa Barbara, CA: Abc-Clio Incorporated.

Larsen, S., & Jenssen, D. (2004). The school trip: Traveling with, not to or from. *Scandinavian Journal of Hospitality and Tourism*, 4(1), 43–57.

Lavie Alon, N., & Tal, T. (2015). Student self-reported learning outcomes of field trips: The pedagogical impact. *International Journal of Science Education*, 37(8), 1279–1298. doi:10.1080/09500693.2015.1034797.

McManus, P. M. (1993). Memories as indicators of the impact of museum visits. *Museum Management and Curatorship*, 12(4), 367–380. doi:10.1016/0964-7775(93)90034-G.

Millan, D. A. (1995). In: 'Experience and the Curriculum Report', *Field Trips: Maximizing the Experience*. Educational Resources Information Centre (ERIC), U.S. Department of Education, Washington DC, pp. 123–144.

Munday, P. (2008). Teacher perceptions of the role and value of excursions in years 7–10 geography education in Victoria, Australia. *International Research in Geographical and Environmental Education*, 17(2), 146–169. doi:10.2167/irgee233.0.

Nespor, J. (2000). School field trips and the curriculum of public spaces. *Journal of Curriculum Studies*, 32(1), 25–43. doi:10.1080/002202700182835.

Pace, S., & Tesi, R. (2004). Adult's perception of field trips taken within grades K-12: Eight case studies in the New York metropolitan area. *Education*, 125(1), 30–41.

Ritchie, B. W. (2003). *Managing Educational Tourism*. Bristol, UK: Channel View Publications.

Ritchie, B. W., Maitland, R., & Ritchie, B. (2009). School excursion management in national capitals cities. *City Tourism: National Capital Perspectives*, 185–200. doi:10.1079/9781845935467.0000.

Roche, D., & Quinn, B. (2016). Heritage sites and school children: Insights from the Batle of the Boyne. *Journal of Heritage Tourism*. doi:10.1080/1743873X.2016.1201086.

Scarce, R. (1997). Field trips as short-term experiential education. *Teaching Sociology*, 25(3)219–226.

Werry, M. (2008). Pedagogy of/as/and tourism: Or, shameful lessons. *The Review of Education, Pedagogy, and Cultural Studies*, 30(1), 14–42.

Wolins, I. S., Jensen, N., & Ulzheimer, R. (1992). Children's memories of museum field trips: A qualitative study. *The Journal of Museum Education*, 17–27. doi:10.1080/10598650.1992.11510204.

14 Young People and Dark Commemorative Events

The Centenary of World War One in Australia

Jennifer Frost and Warwick Frost

Introduction

Commemorative events aimed at remembrance may incorporate stirring or emotion-laden music or strong visual cues to engage people and gain their attention. In the case of events connected to World War One, the scarlet poppy is generally used as the dominant symbol of remembrance, particularly in the UK and Commonwealth nations. This inspired the art installation *Blood Swept Lands and Seas of Red*, created by Paul Cummins and Tom Piper in 2014, to commemorate the centenary of the outbreak of World War One. It comprised 888,246 ceramic poppies, which were placed in the former moat of the Tower of London in London, one for each British or Colonial military fatality during the Great War. While there were criticisms that its beauty was a travesty, given the horrific reality of war (Jones, 2014), many saw it as a powerful symbol of the sacrifice given by these individuals, as well as a reminder of the extent of human loss. It seems to have specifically helped younger people understand the breadth of the losses that were suffered. As one person commented on the British *Guardian* newspaper website: 'The poppy display is a moving, thoughtful experience for EVERYONE, and particularly children who are amazed by the fact that a poppy represents a death' (Jones, 2014: online).

Previous work on commemorative events, held to mark the anniversary of a significant moment in history, has tended to focus on controversy and contestation, partly because cultural heritage is intrinsically dissonant, but also because these events are often linked to dark or disturbing moments in our past (Frost and Laing, 2013; Laing and Frost, 2019). In this way, they can be characterized as a form of 'dark tourism'. While we refer here to dark tourism in its broadest sense as travel 'associated with death, suffering and the seemingly macabre' (Stone, 2006: 146), we acknowledge that this umbrella term can encompass different 'intensities of darkness' (Stone, 2009: 35) and experiences, both from a demand and supply perspective (Sharpley, 2009; Stone, 2009). Thus, it is important to study dark tourism across a variety of contexts to understand the phenomenon more fully (Stone, 2006, 2009). Only a few studies to date have considered how younger people engage with commemorations. For instance, Bird (2016), in a Canadian context, surveyed young people visiting a World War One battlefield in France and an

DOI: 10.4324/9781003032199-18

associated commemoration. He noted students found interacting with the landscape where original events took place provided 'opportunities to gain emotional and even transformative insight that is more resonant than other commemorative experiences' (Bird, 2016: 49). Meanwhile, Laing and Frost (2017), in a study of visits to two Centenary-themed exhibitions staged by Australian museums, noted that social media postings from both teachers and students emphasized their positive experiences; suggesting that these visits may enhance well-being. However, limited work has been done to explore the organizers' perspective of staging commemorative events to appeal to a younger demographic.

This gap in the literature with respect to young people and commemorations is surprising, given a growing body of research suggests that younger people are increasingly making pilgrimages to World War One battle sites (Bird, 2016; Dore, 2006) and evidence of their interest in commemorative activities (see, for example, Jones, 2014). Educational programs have also been set up to take young people to battlefields or museums connected to commemorating warfare, with Pennell (2018: 84) noting that the recent Centenary of World War One commemoration has seen, 'a good degree of state-sponsored centenary activity in Britain [which] has actively targeted young people, for it is they – as the "next generation" – who have to bear the responsibility of carrying memory forward'.

In this chapter, therefore, we examine how Centenary of World War One activities and exhibitions in Australia were structured for children and families. Part of our research strove to understand how these organizers sought to include children and families in their programs, which sought to recognize sacrifice and service, evoke empathy and stimulate discussion about the nature of military conflict.

The Centenary of World War One: An Australian Perspective

Between 2014 and 2018, Australia commemorated the Centenary of World War One through a diverse program of ceremonies, re-enactments, cultural events, and historical exhibitions. In addition to commemorating World War One, the program also included events related to the 75th Anniversary of World War Two and the 50th Anniversary of the Vietnam War under the umbrella of a 'Century of Service'. The result was one of the largest event programs in Australian history and certainly the largest commemorative event since the controversial 1988 Bicentenary of European Settlement.

The Centenary was widely referred to as the Anzac Centenary, with Anzac referring to the Australian and New Zealand Army Corp which served in World War One. Under the aegis of the Australian Department of Veterans' Affairs, a National Commission on the Commemoration of the Anzac Centenary was initially formed to scope out the potential range of activities. As nearly all of the key military moments involving Australian forces in World War One occurred in Europe and the Middle East, travel to 'on-site' commemorations was limited by the cost and time involved. Even the largest overseas event – the 2015 Anzac Day Service at Gallipoli – only had the capacity for 10,000 Australians and New Zealanders to participate (ABC, 2015a). Therefore, the program was strongly focused on staging events within Australia, with attendances at the 2015 Anzac Day Services being 120,000 for Canberra and 85,000 for Melbourne (ABC, 2015a). Major historical exhibitions at war memorials, museums, and galleries were also seen as accessible opportunities and were heavily promoted to encourage as many Australians as possible to engage with the Centenary. Children were seen as one of the key markets for the Centenary, which was apparent in the official strategy, and the focus on education accords with the British experience, as identified by Pennell (2018). In scoping out the main themes of the Centenary, the National Commission reported that:

> The Commission received its highest number of submissions in relation to the *Education* theme: more than 350 ideas were submitted … [a focus should be on] the education of school students about our wartime heritage and its importance in the development of our nation. It is thought that, by educating younger generations, national days of remembrance, such as Anzac Day and Remembrance Day, will continue to be recognised and commemorated in an appropriate and respectful manner.
>
> (National Commission, 2011: 6)

This emphasis was later highlighted in the government summary of the commemoration:

> One of the most important legacies that came from the Anzac Centenary was to improve community understanding and awareness of our wartime history, particularly for younger Australians. The Anzac Centenary gave families, schools and communities an opportunity to start important conversations that will continue long after the national program concluded, ensuring an enduring and unifying legacy for current and future generations.
>
> (Department of Veterans' Affairs, 2020)

Media coverage of smaller, local events also noted that children were a key audience. For example, a media report about a one-week exhibition at a suburban return services league (RSL) club reported that:

> Hundreds of Melbourne school children will this week get the chance to imagine what life was like in the trenches of World War I, in an unlikely place. The Caulfield RSL has commissioned a film set designer to create a replica war zone in its car park for Anzac Day commemorations.
>
> (ABC, 2015b)

Ultimately, our research project sought to understand the Centenary as a commemorative event combining the construction and reinforcement of national identity with darker elements of tragedy and conflict (Frost and Laing, 2013; Viol *et al.*, 2018). We sought to examine objectives and meanings of the Centenary through qualitative interviews with major organizers, including representatives of the Department of Veterans' Affairs, war memorials, museums, and galleries. We now briefly outline our methodology.

Methodology

A hermeneutic phenomenological approach was adopted, focusing on the subjective experience of participants (Szarycz, 2009). Long in-depth semi-structured interviews were conducted with seven individual representatives of organizations involved in the Centenary, to elicit detailed accounts of their strategies and experiences as organizers. A sample of this size is not unusual in that 'the number of people interviewed in phenomenological studies is usually small, and often case-specific' (Szarycz, 2009: 54). They were conducted face-to-face, at a venue chosen by the participants – usually their office. The interviews were recorded and transcribed, with a copy of the transcript subsequently sent to each participant, to allow for member reflections (Tracy, 2010).

Table 14.1 provides a summary of the interview participants, their position or role, and the organizations they represented. Interviewees were given a pseudonym of a first name to maintain confidentiality. This approach was taken in order to encourage participants to be open and honest with their responses and opinions. A thematic analysis was then carried out of the interviews, with a number of themes identified that related to children and families.

Findings and Discussion

A National Responsibility to Educate Youth

Participants felt that there was an important responsibility to educate young people, which was in line with government policy and linked to narratives of identity and a sense of nationhood (Laing and Frost, 2019). For example, Jane spoke of how she saw it as her duty to promote a 'national narrative'

Table 14.1 Interview Participants

Participant	Type of Organization	Role/Position
Jane	Museum	Director/Manager
Kate	Museum	Curator
Liz	Museum	Director/Manager
Penny	Museum	Curator
Tom	Museum	Curator
John	Government Department	Manager
Helen	Destination Marketing Organization	Manager

and 'to continue to educate and make younger generations aware'. She thought it was clearly stated by successive governments that her public funding was for, 'educating younger people about various conflicts that Australia's been involved in, in a language that's appropriate and does have a fit within either history ... [or] citizenship'. Similarly, John saw that he had a duty to 'meet the national responsibility to provide a story of those people who served and to help us put that in context for today, and help our youth take that understanding forward'. Such sentiments paralleled those found by Pennell (2018), where the British Government has funded tours by young people to battlefields on the Western Front based on a belief in their future role as the custodians of memory about World War One. While the idea of a 'master narrative about the past' runs counter to more recent scholarship on history education for the young, which is based on developing an understanding of multiple voices and perspectives (Fournier *et al.*, 2012; Pennell, 2018), participants did perceive that they played a role in providing a more nuanced coverage of the Centenary. For example, Kate noted, '...what we did was tell personal stories that we hoped kind of gave, were a window onto a broader Australian experience, and I think to show that kind of diversity of experience and what was going on at the time'. This focus on multivocality is also linked to new directions in museology (Arnold-de-Simine, 2013; Laing and Frost, 2019), which favor a series of counter-narratives over one grand narrative.

The responsibility to educate was often formalized through school programs. Participants commented that their efforts were often matched to school curriculum, which ensured visitation. As John commented '...school groups are certainly increasing. The national curriculum is very much bedding in. We're, I suppose, a mandatory visited institution'. Similarly, Kate explained that 'school visitation is an incredibly big part of our business, and it's a subject that sits very neatly in the curriculum'. The emphasis placed on learning about the Centenary in school was seen as heightening interest among young people in their military history. According to Helen,

> I've got a five year old and a seven year old boy. This year in particular, they've become more interested in the war and they've sought out more information about the war, because they're learning about it at school. So commemorative events have an impact on our future generations because they start to continue to keep the message alive, and keep it alive in our minds.

Some of those interviewed however were concerned that these educational programs should not be overly militaristic and patriotic. In this sense, they sought to provide young people with a more critical perspective about the events surrounding World War One than they felt had been the case before. Penny observed, 'for me it might have taken 100 years, but now we're going to be more honest about what this war was like and what it did to people'. Meanwhile, Liz was emphatic that there were important contemporary lessons to be learned, arguing, 'I think the First World War was an absolute atrocity and I still am staggered that it kept going for so long with the most appalling consequences'. She further noted, 'you hope that the more you understand it, and this is very naive of me I know, the less likely you are to do it all over again'. This again reflects both the philosophical underpinnings of the 'new museum' and more recent thinking on history education, where, 'critical questioning and independent thinking' are encouraged (Pennell, 2018: 86).

Catering for Young People

The challenges of catering for children and family groups were highlighted by participants. Most were conscious that while the national Centenary program focused on young people; for museums and galleries, this was generally not their typical market. As Tom explained, 'our museum demographic is probably like most museum demographics, it tends to be an older audience'. Moreover, apart from school groups, there were 'not a huge number of teenagers', which he observed as 'a museum problem everywhere'. This led participants who were curators to think about the content that they were curating and how they might present this to a youthful audience. They were highly conscious that they were commemorating a violent conflict and were also aware of pressure to sanitize the program and exhibition interpretation. This is a common issue for organizers of commemorative events (Frost and Laing, 2013). One criticism of the British *Blood Swept Lands and Seas of Red* art installation, as mentioned earlier, was that making the memorial, 'too sanitized' was a way of forgetting the past. There were also discussions as to whether the displays would be suitable for children and educational specialists were consulted through the process. As Liz explained:

> ...it's certainly something we think about but the view of our education staff here is that we don't not do it ... we asked that question in developing them [exhibits] and people who deal with children every day, or with students every day, said no, you shouldn't not tell the stories.

Tom explained his reasoning for not censoring what he wanted to show in his exhibition:

> I think kids in particular are a lot more resilient than we think they are. One of our other exhibitions ...there was some concern beforehand that, oh, this is going to be hard for kids because there's film in there and gunfire shots in the film. The response is that school teachers love the exhibition and bring their school groups to see that exhibition and most kids; it's probably completely twee compared to what they see elsewhere.

Similarly, Penny recounted how she had underestimated what younger people might cope with when developing her museum's offering on the theme of the Centenary:

> The other thing surprisingly was we thought the exhibitions were going to be too hard hitting for younger children, so we recommended ages 11 or 12 upwards and particularly to schools as well. Because there's no barrier at the door, there's a sign which is fairly small and discreet saying 'confronting content'. But we've found even very young children... if they're pretty literate or even drawing drawings, the parent said the child said [they] realised war is terrible, we must never have wars and that sort of thing.

She explained that her institution had only received one complaint about younger people being upset, where a mother of teenage boys wrote that they were 'unhappy for the rest of the day'. Penny thought this was an appropriate response from the boys, suggesting that the messages that were presented had reached their target. She goes on to state that 'I thought, well, you know, you take your sons to an exhibition about an event in which how many millions of people died in the world. Why on earth would you expect to be feeling good after that?'

In another case, exhibition staff were surprised by a child's personal familiarity with warfare. Liz told the story of how

> ...this boy from Africa, I don't know which country, it might have been Somalia, had come as part of a school group. Someone must have been talking about guns and so forth. This boy spoke up and he knew exactly what it was like to hold a gun.

She later observed, '...what better way for young children to learn about the possibility [of war] than from someone of their own age?'

Another issue related to a young audience was respectful behavior. This was particularly pertinent in the context of planning the Anzac Day services, which were held to commemorate the April 25 landings of the Australian and New Zealand Army Corps at Gallipoli in Turkey in 1915. In the 1990s and early 2000s, the Anzac Day Service at Gallipoli had become very popular with young people; almost a rite of passage for backpackers and students on a gap year (Frost, Wheeler and Harvey, 2008). This had led to criticism of the resultant party atmosphere, which John recounted to us during his interview:

> Australia had had an unhappy experience at the 90th of Gallipoli [in 2005] where the commemoration - the service itself - was conducted appropriately, but the management of the crowd and some aspects of the spirit of place tone were not very good. The outcome was... rock videos playing during the night to entertain the crowd, people were allowed to sleep on graves, people left rubbish everywhere, [and] there was alcohol. It was a service that we wouldn't allow to be conducted now.

Changes in organization have led to a more respectful tone and a 'family-friendly' atmosphere at the service. The centenary in 2015 led to further issues as to how to deal with the expected high demand. For the first time, a ballot was instituted with 42,000 applications received for 4,000 places. As John explained, this led to 'a different crowd than we normally have. The normal demographic is a bit younger, backpackers and so on. Because of the ballot the demographic was older'. As the ballot was for individuals, this ruled out school groups and the high level of planning involved discouraged younger people in Europe from making short-term decisions to visit Turkey as they had done in the past. This is in contrast with the planning for commemorative events connected to the Centenary in Canada and the UK, where there was an emphasis placed on young people visiting the battlefields as a learning experience (Bird, 2016; Pennell, 2018).

Relevance and Connection

Those interviewed were cognizant of the desirability of ensuring that exhibitions and ceremonies were relevant to contemporary Australians, particularly children and families. In planning the centenary, they were aware that they could have interaction with people who had little or no direct connection to World War One. As John explained,

> In a multicultural society today, there are far less people who have a direct connection with Australia of 1914... there are millions of Australians whose ancestry and heritage is rooted elsewhere, but who have either a benign view on the Anzac story, in other words they feel it's part of Australia and it's obviously important to Australians... Our research showed very, very little opposition; [though there was] concern I suppose among some multicultural groups that the commemoration would leave them out.

A number of participants told us that in order to make the War relevant to ordinary people today, they chose to focus on the home front. Kate opined 'if you think about it at the most basic level, the majority of Australians experienced the war from Australia', while Liz explained that 'it is widely acknowledged now as a total war which engaged citizens from countries who were occupied to citizens here who were fervent in putting forward their voluntary effort'. This approach resulted in a number of unusual displays. One exhibit was a series of postcards from a soldier to his young son, which Tom felt had relevance for younger visitors:

> The postcards are quite extraordinary... his son was seven or eight at the time and so these postcards are saying, having great fun here, we're going to shoot some 'Fritz' this afternoon or, when there's bombing it's so much fun, dirt flies over you and wish you were here with me in no-man's land. You read these things and think, my God. Then you realize, after a

while – and there's 80 of these things – you realize after a while that he's trying to play it down to [his son] – so you wonder what he was writing to his wife; probably something completely different or maybe not?

Another exhibit consciously focused on the mundane aspects of civilian life, with the justification being, according to Kate, that

It was important to highlight that life – while it was coloured by war – it wasn't always kind of run at a steady, we're miserable, there is war. People get married. People have babies. People go to the rugby and watch their team win … So all of these things I kind of felt that it was important to show some of the light and shade of life at that time, and maybe you could also argue that the highs or those bright spots are so much brighter in a time when you're kind of almost dreading getting that telegram that brings the bad news.

This again showed a more subtle narrative to the War, than a single focus on warfare and atrocity at the frontline.

The Challenges of Emotions and Confrontations

Participants emphasized the ways that they often sought a strong emotional connection, even empathy, with their modern audiences (Laing and Frost, 2019). This strategic approach was summarized by Tom who said

I think the outcome we would have been looking for is to get an emotional response from people, for people to find it an emotionally fulfilling exhibition… perhaps to make them think about the impact of war on their lives, on the lives of their forbears, on the life of society.

This again conforms to the ideas behind the new museum, with emotional engagement of audiences seen as important (Arnold-de-Simine, 2013; Laing and Frost, 2019).

For some, this required the adoption of new curatorial techniques, often borrowed from cinema or the theatre (Arnold-de-Simine, 2013; Laing and Frost, 2019). Penny explained how

I think also we've started become a bit more daring in what we do. We've also become a little bit more, I can't think of what the word is, but it's like 'filmesque'. Learning what movie makers do to evoke particular responses, which means you have a much stronger engagement, which means that you get much more out of it, because you're emotionally engaged as well as intellectually engaged.

Others focused on the emotional challenges of displays that they felt could be seen as too confrontational. Kate told of making choices about presenting images of soldiers who went through reconstructive facial surgery to avoid those that might be most upsetting and exhibiting them so that people did not stumble across them unaware

> Some of them are pretty graphic images. I did select images that were less graphic. Some of them – even though they are drawings or watercolours – they are pretty awful. The positioning of them in the exhibition too, meant that you kind of came upon them. You didn't see them directly, I guess. So that was probably the one thing where I was very conscious about that.

Penny also struggled with this problem, though in the long run she was pleased that such issues were included in her exhibition:

> What were the stories which then were hard to tell or too hard to tell? Facial wounds, sexually transmitted disease, some of the awful things like tuberculosis that some soldiers brought back with them. Then the psychological impacts that went on for so long afterwards… Everywhere along the line I thought someone's going to say, this is too hard, you know we still can't do it even though it's 100 years later. Too graphic, too confronting, you're not being respectful if you're telling a really graphic story about people having breakdowns at home or in the field or whatever… [surprisingly] everyone said thank goodness a more honest story about the war is being told and thank goodness we've used the centenary as an opportunity to be honest about the war.

Penny reported that these questions became especially important when she considered how the exhibition should conclude

> When we develop exhibitions at museums, we tend to want to end on a high note. We tend to want visitors to walk out thinking yippee or I've got more insights or I'm really relaxed and feeling good about something. But I said to my colleagues, this is going to be an exhibition where we don't – we shouldn't feel we need to end up on a high note. There is a positive in that we see people enduring recovery, the generations go on, [and] we become more insightful. But when people leave this exhibition, I think we understand that it isn't meant to be a happy time.

These highly emotive and confrontation subjects were sometimes linked to the need for providing mental health support to ex-service personnel and their families, which again linked to the idea of allowing a diversity of voices to be heard (Arnold-de-Simine, 2013; Laing and Frost, 2019). John explained how

> The commemorations and the remembrance form an important part of the strategy for dealing with the mental health of existing veterans. So for me, a significant element of the commemorative activity we do is

essentially raising awareness in the community and the acceptance of the community of service and sacrifice. If we're running a World War One commemoration... it helps people understand what someone in Afghanistan has experienced and is doing.

This led to an emphasis on including families more overtly in commemorative ceremonies. According to John,

for example... we might have as part of the service read a letter written by a mother... we had a representative of the Widows' Guild not only lay a wreath but also have a formal speaking part in the service, reflecting on the impact on widows particularly but also on families.

Support for mental health was also recognized by Penny, who followed up on her discussion of graphic imagery and long-term psychological problems (see above), by stating that

There are incredibly important arguments for saying once you've been to war and been through this stuff, that the rest of your life you will need some support. Or you may need some support, so let's recognize it that when conflicts happen, they have longer term impacts. So I've been extremely surprised, but delighted at the readiness of so many different groups and organizations to address this sort of topic.

Others took the view that a centenary provides an emotional distance. As Kate put it, '100 years gives you a certain kind of freedom and distance, and you can look at things much more logically and without that emotion, or without that taking sides kind of thing'. Tom explained that there were potential anniversaries of more recent traumatic incidents that were too 'raw' in the community's memory for it to be appropriate to commemorate with an exhibition. He contrasted these with the Centenary of World War One, which he saw as sufficiently distant in the past:

I don't think the rawness is there, to be quite controversial [laughs]. I think a lot of the feeling on World War I is confected by – politically confected – which has happened in the last 30 years because it suited both political sides to raise that trope of the nation building on the War, et cetera, et cetera.

This equates with the idea that some dark events in our past may become lightened over time (Cheer, Reeves and Laing, 2015; Lennon and Foley, 2000).

Conclusion

The potential for controversy and debate was always going to be apparent in the Australian Centenary commemorations; given their subject matter of a global conflict on this scale, and where a younger audience was encouraged.

In this chapter, we examined how organizers of major exhibitions and ceremonies for the Centenary of World War One planned for a strategic focus on children and families. All of those interviewed indicated that issues of adapting what they presented to suit this market were part of their strategic planning and led to key discussions within their teams. All of them, however, explained that apart from some school tours, their exhibitions and ceremonies were open to all ages. Accordingly, their planning was often about modifications for younger people, rather than separate developments.

A number of those interviewed explained how in planning their exhibitions and ceremonies, they decided not to sanitize war or avoid imagery that was too violent or graphic. Such decisions were made with full cognizance that children would be in attendance and some recalled how they had sought external educational advice as to how children would typically cope with such exposure. Working on the basis that children are often far more resilient than adults give them credit for, decisions were made not to censor or limit exhibitions. While there was some apprehension that this might have led to complaints, those interviewed indicated that complaints were few and that they had received a great deal of positive feedback for how they had staged this dark commemoration.

As the Australian government Centenary program strongly identified education of the young as one of its objectives, those interviewed often stated that they had a responsibility to emphasize education and young people. This was, however, interpreted quite broadly. Most indicated that they were not interested in themes of militarism and nationalism, but rather sought to give prominence to sacrifice, duty, and the futility of war. The rationale for this was that audiences – particularly children and families – could see better parallels if they were considering how everyday activities were affected. Many emphasized the long-term legacy of mental health issues and linked this to the contemporary problems of returned service personnel in recent wars. In this way, they wanted to encourage empathy for current generations, as well as those in our past (Pennell, 2018).

There are several practical implications of these findings for those involved in developing or managing dark tourism attractions or events. The first is that younger age groups should not be forgotten when developing interpretation. The second is that the desire to sanitize what is being presented is not necessarily required or optimal, if the aim is to create emotional engagement and stimulate critical reflection among younger visitors. The third is that by providing a range of voices and narratives in what is being presented, younger people can be presented with a more nuanced picture of what happened in our past, as well as the world in which they currently live.

References

ABC (2015a) 'Gallipoli 2015: Hundreds of Thousands attend Anzac Centenary Services across Australia and Turkey', *ABC News*, 25 April 2015, https://www.abc.net.au/news/2015-04-25/thousands-attend-anzac-centenary-service-at-lone-pine-gallipoli/6421740, [Accessed 15 July 2020].

ABC (2015b) 'Gallipoli 2015: Melbourne School Children Get Anzac Experience, as RSL Car Park is Transformed into Replica War Zone', *ABC News*, 20 April 2015, https://www.abc.net.au/news/2015-04-20/school-children-get-anzac-experience-in-rsl-carpark/6407032, [Accessed 31 July 2020].

Arnold-de-Simine, S. (2013) *Mediating Memory in the Museum: Trauma, Empathy, Nostalgia*. Houndmills and New York: Palgrave Macmillan.

Bird, G. (2016) 'Landscape, Soundscape and Youth: Memorable moments at the 90th commemoration of the Battle of Vimy Ridge, 2007', in K. Reeves, G. Bird, L. James, B. Stichelbaut and J. Bourgeois (eds.), *Battlefield Events: Landscape, Commemoration and Heritage* (pp. 48–63). London and New York: Routledge.

Cheer, J., Reeves, K. and Laing, J. (2015) 'Debunking Pacific Utopias: Chief Roi Mata's Domain and the Re-imagining of People and Place in Vanuatu', in S. Pratt and D. Harrison (eds.), *Tourism in Pacific Islands: Current Issues and Future Challenges* (pp. 85–97). London: Routledge.

Department of Veterans' Affairs (2020) *ANZAC Portal*, https://anzacportal.dva.gov.au/commemoration/commemoration-days/national-program/anzac-centenary, [Accessed 15 July 2020].

Dore, L. (2006) 'Gallipoli: A Visitor Profile', *Historic Environment*, *19*(2), 46–51.

Fournier, G., Loughridge, J., Macdonald, K., Sperduti, V., Tsimicalis, E. and Taber, N. (2012) Learning to commemorate: Challenging Prescribed Collective Memories of War. *Social Alternatives*, *31*(2), 41.

Frost, W. and Laing, J. (2013) *Commemorative Events: Identity, Memory, Conflict*. London and New York: Routledge.

Frost, W., Wheeler, F. and Harvey, M. (2008) 'Commemorative Events: Sacrifice, Identity and Dissonance', in J. Ali-Knight, M. Robertson, A. Fyall and A. Larkins (eds.), *International Perspectives on Festivals and Events: Paradigms of analysis* (pp. 161–172). Amsterdam: Elsevier.

Jones, J. (2014) 'The Tower of London Poppies are Fake, Trite and inward-looking – A Ukip-style Memorial', *The Guardian*, 28 October, 2014, http://www.theguardian.com/artanddesign/jonathanjonesblog/2014/oct/28/tower-of-london-poppies-ukip-remembrance-day [accessed 5 December 2014].

Laing, J. and Frost, W. (2017) 'Dark Tourism and Dark Events: A Journey to Positive Resolution and Well-being', in S. Filep, J. Laing and M. Csikszentmihalyi (eds.), *Positive Tourism* (pp. 68–85). London: Routledge.

Laing, J. and Frost, W. (2019) 'Presenting Narratives of Empathy through Dark Commemorative Exhibitions during the Centenary of World War One', *Tourism Management*, *74*, 190–199.

Lennon, J. J. and Foley, M. (2000) *Dark Tourism*. London: Continuum.

National Commission on the Commemoration of the Anzac Centenary (2011) *How Australia May Commemorate the Anzac Centenary*. Canberra: Government of Australia, https://anzacportal.dva.gov.au/sites/default/files/docs/anzac-centenary-report.pdf [Accessed 21 June 2020].

Pennell, C. (2018) 'Taught to Remember? British Youth and First World War Centenary Battlefield Tours', *Cultural Trends*, *27*(2), 83–98.

Sharpley, R. (2009) 'Shedding Light on Dark Tourism: An Introduction', in R. Sharpley and P.R. Stone (eds.), *The Darker Side of Travel: The theory and practice of dark tourism* (pp. 3–22). Bristol: Channel View.

Stone, P. R. (2006) 'A Dark Tourism Spectrum: Towards a Typology of Death and Macabre related tourist Sites, Attractions and Exhibitions', *Tourism: An Interdisciplinary International Journal*, *54*(2), 145–160.

Stone, P. R. (2009) 'Making Absent Death Present: Consuming Dark Tourism in Contemporary Society', in R. Sharpley and P.R. Stone (eds.), *The Darker Side of Travel: The theory and practice of dark tourism* (pp. 23–38). Bristol: Channel View.

Szarycz, G. (2009) 'Some Issues in Tourism Research Phenomenology: A Commentary', *Current Issues in Tourism*, *12*(1), 47–58.

Tracy, S. (2010) 'Qualitative Quality: Eight "big-tent" Criteria for Excellent Qualitative Research', *Qualitative Inquiry*, *16*(10), 837–851.

Viol, M., Todd, L., Theodoraki, E. and Anastasiadou, C. (2018) 'The Role of Iconic-historic Commemorative Events in Event Tourism: Insights from the 20th and 25th Anniversaries of the Fall of the Berlin Wall', *Tourism Management*, *69*, 246–262.

15 Identity and belonging in a dark heritage destination

Perspectives from local children

Antonia Canosa and Rebecca H. Price

Introduction

This chapter explores the complex politics of identity and belonging at dark heritage sites from the perspective of local children. Given the lack of research in this area, we present a theoretically driven approach to explore how children and young people living in dark tourism destinations negotiate identity and belonging. We propose a 'spectrum of connection' model to identify how local children respond to the changes at a local level post-disaster when tourists start visiting their hometown. Our discussion is guided by two main questions: (a) how might tourists visiting local communities post-disaster change children's perceptions of who they are; and (b) how might children's personal connection to dark tourist attractions complicate their experiences of belonging?

In this chapter, we argue that children's experiences of growing up in a difficult heritage destination are particularly complex and worthy of attention. We build upon key tenets of childhood studies to advance knowledge about children's experiences of living in dark tourism sites. Our underlying philosophical stance augments the key principle that childhood is socially constructed and that children are competent social actors in their lives, thus capable of making a significant contribution to tourism research, policy, and planning (Canosa & Graham, 2020; Woodhead, 2008). Further explored in the following sections, we employ 'identity' and 'belonging' as a theoretical lens through which to explore children's perspectives. We combine our previous research at dark tourism destinations and the complex politics involved in negotiating identity and belonging at these sites (Kerr & Price, 2018; Kerr et al., 2017; Kerr, Stone, & Price, 2021; Price, 2018; Price & Kerr, 2018). Importantly, this chapter contributes to an ongoing effort to foster a child-centered agenda in tourism and hospitality research (Canosa, Graham, & Wilson, 2019; Kerr & Price, 2018; Price, 2018).

Background

Children's experiences of tourism are multidimensional, contextual, and relational. As such, child tourist experiences are particularly complex. Historically, children have been marginalized in social research due to the

DOI: 10.4324/9781003032199-19

conception of childhood as a period of 'human becoming' and their role as future citizens, rather than as 'human beings' and active community members in the present (James & Prout, 1997; Peleg, 2019; Wall, 2019). This is particularly true in business-oriented disciplines like tourism studies where children's perspectives have been largely neglected (Canosa, Moyle, & Wray, 2016; Kerr & Price, 2018; Khoo-Lattimore, 2015; Schänzel & Carr, 2015; Yang et al., 2020), or explored with a future-framed approach (Seraphin & Green, 2019). Citing progressive Polish educator Janusz Korczak, Peleg (2019: 20) argues that 'children are not the people of tomorrow, but are people of today' and need to be afforded the right to participate in their own development including the right to be 'properly researched' (Beazley et al., 2009).

It is important however that effort to build a child-centered scholarship in tourism studies is built on strong foundational work. Moreover, efforts should focus on theoretical and methodological advancements in social research *with* and *by* children, rather than *on* children that have been possible thanks to the radical and innovative cross-disciplinary agenda of childhood studies (Canosa & Graham, 2020; Kellett, 2011; Price, 2018; Prout & James, 1997). Central to any paradigmatic shift is our need to challenge common perceptions that children are immature and unable to make a meaningful contribution to society. Hence, we need to position children as *subjects* rather than *objects* in the research process (Freire, 1970).

Conceptualizing identity and belonging in childhood

Formation of identity is a complex process whereby young people attempt to distance themselves from the world of the child but are inevitably denied access to the adult world (Sibley, 1995). Globalization and hypermobility have added to this complexity making the search for identity and belonging an ongoing project or 'quest' which is actively and reflexively achieved (Rapport & Dawson, 1998). Within postmodern and poststructuralist traditions, identities are highly changeable, contextual, dynamic, and hybrid (Bauman, 2009; Hall, 1990; Lash & Friedman, 1992; Olwig, 1997; Poole, 1994; Rapport & Dawson, 1998). Indeed, identity has become a key issue of contemporary times and, subsequently, there have been a growing number of studies exploring the concept. The subject field of 'Identity Studies' is as broad as the meanings attached to the term 'identity' which have become quite 'fuzzy' and abstract (Côté, 2006: 7). Irrespective of the disciplinary approach, modern ideas about the development of identity in young people share some core assumptions which are outlined by Furlong (2013: 125):

- Identity develops in the social context in which the individual grows up and is embedded in the culture of the time and place.
- Identity is a lifetime project.

- Identities are multiple and relational and they often overlap or conflict.
- The protraction of the youth phase and the increased complexity of the socio-economic environment have implications for the development of identity among youth.

Like identity, lived experiences of belonging are multidimensional (Antonsich, 2010). Among the multiple 'modes of belonging', Antonsich (2010) argues there are two analytical dimensions: 'place-belongingness and 'the politics of belonging'. Place-belongingness is described as a 'personal, intimate, feeling of being 'at home' in a place'. Conversely, politics of belonging refers to social dimensions and boundary practices that shape understandings of social inclusion and/or exclusion. In short, place-belongingness amounts to the 'us' versus 'them' (Antonsich, 2010: 645). Emotional feelings of attachment to place have often been discussed in relation to notions of place attachment, place identity, and sense of place (Antonsich, 2010). These are particularly prevalent conceptualizations in the tourism literature (Carter, Dyer, & Sharma, 2007; Cui & Ryan, 2011; Dredge, 2010; Kerstetter & Bricker, 2009; Williams, 2002). Nevertheless, *belonging* as an all-encompassing concept has rarely been adopted as a theoretical concept in tourism research (Buzinde & Manuel-Navarrete, 2013; Canosa, Graham, & Wilson, 2018). According to Antonsich (2010), it is important to explore both the personal (place-belongingness) and the social (politics of belonging) dimensions of belonging in order to have a more nuanced understanding of this ephemeral but important concept.

In premodern societies, belonging was a phase which all adolescents experienced, whereas today, 'the search for belonging becomes a generalised condition' (Thomson, 2007: 152). Belonging is understood as a process of 'becoming' rather than a status of 'being' (Antonsich, 2010: 652), and similar to identity it is often performed, enacted, and displayed (Bell, 1999). Belonging and connection become matters of choice which are inscribed in the biographical narratives told by young people (Giddens, 1991; Yuval-Davis, 2006). Young people may understand belonging as either 'group membership, identity, identification and recognition' or as 'dis-identification and exclusion' (Thomson, 2007: 147). Thus, identity and belonging are subtly interwoven and define 'what you have in common with some people and what differentiates you from others' (Weeks, 1990: 88). Ultimately, a sense of belonging is an important factor in the wellbeing of young people (Thomson, 2007).

Identity and belonging are adopted as key constructs in this chapter to explore the lived experiences of children and young people living in dark tourism destinations. In this study, the terms 'children', 'young people', and 'youth' are used interchangeably and refer to those under 18 years of age consistent with definitions from the United Nations Convention on the Rights of the Child (United Nations General Assembly, 1989). Nevertheless, we acknowledge that childhood cannot be understood just as a stage in the life-cycle of the individual but as a 'vital conjuncture' or critical period of choices, opportunities, and possibilities that might have important consequences for the individual's life course and for the formation of identities (Johnson-Hanks, 2002: 865). Building on these principles, we conceptualize identity as 'situated, multiple and relational' (Brandth & Haugen, 2011: 37): a dynamic process which is explored, negotiated, and affirmed throughout the individual's life span (Hall, 1996).

Identity and belonging in a dark heritage destination

While the intersection of local children with dark tourism has yet to be studied in the scholarly literature, one might find clues in popular literature and the media where anecdotes of children and memorials occur (see, for example, Associated Press, 2017; Mercier, 2019; Peltz, 2016; Price & Kerr, 2018). For instance, any visitor might notice children present at dark tourism destinations, as shown in Figure 15.1. Likewise, readers might draw on recent research on local children's tourism experiences (Canosa, Graham, & Wilson, 2018, 2020) and children's experiences of terrorism, mass trauma, and natural disasters (Gibbs et al., 2015; Kerr, Stone, & Price, 2021).

When considering local children at a dark tourism destination, we must consider how their nascent identities are shaped by the influx of tourists into their homes. Personal memories of loss or distress become entangled with the experiences of being 'on display' for visiting tourists. This can evolve into

Figure 15.1 Children pose for a photo at Dealey Plaza – site of US President John F. Kennedy's assassination.

Source: Author R.H. Price.

children's outspoken pride in their community and its history of resilience post-disaster or conversely jeopardize their connection to place. Therefore, in the subsequent sections, we explore identity and belonging in the context of children growing up in dark heritage destinations.

Negotiating identity and belonging in a dark heritage destination

When unforeseen events cause the formation of a new dark tourism destination, children may be particularly susceptible to the changes brought about by tourism activity in the community where they grow up. For example, archival research into local children's recollections of the United Flight 93 plane crash in Pennsylvania, USA, on September 11, 2001, shows that children wanted to feel helpful and 'welcoming' toward visitors and, in their letters to families of the victims, hoped to be recognized as such (Kerr et al., 2017). This tragic event led children to re-evaluate their role in their community, welcoming outsiders in ways that they had never imagined. Their identities were likely shaped by their need to show empathy toward the visiting families of the victims.

We do not know whether these children who lived near the Flight 93 crash came to experience mixed feelings toward dark tourists. However, our prior research has indicated that overcrowding due to tourism influx can jeopardize children's connection to place and their sense of belonging due to towns primarily accommodating the needs of tourists (Canosa et al., 2018). This is further compounded in dark tourism sites if local children have witnessed tragic events or are directly related to the victims.

When discussing the experiences of local children who reside at or near dark tourism destinations, we cannot overlook the complicating effect of personal connection. As the model in Figure 15.2 illustrates, a local child's connection to a dark tourism site can range across a broad 'spectrum' (after Stone, 2006): from simply having a dark tourism destination situated within

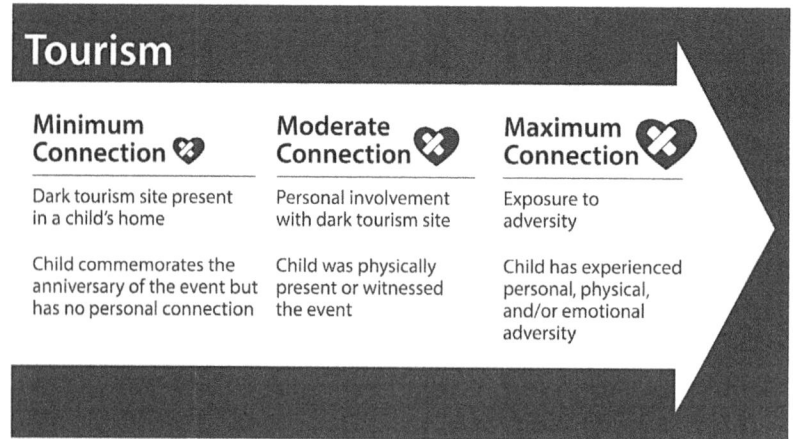

Figure 15.2 Spectrum of local children's connection to dark tourism destination.

their home town, to tragic and very personal connection with the site. This personal connection and the possible memories of the event can feed grief, distress, bereavement, and even post-traumatic stress (see Kousky, 2016; McLaughlin et al., 2009; Pfefferbaum, 2020). What might examples of local children's connection entail? While rarely studied in the scholarly arena, media reports provide some insight into what the varying degrees of connectedness might look like in different situations.

Minimum connection

What might be considered 'minimum connection' can be found in children who live in Tulsa, USA, at the 100th anniversary of the Tulsa Race Massacre. In 2021, the people of Tulsa commemorated the 100th anniversary of the massacre in which hundreds of Black residents were murdered. A few surviving victims remained however local children only know of the event through history lessons and the commemoration ceremonies.

One hundred years later, the city has turned the mass murders into a tourism attraction. According to reports, a local commission raised millions of dollars through tourist-focused events, while neglecting to pay reparations to the three remaining survivors or their descendants (Jamieson & Felton, 2021). Meanwhile, children of 2021 Tulsa still experienced connection, although time diminished its effects. Though tourists flocked to the remembrance events, 'in the decades since [the 1921 massacre], Black children in north Tulsa have been told not to play on what's become known to locals as the "valley of the dry bones"' (Jamieson & Felton, 2021).

Moderate connection

To illustrate 'moderate connection', it is useful to consider children who lived near the 9/11 United Airlines Flight 93 crash in rural Pennsylvania, USA. While these children were neither victims nor survivors they have personal memories of the event and have witnessed the changes in their communities. Our study (Kerr et al., 2017: 5–6) analyzed letters from sixth-grade students (ages approximately 11–12 years) who lived nearby the 9/11 crash.

One girl recalled that 'Our whole school shook', while a boy mentioned 'It shook us all up and I still am'. Other children wrote letters to the families who visited the crash site. One boy wrote, 'I feel very sorry for your Innocet[sic] family members that died but I also would call them heroes for landing it in a field and not on our houses. Because it was very close to my house'. In these examples, children remember the fear they felt as the event happened, and they connect those feelings to the visitors who flooded the scene shortly after. Arguably, this may result in feelings of distress, grief, and post-traumatic stress which may be exacerbated by the visiting tourists.

Unlike child survivors, these children with moderate connections remembered the events and witnessed the changes to their communities. As a temporary, then a permanent memorial was constructed at the crash site, what was once these children's pastoral playgrounds became a dark tourism destination. Yet they did not report experiencing personal danger or loss.

Maximum connection

'Maximum connection' may include factors like the death of a loved one or exposure to physical adversity, factors which can also cause long-lasting physical and emotional impacts (Kousky, 2016; McLaughlin et al., 2009). Considering maximum connection to a dark tourism event or site, one finds many examples in the children of Hurricane Katrina. In 2005, New Orleans, USA, was forever changed by the devastating impact of Hurricane Katrina, which killed more than 1,500 and displaced a million people, and caused over 161 billion dollars in damage ('Hurricane Katrina Statistics Fast Facts', 2020). Lacey Lawrence experienced both loss and personal danger as a 10-year-old in the floodwaters:

> [Lacey] escaped Katrina's waters on an air mattress, as police officers shoved away bodies with oars, and some proprietors guarded swamping businesses with shotguns. An uncle disappeared, probably drowned. A 12-year-old cousin got lost, alone, and wasn't heard from for hours.
>
> (Carey, 2017)

Twelve years later, Lacey told the *New York Times*, 'You lose everything and you don't know how to deal with it – no one prepares you for that' (Carey, 2017). Shaysa Sheif, another child of Katrina, experienced personal danger as well, trapped with her family for days after Katrina, with no power, little food or water, and no one coming to their rescue (Carey, 2017). In Glenn Sullivan's case, as a 10-year-old evacuating with his family as the storm hit, he experienced years of homelessness in the wake of the devastation. Later, he returned home to the same, yet unrecognizable neighborhood and schools (Sullivan, 2015). Hartnell (2009: 725) noted that:

> [The] struggle experienced by former New Orleans residents - mostly poor and predominantly black – [is] to reclaim and rebuild their homes. Instead of supporting these residents, the authorities are championing... New Orleans's makeover into a playground for wealthy tourists.

Is it any wonder, then, that local children might experience feelings of resentfulness toward dark tourists? As child survivor Glenn Sullivan (2015) notes in his autobiographical essay: 'Tourism is thriving but kids and adults in poor neighborhoods aren't'. Maximum connection refers, simply, to local children who remember their losses. Whether the loss be of a day or of years, of homes or of loved ones, these children witness a tourist influx that goes hand in hand with the commercialization and commodification of their still-hurting communities.

This section has presented a spectrum of connections that describes how local children might experience their community becoming a tourist attraction after a disaster or terror attack. When local children encounter tourists – or when their homes become dark tourism destinations, how might their levels of connectedness to the tragic events influence their responses? What are the implications for their felt sense of identity and belonging, including their experiences of grief and their attitudes toward dark tourists? How might local children feel when watching something of theirs – at minimum their home, at maximum a personal tragedy – be the focus of smiling selfies and other tourist behaviors of questionable etiquette? (see Compton, 2019). It is to some of these issues that we now turn.

Children's participation in processes of change

A sense of belonging and connection to place is an important factor in the social and emotional wellbeing of children and young people, particularly for those growing up in tourism destinations. Antonsich (2010) argues that the personal and intimate feeling of being 'at home' somewhere (place-belongingness) is further complicated by the social dimension and the boundary practices that shape understandings of social inclusion and exclusion (politics of belonging). At difficult heritage sites, particularly those that are newly formed due to sudden terror attacks or disasters, children's lived experiences of belonging are complex due to the influx of visiting dark tourists. Although children may be welcoming at first, research shows that they often engage in practices of inclusion/exclusion to create 'locals only' spaces where they feel 'at home' away from the visiting tourists (Canosa et al., 2018). The feelings of being 'on display' may jeopardize the connection to place and sense of belonging. Children have commented on these feelings as their home turns into a makeshift memorial at, for example, the site of the plane crash in Pennsylvania, USA (Kerr et al., 2017; Kerr & Price, 2018).

However, when children are actively involved in processes of change following terror attacks or other mass trauma events, research suggests that they achieve a sense of belonging which is important for their mental health and wellbeing (Dyregrov, Salloum, Kristensen et al., 2015). For instance, a recent report on children and young people's experiences of disaster in Australia following the 2020 catastrophic bush fires, shows that they often 'feel invisible, forgotten and unable to influence the world around them' and that they wish they had a 'voice' and involvement in the disaster recovery process (Office of the Advocate for Children and Young People, 2020: 30). This is important in improving their mental health and wellbeing, and in fostering a connection to land and community. In Japan, for example, local children were involved in actually creating a memorial for the school children victims of the tsunami in 2011 (Parry, 2017). Their involvement was significant and highlights their role as agents of change in the community. This process would have likely facilitated their connection to place by identifying with the memorial constructed. It may have even mitigated the negative impacts of tourism influx on their felt sense of identity and belonging.

Community pride has been evidenced in previous research with children living in difficult heritage sites. As previously mentioned, children in the town nearest to the 9/11 incident wrote in their letters to families of victims and visitors: 'Good ol'[sic] [Stemsburg] P.A. welcomed visitors' (Kerr et al., 2017: 6). One boy mentioned 'You know you are always welcome here…I hope you felt welcome here…Before this all happened, you probably never heard of [Stemsburg]'; and another wrote 'When you need a home away from home, you can always look to good ol'[sic] [Stemsburg] P.A.' (Kerr et al., 2017: 6). Children in this study also expressed pride in the resilience or 'coping' of their community, and how they helped by collecting money and supporting visitors (Kerr et al., 2017). In one letter, a local girl asked visitors to join her community: 'Now that you have my words, may I please have some words from you to bring my spirits up' (Kerr et al., 2017: 6). Thus feelings of 'not belonging' can be mitigated by actively involving children in processes of change post-disaster which may result in community pride.

Conclusion

In this chapter, we argue the perspectives of local children living in dark tourism destinations deserve special consideration given the complexity of lived experiences. These communities hold particular memories and personal/family stories which further complicate children's evolving process of identity formation and the connections to place and community that are so important for their social and emotional wellbeing (Graham, 2004). The chapter also highlights the evident neglect of this area of inquiry and the absence of research specifically focused on local children's lived experiences of identity and belonging at difficult heritage sites.

Thus we have drawn on children's perspectives from our previous research at dark tourism destinations to explore the complex politics of identity and belonging at play when growing up in these communities. Personal accounts of loss or distress presented in this chapter become entangled with the experiences of being 'on display' for visiting tourists. We have argued that this may further complicate children's experiences of loss and grief. In turn, these may require interventions that help local children feel actively included in change processes happening in their community as a result of sudden and tragic events, such as terror attacks or natural disasters. Their 'voice' and involvement in the disaster recovery process is important given children often feel invisible and neglected in these circumstances. In addition, evidence shows their involvement improves mental health and wellbeing and fosters connection to land and community.

Ultimately, this chapter has sought to contribute to building a child-centered scholarship in tourism and hospitality research (Canosa, Graham, & Wilson, 2019) and more specifically to address the dearth of research on children's experiences of life at dark heritage sites (Kerr & Price, 2018).

The theoretical lens presented in this chapter may provide a roadmap for future research in this space, which will clearly need to be epistemologically and ontologically founded on conceptions of children as active social agents in the communities where they live. This can only be achieved with a reflexive, ethical and participatory approach to engaging children in the research process.

References

Antonsich, M. (2010). Searching for belonging–an analytical framework. *Geography Compass*, *4*(6), 644–659.

Associated Press. (2017). Youngest victim of marathon bombing has park named after him. *Denver Post*. [Online]. 16 August. [Accessed 31 July 2020]. Available from: www.denverpost.com/2017/08/16/youngest-victim-boston-marathon-bombing-has-park-named-after-him/

Bauman, Z. (2009). Identity in the globalizing world. In A. Elliott & P. du Gay (Eds.), *Identity in question* (pp. 1–12). London: Sage.

Beazley, H., Bessell, S., Ennew, J., & Waterson, R. (2009). The right to be properly researched: Research with children in a messy, real world. *Children's Geographies*, *7*(4), 365–378.

Bell, V. (1999). Performativity and belonging an introduction. *Theory, Culture & Society*, *16*(2), 1–10.

Brandth, B. & Haugen, M.S. (2011). Farm diversification into tourism – implications for social identity? *Journal of Rural Studies*, *27*(1), 35–44.

Buzinde, C.N. & Manuel-Navarrete, D. (2013). The social production of space in tourism enclaves: Mayan children's perceptions of tourism boundaries. *Annals of Tourism Research*, *43*, 482–505.

Canosa, A. & Graham, A. (2020). Tracing the contribution of childhood studies: Maintaining momentum while navigating tensions. *Childhood*, *27*(1), 25–47.

Canosa, A., Graham, A., & Wilson, E. (2018). Growing up in a tourist destination: Negotiating space, identity and belonging. *Children's Geographies*, *16*(2), 156–168.

Canosa, A., Graham, A., & Wilson, E. (2019). Progressing a child-centred research agenda in tourism studies. *Tourism Analysis*, *24*(1), 95–100.

Canosa, A., Graham, A., & Wilson, E. (2020). Growing up in a tourist destination: Developing an environmental sensitivity. *Environmental Education Research*, 1–16. DOI:10.1080/13504622.2020.1768224.

Canosa, A., Moyle, B., & Wray, M. (2016). Can anybody hear me? A critical analysis of young residents' voices in tourism studies. *Tourism Analysis: An Interdisciplinary Journal*, *21*(2), 325–337.

Carey, B. (2017). Life after the storm: Children who survived Katrina offer lessons. *The New York Times*. [Online]. 8 September. [Accessed 28 June 2021]. Available from: https://www.nytimes.com/2017/09/08/health/katrina-harvey-children.html

Carter, J., Dyer, P., & Sharma, B. (2007). Dis-placed voices: Sense of place and place-identity on the Sunshine Coast. *Social & Cultural Geography*, *8*(5), 755–773.

Compton, N.B. (2019). How to navigate the etiquette of dark tourism. *Washington Post*. [Online]. 17 October. [Accessed 13 July 2020]. Available from: https://www.washingtonpost.com/travel/tips/how-navigate-etiquette-dark-tourism/

Côté, J. (2006). Identity studies: How close are we to developing a social science of identity?—an appraisal of the field. *Identity*, *6*(1), 3–25.

Cui, X. & Ryan, C. (2011). Perceptions of place, modernity and the impacts of tourism - differences among rural and urban residents of ankang, china: A likelihood ration analysis. *Tourism Management, 32*, 604–615.

Dredge, D. (2010). Place change and tourism development conflict: Evaluating public interest. *Tourism Management, 31*(1), 104–112.

Dyregrov, A., Salloum, A., Kristensen, P., & Dyregrov, K. 2015. Grief and traumatic grief in children in the context of mass trauma. *Current Psychiatry Reports, 17*(48), 6.

Freire, P. (1970). *Pedagogy of the oppressed* (2nd ed.). New York: Continuum International Publishing Group.

Furlong, A. (2013). *Youth studies: An introduction.* New York: Routledge.

Gibbs, L., Block, K., Harms, L., MacDougall, C., Baker, E., Ireton, G., ... Waters, E. (2015). Children and young people's wellbeing post-disaster: Safety and stability are critical. *International Journal of Disaster Risk Reduction, 14*, 195–201.

Giddens, A. (1991). *Modernity and self-identity: Self and society in the late modern age.* Stanford: Stanford University Press.

Graham, A. (2004). Life is like the seasons: Responding to change, loss, and grief through a peer-based education program. *Childhood Education, 80*(6), 317–321.

Hall, S. (1990). Cultural identity and diaspora. In J. Rutherford (Ed.), *Identity: Community, culture, difference* (pp. 222–237). London: Lawrence and Wishart Ltd.

Hall, S. (1996). Introduction: Who needs 'identity'. In S. Hall & P. Du Gay (Eds.), *Questions of cultural identity* (pp. 1–17). London: Sage.

Hartnell, A. (2009). Katrina tourism and a tale of two cities: Visualizing race and class in New Orleans. *American Quarterly, 61*(3), pp. 723–747.

'Hurricane Katrina Statistics Fast Facts'. 2020. *CNN.* [Online]. Updated 12 August. [Accessed 2 July 2021]. Available from: https://www.cnn.com/2013/08/23/us/hurricane-katrina-statistics-fast-facts/index.html

James, A. & Prout, A. (Eds.). (1997). *Constructing and reconstructing childhood: Contemporary issues in the sociological study of childhood.* London: Farmer Press, Taylor and Francis.

Jamieson, A. & Felton, E. (2021). People in Tulsa are marking 100 years since the massacre that killed scores of Black residents. Some survivors are still alive. *Buzzfeed News.* [Online]. 31 May. [Accessed 10 June 2021]. Available from: https://www.buzzfeednews.com/article/amberjamieson/tulsa-massacre-1921-100-year-anniversary

Johnson-Hanks, J. (2002). On the limits of life stages in ethnography: Toward a theory of vital conjunctures. *American Anthropologist, 104*(3), 865–880.

Kellett, M. (2011). Empowering children and young people as researchers: Overcoming barriers and building capacity. *Child Indicators Research, 4*(2), 205–219.

Kerr, M.M., Fried, S, Price, R.H., Cornick, C., & Dugan, S. (2017). Rural children's responses to the Flight 93 crash on September 11, 2001. *Journal of Rural Mental Health, 41*(3), 176–188.

Kerr, M.M. & Price, R.H. (2018). 'I know the plane crashed': Children's perspectives in dark tourism. In P.R. Stone, R. Hartmann, T. Seaton, R. Sharpley, & L. White (Eds.), *The Palgrave handbook of dark tourism studies.* London: Palgrave Macmillan, pp. 553–584.

Kerr, M.M., Stone, P.R., & Price, R.H. (2021). Young tourists' experiences at dark tourism sites: Towards a conceptual framework. *Tourist Studies, 21*(2), 198–218.

Kerstetter, D. & Bricker, K. (2009). Exploring Fijian's sense of place after exposure to tourism development. *Journal of Sustainable Tourism, 17*(6), 691–708.

Khoo-Lattimore, C. (2015). Kids on board: Methodological challenges, concerns and clarifications when including young children's voices in tourism research. *Current Issues in Tourism*, *18*(9), 845–858.

Kousky, C. (2016). Impacts of natural disasters on children. *The Future of Children*, *26*(1), 73–92.

Lash, S. & Friedman, J. (1992). Introduction: Subjectivity and modernity's other. In S. Lash, J. Friedman, & N. Abercrombie (Eds.), *Modernity and identity*. Oxford: Blackwell, 1–29.

McLaughlin, K.A., Fairbank, J.A., Gruber, M.J., Jones, R.T., Lakoma, M.D., Pfefferbaum, B., Sampson, N.A., & Kessler, R.C. (2009). Serious emotional disturbance among youths exposed to Hurricane Katrina 2 years postdisaster. *Journal of the American Academy of Child & Adolescent Psychiatry*, *48*(11), pp. 1069–1078.

Mercier, J. (2019). Visiting the 9/11 Memorial & Museum: Tips for going to Ground Zero with kids. 06 September. *Mommy Poppins*. [Online]. [Accessed 31 July 2020]. Available from: https://mommypoppins.com/9-11-memorial-museum-kids-visiting-families-tips

Office of the Advocate for Children and Young People. (2020). Children and young people's experience of disaster. Available from: https://www.acyp.nsw.gov.au/disaster-report-2020

Olwig, K.F. (1997). Cultural sites: Sustaining a home in a deterritorialised world. In K.F. Olwig & K. Hastrup (Eds.), *Siting culture: The shifting anthropological object* (pp. 17–38). London: Routledge.

Parry, R.L. 2017. *Ghosts of the Tsunami: Death and life in Japan's disaster zone*. New York: MCD.

Peleg, N. (2019). *The child's right to development*. Cambridge: Cambridge University Press.

Peltz, J. (2016). Now adults, children of 9/11 attacks draw inspiration from tragedy. 4 New York. [Online]. 8 September. [Accessed 31 July 2020]. Available from: www.nbcnewyork.com/news/local/nyc-children-911-attacks-draw-15-years-tragedy-sept-11-victims/963323/

Pfefferbaum, B. (2020). Children's exposure to single incidents of terrorism: Perspectives over 25 years since the Oklahoma City bombing. *Current Psychiatry Reports*, *22*(8), pp.39.

Poole, P. (1994). Socialisation, enculturation and the development of personal identity. In T. Ingold (Ed.), *Companion encyclopaedia of anthropology* (pp. 831–860). London: Routledge.

Price, R.H. (2018). *Expectations and revelations: Children discuss conducting research during a multi-day school excursion*. Ph.D. thesis, University of Pittsburgh.

Price, R.H. & Kerr, M.M. (2018). Child's play at war memorials: Insights from a social media debate. *Journal of Heritage Tourism*, *13*(2), pp. 167–180.

Prout, A. & James, A. (1997). A new paradigm for the sociology of childhood? Provenance, promise and problems. In A. James & A. Prout (Eds.), *Constructing and reconstructing childhood: Contemporary issues in the sociological study of childhood* (pp. 7–32). London: Farmer Press, Taylor and Francis.

Rapport, N. & Dawson, A. (1998). Home and movement: A polemic. In N. Rapport & A. Dawson (Eds.), *Migrants of identity: Perceptions of home in a world of movement* (pp. 19–38). Oxford: Berg.

Schänzel, H. & Carr, N. (2015). Special issue on children, families and leisure – first of two issues. *Annals of Leisure Research*, *18*(2), 171–174.

Seraphin, H. & Green, S. (2019). The significance of the contribution of children to conceptualising the destination of the future. *International Journal of Tourism Cities*, 5(4), pp. 544–559.

Sibley, D. (1995). *Geographies of exclusion*. London: Routledge.

Stone, P.R. (2006). A dark tourism spectrum: Towards a typology of death and macabre related tourist sites, attractions and exhibitions. *Tourism: An Interdisciplinary International Journal*, 54(2), 145–160.

Sullivan, G. (2015). Katrina might as well have hit New Orleans a day ago if you're young, male and black: Tourism is thriving but kids and adults in poor neighborhoods aren't. *The Hechinger Report*. [Online]. 25 August. [Accessed 2 July 2021]. Available from: https://hechingerreport.org/author/glenn-sullivan/

Thomson, R. (2007). Belonging. In M.J. Kehily (Ed.), *Understanding youth: Perspectives, identities and practices* (pp. 147–179). London: Sage/The Open University.

United Nations General Assembly. (1989). Convention on the rights of the child. [Online]. 20 November. [Accessed 5 July 2021]. Available from: http://www.ohchr.org/Documents/ProfessionalInterest/crc.pdf

Wall, J. (2019). From childhood studies to childism: Reconstructing the scholarly and social imaginations. *Children's Geographies*, 1–14, DOI: 10.1080/14733285.2019.1668912.

Weeks, J. (1990). The value of difference. In J. Rutherford (Ed.), *Identity, community, culture, difference*. London: Lawrence and Wishart, 98–112.

Williams, D.R. (2002). Leisure identities, globalization, and the politics of place. *Journal of Leisure Research*, *34*, 351–367.

Woodhead, M. (2008). Childhood studies: Past, present and future. In M.J. Kehily (Ed.), *An introduction to childhood studies* (pp. 17–31). Maidenhead: Open University Press.

Yang, M.J.H., Chiao Ling Yang, E., & Khoo-Lattimore, C. (2020). Host-children of tourism destinations: Systematic quantitative literature review. *Tourism Recreation Research*, *45*(2), 231–246.

Yuval-Davis, N. (2006). Belonging and the politics of belonging. *Patterns of Prejudice*, *40*(3), 197–214.

Part V

Dark Tourism Research and Children

Methodological Approaches

16 Ethical Research with Children and Young People

Addressing Complexities in (Dark) Tourism

Rebecca H. Price

Research with children and young people at dark tourism destinations is a complex endeavor, as illustrated in other chapters in this book. In this chapter, however, the main focus is on *ethical* practices in such research. This includes issues unique to children and those unique to the dark tourism context. Situated within the adult–child power dynamic, ethical complexities in research with children include consent, reward and compensation, autonomy, and confidentiality. This chapter explores each of these issues and highlights how dark tourism contexts might add a unique variation. Each section concludes with recommendations for tourism researchers for maintaining ethical research relationships with their child counterparts, especially in a dark tourism context.

Here, the terms 'child' and 'young people' refer to children from elementary to high school age, or generally aged five to mid-teens. Clearly, such an age range will include developmental differences, and these are noted where important. Also, studies specific to adolescents are noted with the term 'youth'.

Adult–Child Power Dynamic

It is obvious that research involving children should prevent harm from coming to them. To that end, there are legal, institutional, and other guidelines established for this purpose – to protect the vulnerable research participant (see, for example, Thompson et al., 2020; Health and Human Services, 2016; Powell et al., 2013). Yet ethical research with children involves more than preventing actual harm. This complicated construction of adult-directed, child-focused research takes place in a setting in which adults are in control – which is to say, everyday life.

Power relationships in research cannot be examined without first acknowledging those between adults and children in society at large. Adults make decisions and children are often 'along for the ride'. This applies to tourism in many cases, as well (Kerr and Price, 2018). Every interaction is situated within the adult–child power dynamic. To maintain ethical relationships with child counterparts, adults can empower children to take more control of the research study. This co-research methodology is rooted in the perspective

DOI: 10.4324/9781003032199-21

that research can embrace children as social actors – meaning that adults take responsibility for, rather than away from, children (Christensen and Prout, 2002; Price, 2018). This perspective acknowledges that children are able to meaningfully contribute to the study in a variety of ways, including idea generation, data collection, analysis, interpretation, or dissemination. Specific methods for co-research with children are explored in Chapter 19.

From an ethical standpoint, if we adults take responsibility *for* children in a study, then we also have to take responsibility for the ways in which their experiences differ from those of adults. For example, to take responsibility for obtaining children's consent, ethically we must acknowledge that children's consent is different from adult consent. In a child's mind, an adult researcher holds much power, possibly including school success (Freeman and Mathison, 2009). The adult holds power even over the child's basic needs: this can include whether, when, and what the child eats or has a drink of water, whether and when the child is permitted to use the restroom, and other things taken for granted by adult research participants. To ethically research with children, adults should employ a mindset that uses their power to compensate for the inherent inequalities differentiating adult and child participant experiences.

Dark tourism research may increase the complexities within adult–child power dynamics. Dark destinations often involve evocative experiences, whether they be exhilarating, terrifying, sad, or some combination (Kerr et al., 2021). Yet, who better to analyze and interpret children's experiences at a dark tourism destination than children themselves? In prior research, our team studied youth experiences on school excursions to a 9/11 memorial, a Holocaust museum, and other destinations (see Burns, 2018; Croom et al., 2018; Price, 2018). With backgrounds in education and child mental health, our adult team members were well aware of the power dynamic that was weighted in our favor. We attempted to face it head-on by adopting co-research practices. We acknowledged that children are capable of expressing agency, forming opinions, and creating their own meanings (Price, 2018; see also Christensen and Prout, 2002; Darbyshire et al., 2005; Pinter, 2014). We viewed them as autonomous actors with the freedom to opt in or out of any part of the study at any time. We viewed their contributions as we would those of adult 'consultants'; we valued their opinions and gave them multiple means by which to share their voices.

It is worth mentioning that the practice of 'consulting' is valuable even with the youngest of children. Even the youngest children competently comment about their own experiences; adult researchers can (and should) use this valuable information when designing and interpreting museum exhibits and other tourism experiences (Dockett et al., 2011; Dockett and Perry, 2005).

We discarded some adult research constructs, exchanging interviews for less formal 'research conversations' (see Dockett and Perry, 2011; Pinter, 2014; Pinter and Zandian, 2015). With this less-structured approach, we encouraged child participants to interview each other. We suggested questions, but youth were encouraged to talk about whatever came to mind. These conversations took place in their natural social groupings on a tour bus, which removed the need to remove individuals from the group, thus minimizing part of the 'adult-effect' (Price, 2018). We also employed multiple methods, from which child participants could choose. While many youth joined the research conversations; others chose to share their thoughts and feelings in writing later on.

Other ways in which adult researchers have empowered children have included having children suggest the actual research questions, as in Chen et al.'s (2010) evaluation of a girls' afterschool program. Mayes and Groundwater-Smith (2010) enabled a committee of children to vote on research topics. Other ways to flip the script include encouraging children to interpret the data and to present the final reports, as we are doing in our current research.

Suggestions for Researchers

- Replace formal research constructs with less formal constructs, for example, replace 'interviews' with 'conversations'.
- Simplify data collection to enable young participants to manage it.
- Preview questions with a few children before finalizing the items.
- Have children suggest data collection measures, research topics or questions.
- Offer the opportunity for children to contribute after the study has concluded such as data analysis or dissemination.

Despite one's best efforts to be mindful of adult–child power dynamics, ethical complexities can (and do) emerge with consent, reward, respect, and confidentiality. These issues are further explored in the following sections.

Consent

The issue of children's consent will always be complex. To ethically research with children, adult researchers should take responsibility for children starting as, and remaining, willing participants. However, adult researchers should also acknowledge that children's consent occurs within the aforementioned adult-oriented power dynamic: children may consent because they fear the consequences of saying no to a grown-up.

As previously mentioned, research is often governed by rules and regulations meant to protect the vulnerable participant. Our primary goal as adults should be to protect children from harm, be it physical or emotional. One means by which we protect children is by obtaining consent. Guidelines usually recommend that consent of an adult guardian be obtained prior to a child's participation, whether the consent comes from a parent, classroom teacher, or other guardian. Yet at the same time, it is also ethically important to obtain the consent of a child, whether or not it is 'required'. Why? If we respect children as autonomous beings, then we respect children's right to choose whether or not to participate. Some might argue that children do not or cannot understand enough to grant consent (Einarsdóttir, 2007). One way to manage this is to educate child participants before their roles commence within a study, to help them understand the study goals and their potential roles within it. Additionally, it is the *adult's* responsibility that the child participants understand that they are able to opt out of any part of the study at any time, without consequences. To ensure this, adults should gain and regain children's consent throughout the process (Christensen and Prout, 2002; Einarsdóttir, 2007; Jones, 2004; Merewether and Fleet, 2013; Pinter, 2014; Woodhead and Faulkner, 2008).

During our school excursion research, children visit many potentially evocative dark tourism sites, including the Flight 93 National Memorial, Arlington National Cemetery, the United States Holocaust Museum, and the National 9/11 Pentagon Memorial, among others. Before the excursion, we provide training to students about the study's aims, their research roles, and consent. We also offer multiple opportunities to opt out of research activities, as well as multiple data collection activities (Price, 2018, pp. 8–9).

In one study, for example, my colleagues and I worked diligently to establish research roles and understanding of consent before the excursion began. In our pre-study classroom sessions, we provided information about the data collection activities in which the youth would be involved, and we repeatedly re-stated that they were welcome to opt out of any research activity at any time. We gained consent before the trip took place, and we gained consent before research activities such as the previously mentioned research conversations. Yet, we found ourselves repeating the opt-out invitation throughout the trip after checking in with students and hearing occasional comments such as 'I didn't really want to, but I just did it' and 'Um, I thought that we had to, so I did it'. As one might expect, offering the option to opt out led some children to withdraw from certain research activities.

Worldwide, issues of consent have been a recurring concern across the research literature (see for example, Porter and Abane, 2008). While we provided training before and reminders during the study, some of our young participants still seemed confused about whether or not they 'had to' participate in research activities. This echoes Einarsdóttir (2007), who argued that adults can never really know if children understand enough about research to realize what their consent entails. Although we followed Einarsdóttir's (2007) advice and reminded our young colleagues that they might opt out of any activity, consent and what it means remained unclear for some research participants. This confusion may arise more commonly in school-sponsored trips. Students may assume that research activities are mandatory like their class assignments. School staff, in turn, may want to be sure that the school is fully participating. For these reasons, we make a special effort to remind school staff chaperoning trips that it is fine for their students to opt in and out of research activities.

In a dark tourism research context, consent may hold even greater value. Ethically obtaining consent requires providing children with the option to opt out of research activities, empowering them to further control their own experiences within the context of a site of painful heritage. Simply, recognizing that children may consent 'also requires acknowledging their right to dissent and hence to opt out of the research' (Dockett and Perry, 2011, p. 233). While children may consent to participate in a study, they may not know in advance what kind of feelings an evocative destination might create for them. To ethically manage the issue of consent, researchers might employ the following suggestions.

Suggestions for Researchers

- Follow whatever oversight or guidelines govern your research.
- Even when guardians give their consent, repeatedly remind children that they may opt in or out of any activity.
- Thoroughly explain the research aims and research roles to children, at their comprehension level.
- Explain the concept of consent to children, for example, 'If you don't want to do something, you don't have to. No one will be angry if you don't.'
- Reinforce consent at every juncture, for example, 'Is it ok if we talk about this for a minute? Can I record you? If you don't feel like it, that's completely fine. I'll move on.'

As children consent and participate in a study, issues of reward and compensation become evident.

Reward and Compensation

Reward, or compensation, is an ethical issue often overlooked in research with children. Researchers often reward adults for their research participation, whether with a gift card or a small amount of money, a raffle entry, or something else of minimal value. Doing so recognizes the value of their participation. It may also acknowledge that researchers recognize the effort, or work, extended by participants. Ethical research with children in any setting requires acknowledging their physical, emotional, or time contributions. How might adult researchers compensate child participants? What complexities might a dark tourism destination add?

In the aforementioned study of a children's school excursion that encompassed several dark tourism destinations, we carefully planned training, consent activities, multiple data collection activities, and many opportunities for collaborative efforts. Yet, it was impossible to ensure that every one of our young collaborators would enjoy the experience. A few children voiced negative comments that alluded to the labor-intensive work of conducting research. For example, one youth commented, 'It got annoying sometimes when we had been walking around and all we wanted to do was go back on the bus, but we instead had to do research' (Price, 2018, p. 89). Another wrote, 'i [sic] would not do it for a job' (Price, 2018, p. 90). On the other hand, many young participants expressed positive opinions of the research experience.

As one wrote, 'I had so much fun and would love to Do It agin [sic]' (Price, 2018, p. 90). Research participation as work has been established in the realm of adult research. Comments from our young research collaborators indicate that they, too, experienced the labor of research. As one adolescent reflected after conducting a months-long research study with their classmates, 'we learned how to pace ourselves, and how to walk away to take a breath for a quick second before returning to work, which is a very important thing to do as a young researcher' (Mechanicsburg Exempted Village School District Ninth Grade Research Team, 2020).

Talk of work and labor inevitably leads the conversation to compensation. As Bradbury-Jones and Taylor (2015) asked, what is appropriate compensation for a child researcher? Such compensation is subject to ethical and legal standards imposed by institutions and governing authorities. Moreover, for an adult researcher to decide that a child has been fairly compensated hints at paternalism. Instead, adults might employ strategies that empower child researchers to decide whether compensation is proportionate to the work involved. This is where compensation intersects with consent, fully situated within the adult–child power dynamic. To have power, child participants should understand that if the experience does not meet their expectations, they may stop participating.

To clarify, in order for children to determine whether they are adequately compensated, they need to understand: a) what to expect from the research experience, and b) that if what they gain from the experience does not seem of equal value to the effort that they extend, that they may opt out at any time without repercussions. To empower children to decide whether they are being fairly compensated, adults might convey: (a) realistic descriptions of children's research roles; (b) realistic descriptions of the research experience; and (c) that children may stop participating at any time, without penalties. It remains in their power to quit the 'job'.

While there has been much conversation about children's consent in prior literature, researchers have yet to discuss how closely consent and compensation are related. If children truly understand that they may opt out of any activity at any time without penalty, then perhaps they can decide whether what they receive from the experience is worth the effort that they extend. In order to address the problem of compensating children, adults might instead empower them to make their own decisions about whether the compensation that they receive is fair.

Dark tourism contexts add their own complications to this experience. An experience that is emotionally fraught may be considered too dear for a young participant to fully engage. Yet an experience that is exhilarating may be its own reward. It is up to the child to determine the value of their experience, just as it is up to an adult. It is the adult's responsibility, however, to ensure that child participants have a continuous and recognizable freedom to exercise their power to evaluate and maintain, modify, or discard the research relationship. Even the youngest children can (and will) express whether or not they are happy with their current situation.

Suggestions for Researchers

- Pay attention to children's demeanor throughout the research activity.
- Check with child participants and offer opportunities for fun, for example, 'How are you feeling about…? Would you like…? Want to play with…?'
- Offer opportunities for children to have fun and to act like children.
- Before a child participates, provide a realistic description of the upcoming research experience and the role that the child will play, in language that the child understands.
- Reinforce throughout the study that a child can opt out at any time.
- Reinforce before and during each activity that a child may choose to stop participating without consequences.

As noted above, adults can potentially improve the research experience for children by offering options and opportunities for fun. This further reinforces the idea of respecting the children's autonomy in research, which is explored in the following section.

Respecting Children's Autonomy

As noted previously, ethical research with children at any destination requires that adults respect children's autonomy: that they have unique and individual thoughts, opinions, and capabilities. In practice, this may look different depending on the age of the child, but even very young children will be able to express whether or not they are having a good time. For researchers, autonomy means allowing children to have some control over their research experiences. For researchers at dark tourism destinations, embracing children's autonomy can allow them to manage their exposure to items and experiences that are potentially distressing.

In prior articles, colleagues and I have noted that dark tourism contexts in particular are not always child-friendly (Kerr and Price, 2016, 2018; Price and Kerr, 2018). Our prior work suggested that children's tourism experiences are affected by factors that differ from those of adult visitors, including a) incomplete understanding of death; b) lack of agency in travel and destination choice; c) youthful exploratory behavior; and d) emotional vulnerability (Kerr and Price, 2018). The ways in which context influences a child's visit to such a site are included in our conceptual model (Kerr et al., 2021). Savage (2009) pondered whether a Holocaust site's context might cause vicarious trauma in children. Similar theories have emerged from studies in psychiatry and related fields, which indicate that children may become distressed even when only indirectly exposed to human suffering (Burnham, 2005; Pfefferbaum et al., 2000; see also Kerr and Price, 2016, 2018).

This indirect exposure might include multimedia presentations, exhibits, interpretation, and other 'dark' experiences (Kerr and Price, 2018). Respecting children's autonomy helps adult researchers to ethically research with children in places like this. Examples include employing co-research methods as explored in Chapter 19. Respecting children's autonomy entails giving them some control over their research experiences.

For dark tourism researchers, it could be fruitful to empower children to mold research questions to meet their own needs, perhaps negating potential distress. Young researchers might ask questions and explore topics that feel both comfortable and important to them. As noted in Price (2018) young co-researchers were able to self-regulate their levels of involvement with potentially disturbing aspects of the site, without feeling that they had failed to complete a task. In this way, children contributed to the research while choosing how they might like to be involved, according to their own comfort levels with the site content.

In addition, when adults hand over some research tasks to children, they encourage children's positive engagement with the site and interpretation. Tourism researchers have continued to call for a greater use of actively engaging interactive and participative experiences (Campos et al., 2018; see also Azevedo, 2010; Buhalis, 2001; Eraqi, 2011; Mathisen, 2013; Morgan et al., 2009). For younger children, this active engagement may include physical activity.

When children visit tourist sites, they desire to be physically active (Khoo-Lattimore, 2015; Price and Kerr, 2018; Rhoden et al., 2016; Small, 2008). Both Roche and Quinn (2016) and Small (2008) noted that children's physical activity during a visit is a strong factor in shaping their memories. Positive physical experiences may lead to happy memories; the opposite may also be true (Small, 2008). Promoting children's physical activity by involving them in research may empower them to engage with the site comfortably and on their own terms, and to create happy memories about their experiences there.

Finally, choice is tied to consent. Inviting young tourists to opt out of exhibits or research activities allows them to further control their own experiences with the context of a site of painful heritage. In our experience, when children opt *in* to evocative exhibits, coupling their participation with research conversations provides opportunities to share their feelings with their classmates and with adults who value what they share.

Suggestions for Researchers

- Offer children more involvement, not less, in research tasks. Allow children to manage their levels of involvement during the course of the study.
- Allow children the freedom to engage with the site in ways that make sense to them, including play and exploration.
- Encourage children to choose research activities from a menu. Allow children to change their minds about their levels and type of involvement as the research activities progress.

A final consideration when wanting to ethically research with children in dark tourism destinations is confidentiality, as further explored in the following section.

Confidentiality, Anonymity, and Privacy

Consent and confidentiality are often emphasized in childhood studies research (Canosa and Graham, 2016; Pinter and Zandian, 2015). What has sometimes been called 'confidentiality' in prior literature has more to do here with anonymity and privacy, that is, the freedom of young participants to express their thoughts and opinions without fear of judgment, and without having their contributions traced back to themselves. Evocative dark tourism contexts can provide a unique challenge. When students have feelings they do not wish to share, they may feel vulnerable expressing those feelings. That is why offering private options for contributing might help (Hill, 2006). Such was the case in one study when adolescents were unwilling to talk in front of their peers about their reactions to a dark site. But later, when they could write about how they felt on anonymous comment cards, many more participated, sharing both positive and negative experiences. For others, talking it out with friends in informal research conversations seemed to provide an opportunity to express and unburden themselves of uncomfortable emotions (Price, 2018).

It is impossible to determine whether or how children will want to express their feelings regarding a dark tourism research situation. Child participants, themselves, may be unable to predict how they will feel, until the time at which they are asked to share. One way to counter this problem of conflicting preferences is to offer multiple data collection methods from which children might choose.

It remains important to increase the young tourist's comfort level within the research situation. We as adults need to clearly and repeatedly convey to children that their names and personal identifying information will not be linked to their comments in a way that can be traced back to them by school, family, or others. It is our job as adult researchers to make sure that this remains true. Like any other learning experience, it is ideal to design some form of education regarding privacy, anonymity, and confidentiality.

Chapter 18 explores in-classroom preparation before a research study commences. As noted in Chapter 18, it is important for researchers to spend time in classrooms when initiating a school partnership. In this type of research scenario, this time would be well-used in explaining the goals and purpose of the study, how the children can help researchers, and what exactly will happen to their contributions. Methods might include multiple learning opportunities including handouts, video, and crucially, role-play (see, for example, Filip, Singer, and Olubowale, n.d.-a; n.d.-b). Ideally, the opportunity to act out situations in which privacy and consent come into play allows them to learn what these terms will mean in the research context.

Suggestions for Researchers

- Before the study begins, educate young participants about confidentiality, privacy, and anonymity, and how each idea will explicitly apply to them in this research context.

- Identify pseudonyms and explain how they work.
- Explain how participant data will be anonymized and protected.
- Use multiple methods to teach these concepts, possibly including instructional videos.
- Repeatedly emphasize that participants' names and other identifying information will not be linked to their comments in a way that can be traced back to them by school, family, or peers. And – make sure that this happens!
- Offer a menu of data collection activities, including private methods: written, recorded, or drawn.

Conclusion

This chapter explores the complexities of ethical research with children in dark tourism settings, and it provides practical suggestions for researchers to address these issues. Acknowledging that adults 'hold all the cards' in a research situation, researchers can address issues of consent, compensation, autonomy, and confidentiality as they arise. Yet certain factors, like consent, remain problematic. By considering the experiences of prior researchers across the social sciences, and by empowering children to act within the adult-oriented world of the research study, adults can work together with child counterparts to reach an understanding of children's experiences at dark tourism destinations.

References

Azevedo, A. 2010. Designing unique and memorable experiences: Co-creation and the 'surprise' factor. *International Journal of Hospitality & Tourism Systems.* **3**(1), pp. 42–54.
Bradbury-Jones, C. and Taylor, J. 2015. Engaging with children as co-researchers: Challenges, counter-challenges and solutions. *International Journal of Social Research Methodology.* **18**(2), pp. 161–173.
Buhalis, D. 2001. The tourism phenomenon: The new tourist and consumer. In: Wahab, C. and Cooper, S. eds. *Tourism in the age of globalisation.* London: Routledge, pp. 69–96.
Burnham, J.J. 2005. Fears of children in the United States: An examination of the American Fear Survey Schedule with 20 new contemporary fear items. *Measurement and Evaluation in Counseling and Development.* **38**(2), pp. 78–91.
Burns, L.M. 2018. *'They're just fun to be with': Building a community of learners through overnight school trips.* Ph.D. thesis, University of Pittsburgh.
Campos, A.C., Mendes, J., Oom do Valle, P., and Scott, N. 2018. Co-creation of tourist experiences: A literature review. *Current Issues in Tourism.* **21**(4), pp. 369–400.
Canosa, A. and Graham, A. 2016. Ethical tourism research involving children. *Annals of Tourism Research.* **61**, pp. 213–267.
Chen, P., Weiss, F.L., Johnston Nicholson, H., and Girls Incorporated®. 2010. Girls study Girls Inc.: Engaging girls in evaluation through participatory action research. *American Journal of Community Psychology.* **46**, pp. 228–237.

Christensen, P. and Prout, A. 2002. Working with ethical symmetry in social research with children. *Childhood.* **9**(4), pp. 477–497.

Croom, A.R., Squitiero, C., and Kerr, M.M. 2018. Something so sad can be so beautiful: A qualitative study of adolescent experiences at a 9/11 memorial. *Visitor Studies.* **21**(2), pp. 157–174.

Darbyshire, P., MacDougall, C., and Schiller, W. 2005. Multiple methods in qualitative research with children: More insight or just more? *Qualitative Research.* **5**(4), pp. 417–436.

Dockett, S., Main, S., and Kelly, L. 2011. Consulting with young children: Experiences from a museum. *Visitor Studies.* **14**(1), pp. 13–33.

Dockett, S. and Perry, B. 2005. Researching with children: Insights from the Starting School Research Project. *Early Childhood Development and Care.* **175**, pp. 507–521.

Dockett, S. and Perry, B. 2011. Researching with young children: Seeking assent. *Child Indicators Research.* **4**, pp. 231–247.

Einarsdóttir, J. 2007. Research with children: Methodological and ethical challenges. *European Early Childhood Education Research Journal.* **15**(2), pp. 197–211.

Eraqi, M.I. 2011. Co-creation and the new marketing mix as an innovative approach for enhancing tourism industry competitiveness in Egypt. *International Journal of Services and Operations Management.* **8**(1), pp. 76–91.

Filip, A., Singer, C., and Olubowale, O. (n.d.-a). *Informed consent.* Unpublished.

Filip, A., Singer, C., and Olubowale, O. (n.d.-b). *Pseudonyms.* Unpublished.

Freeman, M. and Mathison, S. (2009). *Researching children's experiences.* New York: The Guilford Press.

Health and Human Services. (2016). *Special protections for children as research subjects.* [Online]. [Accessed 12 January 2021]. Available from: www.hhs.gov/ohrp/regulations-and-policy/guidance/special-protections-for-children/index.html

Hill, M. 2006. Children's voices on ways of having a voice: Children's and young people's perspectives on methods used in research and consultation. *Childhood.* **13**(1), pp. 69–89.

Jones, A. 2004. Involving children and young people as researchers. In Fraser, S., Lewis, V., Ding, S., Kellett, M., and Robinson, C. eds. *Doing research with children and young people.* London: Sage, pp. 113–130.

Kerr, M.M. and Price, R.H. 2016. Overlooked encounters: Young tourists' experiences at dark sites. *Journal of Heritage Tourism.* **11**(2), pp. 177–185.

Kerr, M.M. and Price, R.H. 2018. 'I know the plane crashed': Children's perspectives in dark tourism. In Stone, P.R., Hartmann, R., Seaton, T., Sharpley, R. and White, L. eds. *The Palgrave handbook of dark tourism studies.* London: Palgrave Macmillan, pp. 553–584.

Kerr, M.M., Stone, P.R., and Price, R.H. 2021. Young tourists' experiences at dark tourism sites: Toward a conceptual framework. *Tourist Studies.* **21**(2), pp. 198–218.

Khoo-Lattimore, C. 2015. Children on board: Methodological challenges, concerns and clarifications when including young children's voices in tourism research. *Current Issues in Tourism.* **18**(9), pp. 845–858.

Mathisen, L. 2013. Staging natural environments: A performance perspective. *Advances in Hospitality and Leisure.* **9**, pp. 163–183.

Mayes, E. and Groundwater-Smith, S. 2010. Year 9 as co-researchers: "Our gee'd-up school". *AARE Annual Conference, December 2010, Melbourne.* [Online]. [Accessed 27 July 2020]. Available from: www.aare.edu.au/data/publications/2010/1723bMayesGroundwaterSmith.pdf

Mechanicsburg Exempted Village School District 9th Grade Research Team. 2020. *The National 9/11 Pentagon Memorial: School trip visitor study.* Unpublished.

Merewether, J. and Fleet, A. 2013. Seeking children's perspectives: A respectful layered research approach. *Early Childhood Development and Care.* **184**(6), pp. 897–914.

Morgan, M., Elbe, J., and Curiel, J.E. 2009. Has the experience economy arrived? The views of destination managers in three visitor-dependent areas. *International Journal of Tourism Research.* **11**, pp. 201–206.

Pfefferbaum, B., Seale, T. W., McDonald, N. B., Brandt, Jr., E. N., Rainwater, S. M., Maynard, B. T., Meierhoffer, B., and Miller, P.D. 2000. Post-traumatic stress two years after the Oklahoma City Bombing in youths geographically distant from the explosion. *Psychiatry.* **62**(4), pp. 358–370.

Pinter, A. 2014. Child participant roles in applied linguistics research. *Applied Linguistics.* **35**(2), pp. 168–183.

Pinter, A. and Zandian, S. 2015. "I thought it would be tiny little one phrase that we said, in a huge big pile of papers:" Children's reflections on their involvement in participatory research. *Qualitative Research.* **15**(2), pp. 235–250.

Porter, G. and Abane, A. 2008. Increasing children's participation in African transport planning: Reflections on methodological issues in a child-centered research project. *Children's Geographies.* **6**(2), pp. 151–167.

Powell, M.A., Taylor, N., Fitzgerald, R., Graham, A., and Anderson, D. 2013. *Ethical research involving children.* [Online]. Florence, Italy: UNICEF Office of Research – Innocenti. [Accessed 12 January 2021]. Available from: www.unicef-irc.org/publications/706-ethical-research-involving-children.html

Price, R.H. 2018. *Expectations and revelations: Children discuss conducting research during a multi-day school excursion.* Ph.D. thesis, University of Pittsburgh.

Price, R.H. and Kerr, M.M. 2018. Child's play at war memorials: Insights from a social media debate. *Journal of Heritage Tourism.* **13**(2), pp. 167–180.

Rhoden, S., Hunter-Jones, P., and Miller, A. 2016. Tourism experiences through the eyes of a child. *Annals of Leisure Research.* **19**(4), pp. 424–443.

Roche, D. and Quinn, B. 2016. Heritage sites and schoolchildren: Insights from the Battle of the Boyne. *Journal of Heritage Tourism.* **12**(1), pp. 7–20.

Savage, K. 2009. *Monument wars: Washington, D.C., the National Mall, and the transformation of the memorial landscape.* Berkeley: University of California Press.

Small, J. 2008. The absence of childhood in tourism studies. *Annals of Tourism Research.* **35**(3), pp. 772–789.

Thompson, S., Cannon, M., and Wickenden, M. 2020. *Exploring critical issues in the ethical involvement of children with disabilities in evidence generation and use, Innocenti working papers no. 2020-04.* [Online]. Florence, Italy: UNICEF Office of Research – Innocenti. [Accessed 12 January 2021]. Available from: www.unicef-irc.org/publications/1110-exploring-critical-issues-in-the-ethical-involvement-of-children-with-disabilities.html

Woodhead, M. and Faulkner, D. 2008. Subjects, objects, or participants? Dilemmas in psychological research with children. In Christensen, P. and James, A. eds. *Research with children: Perspectives and practices.* 2nd edition. Abington: Routledge, pp. 10–39.

17 Research Methods for Studying Young Tourist Experiences

Mary Margaret Kerr, Rebecca H. Price, and Gopika Rajanikanth

Introduction

This chapter takes the reader on a journey through research methods that can be adapted to accelerate children's inclusion in tourism research generally and dark tourism research specifically. We draw on methods not only from museum studies and tourism (Veal, 2017) but also from childhood studies, psychology, and education. Our goal is to inspire the reader to explore these methods in more depth (see Chapters 16, 18, and 19 in this volume) and identify approaches suitable for their inquiries. In particular, Chapter 18 describes the logistics for studies with school groups, including recruitment, advance planning, ethical safeguards, scheduling, and researcher-student interactions for conducting observations, surveys, and interviews.

We begin our journey with methods requiring no interactions with young tourists, then move on to those methods increasingly dependent on such interactions. We first share methods used to analyze children's archived artwork and letters brought to dark tourism sites as memorial tributes. We then explain how to study youth authored entries in visitor books and comment cards, before stopping to consider social media comments about young tourists. Our journey continues with observations, surveys, interviews, and photo-elicitation. To illustrate the need for developmentally suitable protocols, we pause along the way to share age-related examples and considerations. As Poria and Timothy (2014) cautioned:

> Researchers should consider becoming familiar with developmental psychology and specifically social developmental theories. … As far as the research method is concerned, scholars should consider children's age and other demographic differences including their cultural background, not approaching them as a homogenous group. Specifically, children's developmental age is critical when designing the methodology or framing the research question.
>
> (p. 95)

Let us now start with an exploration of the most socially distant approaches to studying young tourist experiences: examining the items that they leave behind.

DOI: 10.4324/9781003032199-22

Young tourists' artifacts

To study children's experiences without interacting with children, a researcher might gain some benefit from examining objects left behind or letters children send. This kind of artifact study allows the researcher to hear the 'echo' of children's voices – what they said once upon a time, yet it lacks the influential presence of the children themselves and the elaboration that they can provide. Artifact analysis may benefit the very nascent stages of research, with later studies building upon what is discovered through this method. Additionally, artifact analysis can be used to inform interpretation for children at dark tourism destinations, as we found to be the case at a 9/11 memorial (Kerr and Price, 2018).

Our methods chapter (Kerr and Price, 2018) provides an introductory overview of artifact analysis methods, in addition to examples of how such research can influence interpretation. Studying children's artifacts can provide a window through which the researcher can spy on children's thoughts and feelings at a point in time. Artifacts can include written comments, artwork, videos, and photographs created while touring (Kerr and Price, 2018, p. 561). Additionally, we found (similar to Doss, 2010 and Sturken, 2007) that children as well as other visitors often bring memorial tribute objects with them, or create them on site. This introduction to children's artifact analysis includes illustrations of our own work at a dark tourism destination, including its influence on designing interpretation for children, examples which are worth reviewing for researchers attempting to analyze children's artifacts (Kerr and Price, 2018, pp. 562–574).

Letters from children also provide valuable insight into their feelings and experiences related to a dark tourism destination. Kerr et al. (2017a) provide a detailed description of the methods our multi-disciplinary research team employed to analyze letters sent by young adolescents, which are stored in the Flight 93 National Memorial (a 9/11 terrorism site). We developed an eight-step data analysis process that included examination of the artifacts and iterative coding which identified emergent themes. This method is described in detail below: (Kerr et al., 2017a, pp. 179–180).

First, two analysts read and photographed the letters, then transcribed the images. Next, the lead author verified the transcribed wording of each letter by reexamining photographs of the original letters. Then, these three individuals read each letter multiple times to become familiar with its contents. After that, two of the analysts independently used existing codes (established from the ongoing review of the collection archives) and codes for acute stress (since the letters were written within a few days of the event). To analyze the letters, we followed accepted coding procedures (Miles et al., 2013; Palinkas, 2006).

After applying the pre-established codes, the two analysts independently coded the letters, and then compared their coding, achieving 90% agreement. They then resolved any discrepancies and proceeded to identify new codes, capturing important emerging concepts in the letters. They compared their coding of 10% of the letters with two other analysts and resolved any discrepancies through discussion. After coding, these four team members identified themes, examining specific words, patterns of words used together, and meanings of the phrases. They then met to review the themes, which resulted in 100% agreement. Finally, the team verified our process, results, and conclusions with the fifth team member and with an interdisciplinary group of other researchers who attended a public poster presentation. These researchers consisted of faculty members and graduate student researchers in psychology and education who offered their comments during the presentation. Following the conference, two graduate students specializing in qualitative research also offered comments via sticky notes on the posters. This thorough account of our methods and the resulting findings provide a starting point for researchers looking to conduct similar analyses of children's missives at a dark tourism destination (Kerr et al., 2017a).

On-site visitor comments

Despite their presence in museums everywhere, "comment books are certainly under-used and under-analyzed" (Coffee, 2006, p. 166). Not surprisingly, adult entries dominate the small visitor comments literature, with only occasional mentions of children (Stone, 2012; Walter, 2004). However, recent studies (Divaker and Kerr, forthcoming; Kerr, Price, and Savine, 2017b) illustrate insights one can glean from this ubiquitous yet overlooked source. For example, consider this description from our work at a 9/11 memorial:

> We allowed children… to "talk with" us through their comment cards. We listened to their voices in order to understand their personal meaning-making of the tragic events commemorated. First, we needed to understand what mattered to children, so that we might design interpretation relevant to them. Second, we studied their word choices and ideas to discern what interpretive concepts and language would, for them, "be easy to understand and process."
>
> (Ham, 2013, p. 124; Kerr et al., 2017b)

Our approach to studying visitor comments adopts the guidance outlined by Coffee (2006) and Macdonald (2005). We first arrange access to the visitor comments, typically through the site's research or visitor services department. The next step requires sorting through boxes of cards, poring over visitor books, or reading files of saved comments. Next, we select only those entries authored by young tourists. Some comments may not reference the authors' age. To address this problem, we look for clues such as references to school or grade, handwriting style, syntax, spelling, and accompanying drawings. For a detailed description of this process, see Kerr et al. (2017b). We then transcribe and upload the comments into a computer-assisted qualitative software program for coding, recording researcher memos, calculating co-coder agreement, and discerning themes (Miles et al., 2013).

Developmental considerations warrant some attention. To encourage young children's participation, one must consider the placement and format of the comment book or cards, as well as the writing implement. If a visitor book rests on an adult-height stand, many children will be excluded from its use. A lower table and chair with cards and markers invites young children to write or draw.

What can we learn from visitor comments? Young tourists' comments not only reveal their emotional reactions, but also their preferred exhibits, evaluations, recommendations, interpretations of artifacts or events, and new realizations. Comments also shed light on family, school, and cultural influences (see Kerr et al., 2017a). Table 17.1 summarizes the uses, benefits, and cautions for visitor comments projects.

Online reviews

Online travel reviews, although authored by adults, nevertheless may include useful observations of young tourists. To find such observations, researchers should think broadly about variations of search terms and synonyms, and about how people talk (or write) in casual conversation and online comments. We found it useful to use terms such as *boy* or *girl*, *child* or *children*, *kid* or *kids*, *son* or *daughter*, *school*, *teenage(er)*, and so forth.

Importantly, online comments also reveal intense adult reactions to child tourists – reactions that, in turn, can influence a young person's visit. Consider this Trip Advisor post about a Halloween haunted tour of Eastern State Penitentiary:

> Literally had a single Dad and his 8 year old daughter behind us in 3/4 attractions and she just kept screaming "MONSTER BE GOOD MONSTER BE GOOD MONSTER BE GOOD" throughout the entire time. We wanted to shove a sock in her mouth and punt her down the stairs.
>
> -xXxMaddiexXx18, October 2014

We certainly found that adults had strong opinions about children's behavior at dark tourism destinations, when we analyzed online comments about an image of children playing on a memorial sculpture (Price and Kerr, 2017). We realized the value of words in betraying emotions and beliefs of the commenters; therefore, we used a stance analysis to find what strongly held beliefs commenters shared (see Kucher et al., 2015). Table 17.1 summarizes the uses, benefits, and cautions for online comments projects.

Observations

Observation studies include (a) informal observations reported anecdotally; (b) structured observations focusing on pre-designated behaviors and talk; and (c) participant observations, in which the researcher joins in the observed

Table 17.1 Overview of Research Methods

Method	Resulting Data	Benefits	Cautions	Illustrative Studies
Archived Tributes	• Children's own language revealed • Expressions of emotion • Expressions of understanding event	• May not require interaction with children and therefore may not require clearances or human subjects	• Unless the child is present, difficult to interpret • Helpful to involve an expert in children's handwriting and art • Archived tributes may not include ages or other identifying information	Kerr and Price, 2018; Kerr et al., 2021
Online Comments	• Adult observations of children's behavior, comments, and reactions to dark tourism sites • Adult perceptions of suitability of a site for young tourists	• May not require interaction with children and therefore may not require clearances or human subjects • Website scraping computer programs can gather these rapidly	• Authors of online comments are typically adults describing children, not children themselves. • The context for the visit may not be shared in the online comment (purpose, length of visit, traveling group)	Abaidoo and Takyiakwaa, 2019; Cui et al. 2020; Price and Kerr, 2017
Visitor Comments	• Children's own language revealed • Expressions of emotion • Expressions of understanding event commemorated • Reveals themes related to nationality and religion	• May not require interaction with children and therefore may not require clearances or human subjects protection review • Youth may be able to collect this data themselves from their peers	• Unless child is present, difficult to interpret • Helpful to involve expert in children's handwriting • Visitor books may not include ages or other identifying information • Comment cards may not be organized and therefore require more time to analyze • Labor intensive if not already transcribed	Coffee, 2006; Kerr, Price, and Savine, 2017b; Macdonald, 2005

(*Continued*)

Table 17.1 (Continued)

Method	Resulting Data	Benefits	Cautions	Illustrative Studies
Drawings	• Expressions of emotion • Expressions of understanding event commemorated • Reveals themes related to nationality and religion	• Can allow younger children or those with special needs to effectively communicate • Computer based qualitative analysis software now available to analyze visual data • Themes may appear in drawings that are not apparent in written text or oral communication • Young tourists may prefer drawing over writing or other methods	• Unless child is present, difficult to interpret • Helpful to involve expert in children's artwork	Hill, 2006; Israfilova and Khoo-Lattimore, 2018a; Kerr and Price, 2018
Surveys	• Demographic Data • Content Knowledge • Asking children to write down their viewpoints	• Easy to administer and produce a lot of data • Surveys may be given online rather than in person • Majority of children don't need help of researcher to participate	• Must be written and formatted for the age group • Should be used in conjunction with other methods, especially with young children • Dependent on whether students are in school the day survey is administered • Harder for individuals who have difficulty expressing their views	Bird, 2016; Croom et al., 2018; Kerr and Price, 2018; Larsen and Jenssen, 2004; Roche and Quinn, 2017

Observations	• Description of how children are reacting to dark site • Learn about tone, body language, interactions with others	• Researchers are experiencing the same events as the children • Beneficial when studying younger children and those with special needs • Making extensive list of what to observe establishes what different researchers should take note of	• Awareness of researchers' biases when recording observations • More planning and expenses in order to accompany the students • Researchers should be experienced in working with children	Burns, 2018; Patterson, 2007; Roche and Quinn, 2017; Sutcliffe and Kim, 2014; Veal, 2017; Waterton and Dittmer, 2014
Interviews	• Children's accounts and perceptions	• If comfortable with researcher and environment, children are more open when talking	• Difference of authority between adult researcher and child • Students may say what they think the adult researcher wants to hear instead of what they actually feel • Translations and analysis shouldn't alter true meaning of response • Researchers should be experienced in working with children	Clark, 2005; Kerr et al., 2021; Larsen and Jenssen, 2004; Mechanicsburg Exempted Village School District 9th Grade Research Team, 2020; Price, 2018; Sutcliffe and Kim, 2014; Wong and Piscitelli, 2019

activity (Veal, 2017). As shown in Table 17.1, researchers choose their methods and observation foci to align with their research interests.

Some dark tourism writers offer informal anecdotes about young tourists playing, socializing, sobbing, or leaving memorial tributes (Baldwin and Sharpley, 2009, Bowman and Pezzullo, 2010; Kerr, Shaffer, and Hartman, 2014; Stevens and Franck, 2015). Veal (2017) defends this practice of "just looking":

> We should not forget how important it is to use our eyes in research, even if the research project does not involve systematic observation data collection. Familiarity with a leisure activity or a leisure or tourism site helps in the design of a good research project and aids in interpreting data.

Building on this data, others rely primarily on more systematic approaches. For example, Patterson (2007) video-recorded and analyzed conversations between parents and young children at the *Mysterious Bog People* exhibit. Roche and Quinn (2017) studied school children on their trip to the Battle of the Boyne Visitor Centre, an Irish heritage site. While observing the children, the researchers took observational notes on children's talk, tone, body language, and what the children seemed to like or find puzzling. Sutcliffe and Kim (2014) watched children at a museum exhibit showing British migrants to Southern Australia in the 1840s. The exhibit featured death and illness, specifically mentioning an infant's death at sea. Observers recorded "the children's physical positions, peer group associations, body language, attentiveness, responses to questions from the Education Officer, and personal conversations" (Sutcliffe and Kim, 2014, p. 6). In her study of adolescent interactions at dark sites, Burns (2018) watched for youth teasing among themselves, assisting others, and telling peers to look at something. These studies offered rich qualitative descriptions, as compared to numerical counts of time spent at an exhibit or engaged in a designated behavior.

Participant-observers join in the dark tourism experience *with* young people, typically on a school visit or longer excursion. In these "mobile ethnography" projects, both anecdotal and formal observations occur. Such child-centered research presumes that children deserve respect as individuals with critical perspectives to share (Canosa et al., 2016; 2019; Kellett, 2005, 2010; Prout and James, 1997; Woodhead, 2008). Eliciting youth views through interviews, focus groups, and participant observation requires rapport with researchers (Burns, 2018; Price, 2018; Tong et al., 2020; Wong and Piscitelli, 2019). Introducing themselves and the project in advance makes young tourists more comfortable, as prior researchers have illustrated (Burns, 2018; Khoo-Lattimore, 2015; Price, 2018). Chapters 16 and 19 offer specific strategies for successful researcher–youth interactions.

Developmental considerations abound during observations. As McCabe (2005: 103) observes, "to develop a meaningful engagement within the sociology of tourism with tourists, we will have to recognize the cultural and interactional contexts in which we engage with our subjects." Lest they misinterpret what they see and hear, observers should have some understanding of the age group, including age-typical behavior, understanding of death concepts, family and school influences, and peer culture (see other sections of this volume). Table 17.1 summarizes the uses, benefits, and cautions for online comments projects.

Surveys

Although common in tourism studies, surveys rarely include those under 18 years, perhaps owing to ethical safeguards (Barker and Weller, 2003). However, surveys and questionnaires involving children can be found in museum studies, and these methods can be used in dark tourism studies (Croom et al., 2018; Kerr and Price, 2018). Researchers should attend to reading and comprehension levels, conceptual understanding of death (see Chapter 5), and cultural factors (Wu et al., 2019).

Using surveys as one of their two measures, Larsen and Jenssen (2004) successfully studied Swedish adolescents' motives for a school trip. The first survey was given four months prior to the students' trip and asked the students why they thought a school trip would be a good idea and what they would do if on a school trip. Four weeks after the trip, students completed 12 more questions. The design assessed "whether the observed reasons (given by the children) occur over the three points of data collection using different methods" (p. 47).

Bird (2016) surveyed Canadian teenagers visiting the WWI Canadian National Historic Site of Vimy Ridge near Arras, France. The pre- and post-trip surveys were given online rather than in a classroom setting. Open-ended questions yielded rich descriptions and insights, such as this one:

> Standing behind the grave of a fallen soldier was a very humbling experience. It was dreadful to think that this man died and they do not know who he is. It made me realize how lucky I am to live a life that is not endangered by war.
>
> (Bird, 2016, p. 55)

Roche and Quinn (2017) also surveyed schoolchildren visiting a battlefield, the Battle of the Boyne. Prior to the trip, they asked 10–12-year-olds about their "prior knowledge, experiences, and opinions" (p. 6) by giving them pictures of heritage sites and asking them to comment on them. After the visit, students took a quiz about the site and could choose to write or draw about the exhibit (Roche and Quinn, 2017).

To understand students' perspectives on a site or some aspect of their trip, we prefer open-ended questions (e.g., "Please share with us about your experience at ____"). We print each question on a half-page card and then clip it to a small clipboard with a pen. Interestingly, youth chose this format after rejecting a page-length survey, complaining that "That looks too much like schoolwork." The clipboards are portable, provide a writing surface while on the bus, and avoid concerns about data usage and privacy on devices connected to the internet. Croom et al. (2018) utilized this method and found that the participants completed their cards within 30 minutes of time on the bus. One obstacle we encountered is motion sickness. Some youth find it uncomfortable to read or write while in motion. They prefer to fill out their cards upon returning to the hotel, or when the motorcoach is parked.

Younger children may prefer drawing their responses (Wu et al., 2019). Clark (2005) has warned against using surveys as the only method when studying young children: "this method runs the risk of being tokenistic if it is used as the only way for young children to convey their views and experiences" (493). Alerby and Kostenius (2011) concluded that a child declining to answer survey items is itself a powerful response showing that survey answer choices are not applicable to them. Table 17.1 summarizes the uses, benefits, and cautions for surveys.

Interviews

Generally, researchers have found that children as young as three can recall their experiences (Docherty and Sandelowski, 1999; Fivush, 2011; Fivush and Haden, 2003; Kerr and Price, 2018). Therefore, interviewing children can be useful for exploring children's experiences at dark sites (Kerr and Price, 2018). Tourism researchers are among the scholars testing new interview protocols to address the developmental, ethical, and logistical considerations we review in this section. For example, children might attempt to reason what they think the adults asking the questions want them to answer, which means what they saw may not accurately represent what they are truly feeling (Clark, 2005; Garbarino et al., 1989 cited in Gollop, 2000). Researchers should also be aware of the difference in authority between an adult and a child, especially in an interview setting. Because of this, some studies have even looked at having children interview other children to minimize this authority imbalance (Clark, 2005; Clark and Moss, 2001; Mechanicsburg Exempted Village School District 9th Grade Research Team; 2020; Price, 2018).

Some have noted that young children will feel more comfortable if the interview is in a familiar place with adults they know (Brooker, 2001; Clark, 2005; Gollop, 2000). It should also be mentioned that certain cultures do not condone children being interviewed individually (Clark, 2005; Gallop, 2000). Making sure the child feels relaxed and protected should be a large priority for researchers and after the interview, there should be an "appropriate debriefing with praise and thanks" (Clark, 2005, p. 493). We now turn to examples of interviewing in the dark tourism literature.

Sutcliffe and Kim (2014) used interviews when they studied 8–10-year-olds. A researcher accompanied the children on their school trip to a museum. The same researcher also went to the students' school a week later to interview them about their visit. The interviewer talked to each student for about fifteen to twenty minutes and asked them different questions. The researchers had some interview guidelines prepared before the actual interview and showed the students images of various places or items from the museum, to aid the children with remembering their visit and direct the discussion. The students were also invited to make further comments on anything they wanted and ask the researchers questions (Kellett and Ding, 2004; Sutcliffe and Kim, 2014). The researchers recorded all the interviews and then transcribed them for their analysis.

In order to study teenagers who had traveled on a school trip, researchers Larsen and Jenssen (2004) used interviews as one of their methodologies. The interviews were recorded with "minidisk recording equipment" (p. 48) and conducted while the students were on their trip. The researchers asked the students questions such as why they wished to go on their trip, what activities they were participating in, what they did or did not enjoy about their trip so far, and what they thought they would recall about their trip after it ended (Larsen and Jenssen, 2004).

Price (2018) details one of our experiences with motorcoach interviews, or "bus conversations," conducted immediately after the students toured the Flight 93 National Memorial. Price's (2018, pp. 50–52) study focuses on the children as researchers and their impressions of the research experience. To that end, we initiated conversations about the experience on the bus afterward, while the bus was en route to Washington, D.C. Like Pinter and Zandian (2015), we invited small groups of two or three students to talk with us on the bus, if they chose. A microphone, attached to a voice recorder, was held by the researcher or students. In addition, students were invited to interview each other by passing the microphone back and forth. We offered a set of questions handwritten on an index card as possible topics, but we explained that they were free to talk about anything that they chose.

These bus conversations have proven so successful that we have used them with over 150 students. Recently, we relinquished the interviewer role to adolescents, who succeeded in gathering data by designing interviews themselves, interviewing their classmates, and analyzing their findings (Kerr et al., 2021; Mechanicsburg Exempted Village School District 9th Grade Research Team, 2020). Table 17.1 summarizes the uses, benefits, and cautions for interviews.

Visual methods: Photo elicitation and drawings

Photo elicitation refers to asking children to take photographs of their surroundings and environment, which can allow researchers to understand what a child might want to discuss or finds most intriguing (Briggs et al., 2014; Cappello, 2005; Dockett et al., 2017; Rhoden et al., 2016). Asking young school students to take and share photographs may require storage on password-protected devices disconnected from the Internet. School concerns about students posting photos on social media drive some of these restrictions, while confidentiality may be the Research Ethics Committee's goal.

To address this problem, researchers can use set up privacy and parental controls or rely on district-issued devices. In an early study, we also tested an app (SonicPics) that would allow youth to type or dictate why they took the picture, without Internet access. Passersby who saw them recording would ask what they were doing. They would explain that they were researchers and why. Students would even come up to their chaperones and say, "I took this picture. Isn't this cool?" "Look here on this wall. Did you see this?" Some students found the iPads cumbersome, however, and suggested smaller devices instead (Price, 2018).

Younger children may want to draw about their visits (Khoo-Lattimore, 2015; Wu et al., 2019). Drawing may be useful in studying the experiences of children of a variety of ages. As a final data collection method in a multi-method study, Israfilova and Khoo-Lattimore (2018a) used drawings as an elicitation technique to help high school students solidify and clarify the feelings that they had expressed in earlier focus group interviews.

Pictures may be useful even when the child does not draw them. Hunter-Jones et al. (2020) detailed their use of the Trajectory Touchpoint Technique, an instrument designed to help understand patient and family experiences in palliative and end-of-life care. This instrument consists of

> clip-art pictures, cartoons, and easy-to-recognize signs and symbols which act as an aide to narrating memoirs…. designed to capture the different tangible and intangible aspects that service users commonly come into contact with during palliative and end-of-life care.
>
> (p. 5)

Participants were asked to tell their stories, choosing images which reflected their stories.

Whether taking photos, drawing pictures, or using others' pictures as touchpoints for their own storytelling, children of a variety of ages can participate in visual methods. As we found, children who may not participate as thoroughly in other methods may share their thoughts with visual methods (Price, 2018; see also Chapter 16).

Table 17.1 summarizes the uses, benefits, and cautions for visual methods.

Combining methods

Poria and Timothy (2014) called for using multiple methods when researching children's experiences (e.g., Larsen and Jenssen, 2004; Roche and Quinn, 2017). Ethically, as noted in Chapter 16, providing children with multiple means by which to share their voices allows more opportunities for them to share their thoughts and feelings. Practically, researchers can benefit from multiple data sources, whether they confirm and "solidify" each other, as Israfilova and Khoo-Lattimore (2018b) noted, or whether offering different methods elicits responses from children who would not feel comfortable participating in one of the methods offered (Price, 2018). Additionally, using multiple measures may allow researchers to achieve responses from children at different developmental stages and ages (Kerr et al., 2021).

Conclusion

This chapter provided various research methods that researchers can use to accelerate research with young tourists. These methods have succeeded in disciplines such as museum studies, childhood studies, psychology, and education and are transferable to the study of children in dark tourism. The variety of methods available suggests that even the most constrained research situation can benefit from children's input. Still, researchers should be aware of children's developmental stages when planning their studies.

The different research methods allow for varying levels of direct contact with children. Studying artifacts or on-site visitor comments left behind at sites will enable researchers to gain insight into young tourists' feelings and thoughts without directly interacting with them. In addition, although usually written by adults, online reviews may include observations of children, also allowing researchers to study young people without direct interactions with them.

Other research methods such as observations, surveys, interviews, and photo elicitations allow increased interactions with young tourists. Researchers can choose between different observational methods based on their research interests. But researchers should have a grasp of typical behaviors for their target age group because a lack of understanding could lead to misinterpretations of observed behavior (McCabe, 2005). When designing surveys, researchers should be mindful of children's reading and comprehension levels (Wu et al., 2019). Interviews also involve increased interaction with children, and researchers should be cognizant of the authority differences between adult researchers and children. To address this issue, some researchers have invited young people to interview one another (Clark, 2005; Clark and Moss, 2001; Mechanicsburg Exempted Village School District 9th Grade Research Team; 2020; Price, 2018). Photo elicitation allows researchers to see what children find most interesting (Briggs et al., 2014; Cappello, 2005; Einarsdóttir, 2005; Rhoden et al., 2016), but researchers need to take special care to address privacy concerns. Inviting children to take photos or draw can help younger children express their thoughts and feelings when speaking or writing might be difficult.

To best understand youth experiences at dark tourism sites, researchers should consider adopting multiple research methods. A range of methods awaits researchers, and combining these measures will provide additional insight into young tourists' experiences.

References

Abaidoo, S., and Takyiakwaa, D. (2019). Visitors' experiences and reactions to a dark heritage site: The case of the Cape Coast Castle (2010–2015). *Visitor Studies*, *22*(1), 104–125.

Alerby, E., and Kostenius, C. (2011). 'Dammed taxi cab' – How silent communication in questionnaires can be understood and used to give voice to children's experiences. *International Journal of Research & Method in Education*, *34*(2), 117–130. DOI: 10.1080/1743727X.2011.578821.

Baldwin, F., and Sharpley, R. (2009). Battlefield tourism: Bringing organised violence back to life. In R. Sharpley and P.R. Stone (Eds.), *The Darker Side of Travel*. Bristol: Channel View Publications, pp. 186–206.

Barker, J., and Weller, S. (2003). Geography of methodological issues in research with children. *Qualitative Research*, *3*(2), 207–227.

Bird, G.R. (2016). Landscape, soundscape and youth: Memorable moments at the 90th commemoration of the Battle of Vimy Ridge. In K. Reeves, G.R. Bird, L. James, B. Stichelbaut, and J. Bourgeois (Eds.), *Battlefield Events: Landscape, Commemoration and Heritage*. London and New York: Routledge, pp. 48–63.

Bowman, M.S., and Pezzullo, P.C. (2010). What's so "dark" about "dark tourism"? Death, tours, and performance. *Tourist Studies*, *9*(3), 187–202.

Briggs, L.P., Stedman, R.C., and Krasny, M.E. (2014). Photo-elicitation methods in studies of children's sense of place. *Children Youth and Environments*, *24*(3), 153–172.

Brooker, L. (2001). Interviewing children. In G.M. Naughton, S.A. Rolfe, and I. Siraj-Blatchford (Eds.), *Doing Early Childhood Research: International Perspectives on Theory and Practice* (pp. 163–177). Philadelphia: Open University Press.

Burns, L. (2018). *Overnight school trips: An overlooked phenomenon* (Unpublished doctoral dissertation). University of Pittsburgh, Pittsburgh, PA.

Canosa, A., Graham, A., and Wilson, E. (2019). Progressing a child-centred research agenda in tourism studies. *Tourism Analysis*, *24*(1), 95–100.

Canosa, A., Moyle, B.D., and Wray, M. (2016). Can anybody hear me? A critical analysis of young residents' voices in tourism studies. *Tourism Analysis*, *21*(2–3), 325–337.

Cappello, M. (2005). Photo interviews: Eliciting data through conversations with children. *Field Methods*, *17*(2), 170–182.

Clark, A. (2005). Listening to and involving young children: A review of research and practice. *Early Child Development and Care*, *175*(6), 489–505.

Clark, A., and Moss, P. (2001). *Listening to Young Children: The Mosaic Approach*. London: National Children's Bureau.

Coffee, K. (2006). Museums and the agency of ideology: Three recent examples. *Curator*, *49*(4), 435–448.

Croom, A.R., Squitiero, C., and Kerr, M.M. (2018). Something so sad can be so beautiful: A qualitative study of adolescent experiences at a 9/11 memorial. *Visitor Studies*, *21*(2), 157–174.

Cui, R., Cheng, M., Xin, S., Hua, C., and Yao, Y. (2020). International tourists' dark tourism experiences in China: The case of the memorial of the victims of the Nanjing Massacre. *Current Issues in Tourism*, *23*(12), 1493–1511.

Divaker, R., and Kerr, M.M. (forthcoming). Close Encounters with death and disease: Young visitors' perspectives at the Mütter Medical History Museum. In Trish Biers and Mary Kate Clary (Eds.), *Routledge Handbook of Museums, Heritage, and Death*. Abington, UK: Routledge.

Docherty, S., and Sandelowski, M. (1999). Focus on qualitative methods: Interviewing children. *Research in Nursing & Health*, *22*(2), 177–185.

Dockett, S., Einarsdottir, J., and Perry, B. (2017). Photo elicitation: Reflecting on multiple sites of meaning. *International Journal of Early Years Education*, *25*(3), 225–240.

Doss, E. (2010). *Memorial Mania: Public Feeling in America*. Chicago: University of Chicago Press.

Einarsdóttir, J. (2005). Playschool in pictures: Children's photographs as a research method. *Early Child Development and Care*, *175*(6), 523–541.

Fivush, R. (2011). The development of autobiographical memory. *Annual Review of Psychology*, *62*, 559–582.

Fivush, R., and Haden, C.A. eds. (2003). *Autobiographical Memory and the Construction of a Narrative Self: Developmental and Cultural Perspectives*. Mahwah, NJ: Psychology Press.

Garbarino, J., Stott, F.M., and Erikson Institute. (1989). *What Children Can Tell Us: Eliciting, Interpreting, and Evaluating Information from Children*. San Francisco: Jossey-Bass.

Gollop, M.M. (2000). Interviewing children: A research perspective. In A.B. Smith, N.J. Taylor and M.M. Gollop (Eds.), *Children's Voices: Research, Policy and Practice*. Auckland: Longman, pp. 18–36.

Ham, S. (2013). *Interpretation: Making a Difference on Purpose*. Golden, CO: Fulcrum.

Hill, M. (2006). Children's voices on ways of having a voice: Children's and young people's perspectives on methods used in research and consultation. *Childhood*, *13*(1), 69–89.

Hunter-Jones, P., Sudbury-Riley, A., Al-Abdin, A., Menzies, L., and Neary, K. (2020). When a child is sick: The role of social tourism in palliative and end-of-life care. *Annals of Tourism Research*, 83. https://doi.org/10.1016/j.annals.2020.102900

Israfilova, F., and Khoo-Lattimore, C. (2018a). Sad and violent but I enjoyed it: Children's engagement with dark tourism as an educational tool. *Tourism and Hospitality Research*. DOI:10.1177/1467358418782736.

Israfilova, F., and Khoo-Lattimore, C. (2018b). Azerbaijan youth culture and its influence on their dark tourism experiences. In C. Khoo-Lattimore and E. Yang (Eds.), *Asian Youth Travellers: Perspectives on Asian Tourism*. Singapore: Springer, pp. 61–78.

Kellett, M. (2005). Children as active researchers: A new research paradigm for the 21st century? Available at: http://oro.open.ac.uk/7539/ (accessed 04 February 2021).

Kellett, M. (2010). Small shoes, big steps! Empowering children as active researchers. *American Journal of Community Psychology*, *46*(1): 195–203.

Kellett, M., and Ding, S. (2004). 'Middle childhood', in S. Fraser, V. Lewis, S. Ding, M. Kellet, and C. Robinson (Eds.), *Doing Research with Children and Young People*. London: The Open University, pp. 161–174.

Kerr, M.M., Fried, S.E., Price, R.H., Cornick, C., and Dugan, S.E. (2017a). Rural children's responses to the Flight 93 crash on September 11, 2001. *Journal of Rural Mental Health*, *41*(3), 176–188.

Kerr, M.M., and Price, R.H. (2018). "I Know the Plane Crashed": Children's perspectives in dark tourism. In P. Stone, R. Hartmann, T. Seaton, R. Sharpley, and L. White (Eds.), *Palgrave Handbook of Dark Tourism Studies* (pp. 553–583) London: Palgrave MacMillan.

Kerr, M.M., Price, R.H., Savine, C.D., Ifft, K., and McMullen, M.A. (2017b). Interpreting terrorism: Learning from children's visitor comments. *Journal of Interpretation Research*, *22*(1). https://lnt.org/wp-content/uploads/2018/10/JIR-v22n1.pdf#page=89

Kerr, M.M., Shaffer, A., and Hartman, M. (2014). Interpreting the Flight 93 crash for children: A collaborative evaluation project. *Legacy: The Magazine of the National Association for Interpretation*, July/August, 2014.

Kerr, M.M., Stone, P.R., and Price, R.H. (2021). Young tourists' experiences at dark tourism sites: Toward a conceptual framework. *Tourist Studies*. *21*(2), 198–218.

Khoo-Lattimore, C. (2015). Kids on board: Methodological challenges, concerns, and clarifications when including young children's voices in tourism research. *Current Issues in Tourism*, *18*(9), 845–858.

Kucher, K., Schamp-Bjerede, T., Kerren, A., Paradis, C., and Sahlgren, M. (2015). Visual analysis ofonline social media to open up the investigation of stance phenomena. *Information Visualization*, *15*(2), 1–24.

Larsen, S., and Jenssen, D. (2004). The school trip: Travelling with, not to or from. *Scandinavian Journal of Hospitality and Tourism*, *4*(1), 43–57.

McCabe, S. (2005). "Who is a tourist?" A critical review. *Tourist Studies*, *5*(1), 85–106.

Macdonald, S. (2005). Accessing audiences: Visiting visitor books. *Museum and Society*, *3*(3), 119–136.

Mechanicsburg Exempted Village School District 9th Grade Research Team (2020). The National 9/11 Pentagon Memorial: School Trip Visitor Study.

Miles, M.B., Huberman, M.A., and Saldaña, J. (2013). *Qualitative Data Analysis: A Methods Sourcebook* (3rd ed.). Thousand Oaks: Sage.

Palinkas, L. (2006). Qualitative approaches to studying the effects of disasters. In F. Norris, S. Galea, M. Friedman, and P. Watson (Eds.), *Methods for Disaster Mental Health Research*. New York: Guilford Press, pp. 158–173.

Patterson, A.R. (2007). "Dad look, she's sleeping": Parent–child conversations about human remains. *Visitor Studies*, *10*(1), 55–72.

Pinter, A., and Zandian, S. (2015). "I thought it would be tiny little one phrase that we said, in a huge big pile of papers:" Children's reflections on their involvement in participatory research. *Qualitative Research*, *15*(2), 235–250.

Poria, Y., and Timothy, D.J. (2014). Where are the children in tourism research? *Annals of Tourism Research*, *47*, 93–95.

Price, R.H. (2018). *Expectations and revelations: Children discuss conducting research during a multi-day school excursion.* University of Pittsburgh. Unpublished dissertation.

Price, R.H., and Kerr, M.M. (2017). Child's play at war memorials: Insights from a social media debate. *Journal of Heritage Tourism*. DOI: 10.1080/1743873X.2016. 1277732.

Prout, A., and James, A. (1997). A new paradigm for the sociology of childhood? Provenance, promise and problems. In A. James and A. Prout (Eds.), *Constructing and Reconstructing Childhood: Contemporary Issues in the Sociological Study of Childhood*. London: Farmer Press, pp. 7–32.

Rhoden, S., Hunter-Jones, P., and Miller, A. (2016). Tourism experiences through the eyes of a child. *Annals of Leisure Research*, *19*(4), 424–443.

Roche, D., and Quinn, B. (2017). Heritage sites and schoolchildren: Insights from the Battle of the Boyne. *Journal of Heritage Tourism*, *12*(1), 7–20.

Stevens, Q., and Franck, K.A. (2015). *Memorials as Spaces of Engagement: Design, Use and Meaning*. New York: Routledge.

Stone, P.R. (2012). Dark tourism and significant other death: Towards a model of mortality mediation. *Annals of Tourism Research*, *39*(3): 1565–1587.

Sturken, M. (2007). *Tourists of History: Memory, Kitsch, and Consumerism from Oklahoma City to Ground Zero*. Durham: Duke University Press.

Sutcliffe, K., and Kim, S. (2014). Understanding children's engagement with interpretation at a cultural heritage museum. *Journal of Heritage Tourism*, *9*(4), 332–348.

Tong, Y., Wu, M.Y., Pearce, P.L., Zhai, J., and Shen, H. (2020). Children and structured holiday camping: Processes and perceived outcomes. *Tourism Management Perspectives*, *35*, 100706.

Veal, A.J. (2017). *Research Methods for Leisure and Tourism*. Essex: Pearson UK.

Walter, T. (2004). Body worlds: Clinical detachment and anatomical awe. *Sociology of Health & Illness*, *26*(4), 464–488.

Waterton, E., and Dittmer, J. (2014). The museum as assemblage: Bringing forth affect at the Australian War Memorial. *Museum Management and Curatorship*, *29*(2), 122–139.

Wong, K.M., and Piscitelli, B.A. (2019). Children's voices: What do young children say about museums in Hong Kong? *Museum Management and Curatorship*, *34*(4), 419–432.

Woodhead, M. (2008). Childhood studies: Past, present and future. In M. Kehily (Ed.), *An Introduction to Childhood Studies*. London: Open University Press, pp. 17–31.

Wu, M.Y., Wall, G., Zu, Y., and Ying, T. (2019). Chinese children's family tourism experiences. *Tourism Management Perspectives*, *29*, 166–175.

18 Research Collaborations with Schools

Mary Margaret Kerr, Cecilia Greene, and
R. Scott Marsh

Introduction

Research involving groups of children typically requires collaboration with child-serving organizations, primarily schools. These partnerships may be necessary to conduct empirical studies, evaluate educational activities developed by the site, or field-test interpretive resources such as children's booklets, audio tours, or activities (Kerr, Dugan, and Frese, 2016; Kerr, Shaffer, Hartman, 2014; Shaffer and Kerr, 2015). However, tourism researchers and professionals may not have ongoing relationships with such organizations. Moreover, researchers may encounter unfamiliar challenges: safeguards for children's participation include securing child protection clearances and research approvals; engaging parents and guardians; promoting children's safety and well-being while traveling; and becoming an unobtrusive participant-observer during children's visits. Logistics include funding, itinerary planning, research equipment management (and instruction, when children collect their own data), and non-disruptive data collection times and spaces.

Not surprisingly, many tourism researchers facing these requirements prefer to limit their studies to adults (Kerr, Stone, and Price, 2021). To encourage more inclusion of children, this chapter outlines proven and practical strategies we have adopted to identify and maintain beneficial partnerships with schools.

Types of school partnerships

School partnerships vary according to the task at hand and the commitment sought. We organize these partnerships here according to the level of commitment required. A minimal commitment partnership is appropriate if one merely wants to sample a group of children (and perhaps their teachers or parents) for a limited period. Examples of this would be a survey or a periodic on-site evaluation of a few hours' duration. These partnerships typically occur through existing networks of local schools. Funding for these partnerships is usually the researcher's responsibility unless the destination site has funding for its evaluation.

DOI: 10.4324/9781003032199-23

Examples of this kind of partnership appear in the literature. Kerr and her colleagues (2021) engaged young students in a one-day field trip to evaluate interpretive materials at a 9/11 Memorial site. Interpreters sought youth perspectives as they unveiled a new memorial and accompanying booklet. This partnership involved 60 students who traveled to the memorial, used the booklet, and gave written feedback on the activities. Their field-test feedback improved the activities.

A multiphase, single-study research project requires a longer partnership such as that described by Roche and Quinn (2017). They discussed the importance of researchers interacting *across time* with students visiting a specific destination site. Similarly, Israfilova and Khoo-Lattimore (2019) spent considerable time at a school to conduct their research with Azerbaijani youth. In their study of a British sailing museum, Sutcliffe and Kim (2014) observed children at the museum before interviewing them back at their school. Lastly, Larsen and Jenssen (2004) studied teenagers attending a school trip and visited the school before and after the trip to give surveys. They also participated in the school trip along with the students. Of note is that these studies involved multiple types of data collection that required time beyond the visit itself. Evaluating a school curriculum (e.g., lessons created by the destination site's education staff) motivates other partnerships. For example, Kucan and Cho (2018) collaborated with a school and a museum to research new culturally relevant pedagogy related to the 1889 Johnstown, PA (US) flood disaster. These curriculum studies (typically found in education journals) contribute to our understanding of children's experiences at dark heritage sites.

Lastly, long-term partnerships may have as their goal the expansion of dark tourism research itself, particularly *with* youth. Our work exemplifies this model (Kerr et al., 2021). However, to develop any collaboration, one must first find schools willing to participate. Therefore, we begin with recruiting school staff and students.

Identifying and engaging school collaborators

Finding school partners can be daunting. We suggest a search of visitor records to identify schools that have visited the site in the past. As a follow-up, school websites often describe the school and its demographic composition—information useful for sampling decisions. Alternatively, university education departments rely on relationships with schools for their teacher training and can introduce researchers to school leaders.

In unexpected ways and places, destination managers and researchers may identify potential school collaborators. This was the case in our partnership. Poring over years of archived children's tributes left at a 9/11 Memorial, our university team came across multiple files attributed to the same school. The curator explained that this school had visited annually for more than a decade. Curious about why a school en route to Washington, D.C. would detour to this remote site year after year, we contacted the school and talked with the two educators who led the trips. That 2014 conversation began a partnership that continues to this day. Crucial to our success is joint planning and openness to one another's perspectives. In the spirit of that collaboration, we co-authored this chapter to share working guidelines for school–research partnerships. We include examples from our own notes to illustrate each guideline.

Guideline 1: Trust and credibility are essential for research involving children

Children participate in research only when educators and parents trust the researchers. Building this trust begins with a known contact in the district (parent or employee) but must quickly expand to those with authority to authorize research activities and travel. In some locales, a citizens' board or government official oversees the school district and its employees. When researchers present the research goals and activities to such a governing body, they can then work out any concerns. Such presentations also allow researchers to consider and highlight the benefits of the research, not only for the tourism arena but also for children and educators. For example, we often highlight the benefits to students of travel, learning history, and understanding something about research methods. Once, while we were conducting a workshop with adolescent co-researchers, a parent stopped by the classroom. Realizing that the students were learning how to code visitor comments, he gave an impromptu talk about the value of coding skills for his company and the job opportunities available for those with this knowledge. He subsequently sought company funding to support one of our trips.

Being physically present in the school builds credibility and a perception that the researchers are trustworthy (Wong and Piscitelli, 2019). Opportunities to build collaboration may include meetings with parents as well as discussions with teachers and trip chaperones. In these meetings, researchers must remember that they are guests of the school. We follow school rules, including where to park, how to sign in, and when and where to eat or drink. We take care to avoid disrupting schedules and to thank those who facilitate our work. Showing this respect to the host organization avoids conflicts and conveys respect for the school's primary mission of educating children.

Rapport- and trust-building activities include live and video classroom visits, informal conversations about the project, shared rituals such as eating (our team always shares cookies), and selecting a fun name for the project. They also include trying out research equipment together, letting students lead researchers on a tour of their school, or jointly visiting the university researchers' campus. We build into our itinerary opportunities for informal conversations. After the trip, we recalled these conversations:

> You ask your research questions first, and that establishes your role. Once that's done, the kids can just talk about their tour site experiences (not their private lives). You are building trust so the next time you talk, the kids feel more comfortable.

There has to be strong ownership on both sides. The researchers made the students feel their opinions mattered. In turn, the students wanted to do the project, wanted to know the next thing to do as student researchers, and they were excited. It was a dynamic relationship of collaboration.

Of course, these activities require professional boundaries, child protection clearances, and research approvals—the topics of our next section.

Guideline 2: Safeguards protect all parties

Researchers, parents, and educators share responsibility for safeguarding children and adolescents. Policies and protocols also protect the research team against claims of unprofessional conduct.

Researchers must identify and comply with child protection clearance requirements. Before approving a study, schools may require copies of these certifications as well as copies of the data collection measures. Two of us recalled this process:

> Any forms we got from the researchers we shared with the Superintendent, and we carry those on the trip. Another safeguard was in case a parent came back later saying, 'I didn't know my kid was doing that.' It was a form they'd sign after the researchers' presentation to parents, saying that the parent had attended the presentation and understood the way kids were participating and that they granted their permission. To reassure parents and ward off problems, the school secretary posts pictures of the trip each day on the school website. Parents can see in those pictures that the researchers are talking to kids, the chaperones are there, kids aren't alone, and it's never just a researcher isolated with a kid. Just posting those alleviates parent concerns.

As Chapter 16 documented, securing young research participants' consent requires their comprehension of the consent form. To ensure that children (and their parents) understand the study, we write family-friendly handouts, checking readability levels. Readability refers to the 'ease with which a reader can read and understand a given text' (Oakland and Lane, 2004, p. 244). Though not a precise measure of a document's readability, one can use the feature in Microsoft Word or use an online program as a starting point. Keep in mind that the reading level may need to be below the students' current grade level, as some will not be proficient readers.

Pseudonyms allow data to be de-identified, thereby reducing risk and facilitating Research Ethics Committee (REC) approval. We introduce the pseudonyms ('research nicknames') before the trip. Aliases that busy youth can easily recall work best. For example, students can each choose the name of a county or state as their pseudonym and then learn to spell it. We also incorporate pseudonyms into a few classroom activities to help students recall them later.

Professional boundaries also protect all participants in a study, especially when young adult research assistants are themselves not much older than adolescent students. We review with each team member the usual research requirements, such as confidentiality and data protection. In addition, we rely on written policies that all researchers must read and sign, including the protocols below:

- We are not chaperones. However, we are sensible adults concerned about others' welfare, so if you have a safety concern, contact a school staff member immediately.
- It is fine to offer general assistance (e.g., answer a child's request for information about where something is, what the schedule is, what time it is, how to use the microphones or record notes). You may help a child find his or her way back to his or her group.
- Unless it is an emergency (e.g., pulling a child away from an oncoming car), refrain from *any* physical contact with students other than shaking hands (if they initiate it).
- Do not share personal information or contact information about yourself or any research team member (e.g., Facebook, Instagram, email, Twitter). Do not communicate with students via social media, phone, or email, before, during, or after the trip.
- Do not take photos of students or agree to be in students' photos.
- Do not break the trip rules set for students. Even when we are not interacting with the students, we are modeling appropriate trip behavior.
- Remember that we are guests of the school. We are not here to *evaluate* others' behavior, school, curriculum, views, conversations, teaching, or the trip.
- Wear your University ID so that students can identify you.
- Do not use alcohol or tobacco while on duty.
- Do not have any drugs or weapons in your possession while on school grounds or during the trip.
- Do not drive students in your car.

In our next section, we introduce the reader to aspects of school culture that can affect collaboration. We highlight successful communication strategies.

Guideline 3: Communication requires an understanding of school culture

Schools are dynamic, busy environments, where hundreds of conversations fill classrooms and corridors every day. Before and after pupils arrive, school staff engage in curriculum planning, parent conferences, continuing education, and co-curricular clubs and sports. During the school day, teachers go from class to class, with few quiet breaks. Voicemails and emails may go unanswered for days. These are not ideal circumstances for communication with outsiders, such as researchers.

Yet, researchers need information and approvals from their school part-ners to meet grant and institutional deadlines, plan with destination manag-ers, and refine their data collection methods. To facilitate communications with educators, we suggest these strategies:

- Be available to talk. Sometimes the nature of the discussion dictates a conversation, not an email or text exchange.
- Have a regular, recurring meeting time via a videoconference or phone call.

These communication strategies help researchers plan, troubleshoot, and evaluate projects with their school collaborators. In the next section, we out-line additional advance planning activities that worked well for us.

Guideline 4: Advance logistical planning identifies both problems and opportunities

Whenever possible, researchers and educators should preview all research activities *in the actual setting*. Doing so allows the team to problem-solve logistical problems such as when, where, and how to collect data without disrupting the naturalistic conditions or compromising the safety of par-ticipants (or research equipment). For example, coordination with tour operators prevents tour guides from talking while students are completing interviews or surveys. Researchers observing students need to know the students' touring plans at a site, to plan where to station themselves unob-trusively. Interviewing requires quiet spaces, which may be hard to locate in crowded venues or busy schools. For example, in our first study, we ini-tially planned to conduct our post-visit student interviews in one of the museum's empty classrooms. We were dismayed when we found out that it was not permitted. With no other options left, we resigned ourselves to conducting the interviews on the bus, while in loud city traffic. To help, our university technology staff found lightweight mics and noise filtering audio recorders. In the end, this plan worked so well that we adopted it on every trip.

Researchers planning surveys, individual and group interviews, or observa-tions with students at school or while traveling may benefit from lessons we and others have learned. We begin with surveys and interviews.

Conducting surveys and interviews

The dark tourism literature mentions interviews with young tourists at sites or at school (see Wong and Piscitelli, 2019; Sutcliffe and Kim, 2014; Larsen and Jenssen, 2004). These research interactions require decisions about times and spaces in the school day or trip that are least disruptive to the school's goals. Further caution applies to individual interviews, as Barker and Weller (2003) warn:

> Finding a quiet and confidential space within a school to conduct interviews is problematic… Researchers must protect both children and themselves by adopting 'cautionary practice' which ensures they are not the sole adult in a closed room with children (Cameron et al., 1999). Thus, an interview may be conducted within view of other members of staff or children.
>
> (p. 214)

Here are some suggestions for researchers at schools:

- Have all clearances, permissions, and communications with the school administration worked out well before the day of the visit.
- Before the visit, coordinate how researchers will communicate with the students involved in the study (i.e., jointly identify with each teacher when to visit their classroom to talk with students).
- Ask about the school calendar to ensure there are no extracurricular activities or other conflicts (e.g., assemblies, pep rallies, school pictures) that will interfere with the project.
- Students and staff need to be informed about the visit in advance, so that they know who is coming and what researchers expect of them.
- Know the logistics of where to park, enter the building, and sign in so that you don't encounter an unexpected delay.
- Class schedules and trip itineraries may not be flexible, so adhere to your agreed-upon schedule.
- Have a detailed process of how you want to proceed with the survey or interview. Know each step of your plan and how you want to enact it.
- Empower the students during the visit by emphasizing that they are colleagues and partners, not merely research subjects.
- Before the visit ends, let the students know what will happen next (e.g., what will happen with the interview notes or surveys, whether there will be follow-up activities).
- Before departing the school, thank the staff and students.

On a trip, these may be opportunities to schedule short surveys or interviews:

- On the bus before tours begin, between sites, or each evening.
- During longer walks to and from sites.
- During a break or lunch stop, if school staff agree.
- During hotel 'hall meeting' times before bed or before breakfast, if you can reserve a suitable space.
- As the visit to a site is finishing, different students can be taken aside if questions are short and to the point.

Conducting observations

Structured observations of young tourists occasionally appear in dark tourism literature (Burns, 2018; Roche and Quinn, 2017; Sutcliffe and Kim, 2014). Classroom observations can reveal how teachers or researchers prepare students for an upcoming trip. Here are some general suggestions:

- Ask the teacher where to *sit*. Standing in a classroom is distracting.
- Ask the teacher how to introduce yourself.
- Maintain confidentiality: do not discuss your impressions of the class with others.
- Turn off your technology notifications.
- Ask the teacher beforehand what to do or say if a student initiates a conversation with you.
- Close your device or notebook after you finish taking notes and keep it with you.

Post-study steps

At the end of the study, researchers and school staff should meet to share their perceptions about how the research went. For example, shortly after a study, teams can discuss issues that arose unexpectedly and problem-solve for the next study.

Feedback from students who participate in the study is also crucial. Consider this teacher's experience:

> After the trip, I led a discussion in my classroom with the students. Our goal was to bring to light problems that might have escaped our notice during the trip. For example, the students wanted more interactions with the research team. They had not wanted to mention this during the trip. This conversation led to our selecting research team members more carefully to ensure that they are comfortable interacting with young adolescents. We also planned more videoconferences with every member of the research team so that the students could get to know the research team members better before the trip.
>
> (Marsh, 2020)

Adolescents may be able to facilitate their own discussions, as evidenced by these recommendations written by 14-year-olds following their study (see Chapter 19).

> It would be very important for teachers to ask students what could be improved on about their experience. This is important because some may not like their experience and then others will. For the people who weren't as interested, they should feel like they have a say about what they think.

Sites should survey the schools to see if people's experience was interesting enough to go back again, and why they chose yes or no. The survey could ask students what kind of emotions they were feeling during and after their experience.

(Mechanicsburg Exempted Village School District
Research Team, 2020)

Post-study opportunities to share findings with the school community may arise during these discussions as well (see Pinter and Zandian, 2014; 2015). Strategies for involving students in dissemination appear in Chapter 19.

Conclusion

Research with young people often involves collaborations with their schools. Such research may run the gamut from brief, unobtrusive observation to dynamic interpersonal interactions over extended periods. Researchers may need to watch classroom lessons, conduct surveys, record interviews, or even travel overnight with young people to visit dark heritage sites. In some cases, researchers may collaborate with schools so that they can teach students to conduct and analyze their own studies, as illustrated in the next chapter. All of these activities demand logistical planning and ethical considerations and safeguards.

Fundamental to these beneficial partnerships is the establishment of credibility and rapport to gain trust. Most importantly, successful partnerships help to ensure the safety, privacy, and well-being of young people. Clear guidelines for individual team member conduct, compliance with human protection research boards, and common-sense planning with school partners create professional boundaries and reassures parents and school officials.

Communications between researchers and school personnel can be challenging because school personnel spend most of their days busy with their students. However, in-person visits, video conferencing, joint planning trips, and thoughtful consideration of the school schedule have enabled us and others to maintain the close communications necessary to a successful research project.

Colleagues in other fields such as childhood studies have outlined helpful methods for working with children in research. The time has now come for tourism researchers to share how they identify, engage, and maintain partnerships with schools. Moreover, our school colleagues have much to teach us about improving our research practices when traveling with children. When researchers and educators share their expertise, young tourists benefit.

References

Barker, J., and Weller, S. (2003) Geography of methodological issues in research with children. *Qualitative Research*, *3*(2): 207–227.

Briggs, L.P., Stedman, R.C., and Krasny, M.E. (2014) Photo-elicitation methods in studies of children's sense of place. *Children, Youth, and Environments*, *24*(3): 153–172.

Burns, L. (2018) *Overnight school trips: An overlooked phenomenon*. PhD Thesis, University of Pittsburgh.

Cameron, C., Moss, P. and Owen, C. (1999) *Men in the Nursery: Gender and Caring Work*. London: Paul Chapman.

Cappello, M. (2005) Photo interviews: Eliciting data through conversations with children. *Field Methods, 17*(2): 170–182.

Einarsdóttir, J. (2007) Research with children: Methodological and ethical challenges. *Early Childhood Education Research Journal, 15*(2): 197–211.

Israfilova, F., and Khoo-Lattimore, C. (2019) Sad and violent but I enjoyed it: Children's engagement with dark tourism as an educational tool. *Tourism and Hospitality Research, 19*(4): 478–487.

Kerr, M.M., Dugan, S.E., and Frese, K.M. (2016, July/August) Using children's artifacts to avoid interpretive missteps. *Legacy: The Magazine of the National Association for Interpretation*.

Kerr, M.M., Shaffer, A., and Hartman, M. (2014) 'Interpreting the Flight 93 crash for children: A collaborative evaluation project'. *Legacy: The Magazine of the National Association for Interpretation*, July/August, 2014.

Kerr, M.M., Stone, P.R., and Price, R.H. (2021) Young tourists' experiences at dark tourism sites: Toward a conceptual framework. *Tourist Studies, 21*(2): 198–218.

Kucan, L., and Cho, B.Y. (2018) "Were There Any Black People in Johnstown?" An investigation of culturally relevant pedagogy in service of supporting disciplinary literacy learning in history. *Urban Education*, doi: 10.1177/0042085918804011.

Larsen, S., and Jenssen, D. (2004) The school trip: Travelling with, not to or from. *Scandinavian Journal of Hospitality and Tourism, 4*(1): 43–57.

Mechanicsburg Exempted Village School District 9th Grade Research Team (2020) *The National 9/11 Pentagon Memorial: School Trip Visitor Study*.

Oakland, T., and Lane, H.B. (2004) Language, reading, and readability formulas: Implications for developing and adapting tests. *International Journal of Testing, 4*(3): 239–252.

Pinter, A., and Zandian, S. (2014) "I don't ever want to leave this room:" Benefits of researching 'with' children. *ELT Journal, 68*(1): 64–74. doi: 10.1093/elt/cct057.

Pinter, A., and Zandian, S. (2015) "I thought it would be tiny little one phrase that we said, in a huge big pile of papers:" Children's reflections on their involvement in participatory research. *Qualitative Research, 15*(2): 235–250.

Roche, D., and Quinn, B. (2017) Heritage sites and schoolchildren: Insights from the Battle of the Boyne. *Journal of Heritage Tourism, 12*(1): 7–20.

Rhoden, S., Hunter-Jones, P., and Miller, A. (2016) Tourism experiences through the eyes of a child. *Annals of Leisure Research, 19*(4): 424–443.

Shaffer, A., and Kerr, M.M. (2015) "Can you tell my child what happened here?" --- Explaining the Story of United Flight 93. *Ranger*. [Magazine of the Association of National Park Rangers].

Sutcliffe, K., and Kim, S. (2014) Understanding children's engagement with interpretation at a cultural heritage museum. *Journal of Heritage Tourism, 9*(4): 332–348.

Tong, Y., Wu, M.Y., Pearce, P.L., Zhai, J., and Shen, H. (2020) Children and structured holiday camping: Processes and perceived outcomes. *Tourism Management Perspectives, 35*: 100706.

Wong, K.M., and Piscitelli, B.A. (2019) Children's voices: What do young children say about museums in Hong Kong? *Museum Management and Curatorship, 34*(4): 419–432.

19 Co-Research with Youth

A Conceptual Model and Case Study

Rebecca H. Price, Mary Margaret Kerr, and Gopika Rajanikanth

Introduction

Our multi-disciplinary research team seeks to understand young tourists' overlooked experiences at dark tourism destinations. When we began this research, we were aware that dark sites have the potential to be powerfully evocative, yet difficult for children to understand. We reasoned that 'co-research' might be especially beneficial for dark tourism research contexts, as it allows children some control over their own experiences.

As Freeman and Mathison (2009) observed, 'research partnerships with youth require specific strategies that level the playing field for adults and youth and create a shared research agenda' (p. 174). However, research methods for documenting these encounters rarely appear in the tourism literature (Kerr and Price, 2016, 2018; Kerr et al., 2017, 2021). One of the few examples available is not specific to dark tourism, yet provides a bit of insight into co-research with adolescents (Leonard, 2019). Notably, adolescents in Belfast took on adult roles by co-constructing a tour route and playing a significant role in the interpretation of visited sites.

This chapter opens with a framework originally introduced in Price (2018) where the systematic review, its methods, and literature are detailed. The second part of the chapter offers a case study that illustrates these co-research strategies with youth at a dark tourism destination. We conclude with implications for adopting co-research paradigms in dark tourism.

Framework for Co-research with Children

To begin, we looked to prior literature for the experiences of past adult–child research collaborations. The method of this systematic review is elaborated in Price (2018). Following each section of this discussion, implications for dark tourism and ethical considerations appear. The conceptual model presented here as Figure 19.1 is divided into areas of the research process: selecting a topic and aims, research design and data collection, data analysis and interpretation, and dissemination.

The following sections provide more detail about the methods in Figure 19.1.

DOI: 10.4324/9781003032199-24

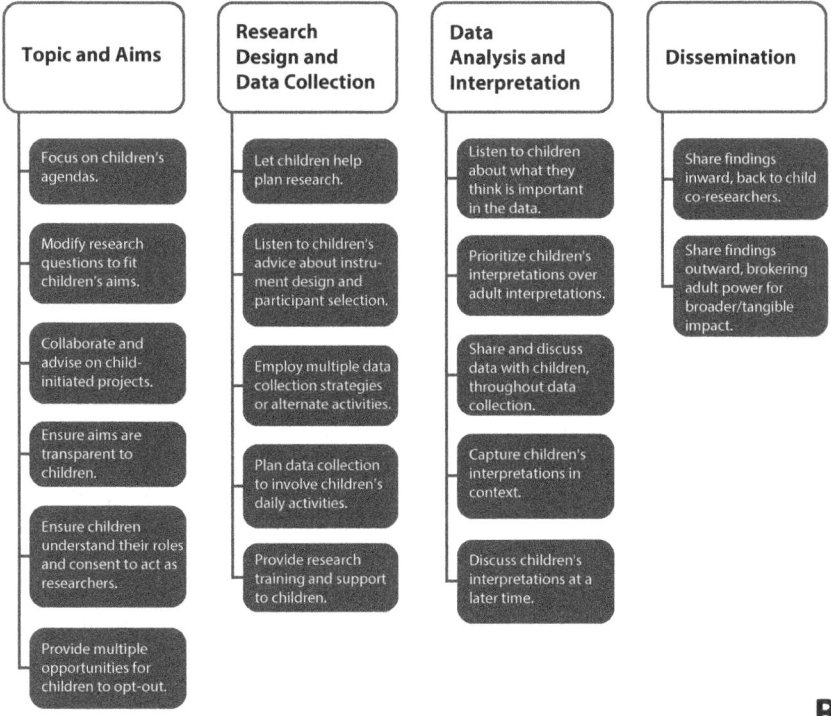

Figure 19.1 Methods of co-research with children and young people (adapted from Price, 2018).

Topic and Aims

Prior researchers have engaged children in this initial phase of research in a number of ways. These include considering children's agendas, children's research roles, and their consent.

Child's agenda

To research with children and embrace their agendas, some adult researchers have modified or adjusted their questions to meet the aims of children or have offered children the freedom to select the topic of research according to their interests (see Chen et al., 2010; Ergler, 2011; Johnson, 2008; Kellett, 2010; Kellett et al., 2004; Mayes and Groundwater-Smith, 2010; Pinter and Zandian, 2014, 2015).

Child's role

While one might label a child as a *co-researcher*, how well might the child understand what that means in context? The systematic review revealed that prior researchers have made considerable effort to ensure that aims remained transparent and that children understood their co-researcher roles (Chen et al., 2010; Coppock, 2011; Kellett, 2010; Kellett et al., 2004; Lundy et al., 2011; Mayes and Groundwater-Smith, 2010; Price and Kerr, 2017). Examples include adult researchers meeting with children (and sometimes parents) to discuss aims and research plans; pre-research training sessions; or the creation of a young researcher steering committee with members deeply involved in every phase of the research and its oversight.

Child's consent

These methods address children's consent to act as researchers, their awareness that they may opt-out, and whether adults involved regain children's consent throughout the research project. While we must presume that adult researchers obtained consent according to the ethical guidelines governing their research, studies included in this section provide a detailed explanation of the efforts made to obtain children's consent (Chen et al., 2010; Coppock, 2011; Ergler, 2011; Kellett, 2010; Kellett et al., 2004; Kirova and Emme, 2008; Mayes and Groundwater-Smith, 2010; Morrow, 2005; Pinter and Zandian, 2015; Smith, Monaghan, and Broad, 2002). These included thorough explanations and discussions (at the children's level of comprehension) of the proposed research and reassurance that children had the opportunity to opt-out of any activity.

Research Design and Data Collection

Prior co-research efforts have engaged children in research design and data collection. Examples include planning, design, and participant selection; data collection methods; and research training and support.

Planning, design, and participant selection

In several studies, adult researchers provided detailed examples of how they collaboratively planned research with children (Chen et al., 2010; Coppock, 2011; Leonard, 2019; Lundy et al., 2011; Porter et al., 2010; Smith et al., 2002). Other studies provided examples of how adult researchers embraced their child colleagues' direct influence on instrument design (Kellett et al., 2004; Pinter and Zandian, 2014, 2015). Young researchers who influenced such designs experienced more engagement with data collection.

Data collection

Several adult researchers detailed how they used multiple collection methods or alternative activities for children (Chen et al., 2010; Cheshire and Edwards, 1991; Clark, 2007; Coppock, 2011; Ergler, 2011; Hunleth, 2011; Johnson, 2008; Kellett, 2010; Kellett et al., 2004; Kirova and Emme, 2008; Lundy et al., 2011; Mayes and Groundwater-Smith, 2010; Morrow, 2005; Pinter and

Zandian, 2014, 2015; Porter and Abane, 2008; Porter et al., 2010; Price and Kerr, 2017). For example, some planned data collection around children's daily activities, rather than in prearranged settings (Coppock, 2011; Hunleth, 2011; Kirova and Emme, 2008; Porter et al., 2010).

Research training and support

Several studies mentioned that adult researchers provided research training and support for the young researchers, both before and during the project (Chen et al., 2010; Clark, 2007; Coppock, 2011; Johnson, 2008; Kellett et al., 2004; Kirova and Emme, 2008; Mayes and Groundwater-Smith, 2010; Pinter and Zandian, 2015; Porter and Abane, 2008; Porter et al., 2010; Price and Kerr, 2017; Schäfer and Yarwood, 2008; Smith et al., 2002). While few adult authors clearly outlined *how* they provided that, it is notable that most studies that included children in the planning phase did provide such training and support (Chen et al., 2010; Coppock, 2011; Porter et al., 2010; Smith et al., 2002).

Data Analysis and Interpretation

Various co-researched studies provided details about at least one method for co-research with children in data analysis and interpretation. These methodological tools include aspects of ideas and agendas, data sharing, and interpretations.

Ideas and Agendas

These studies provided examples of engaging child co-researchers by encouraging children's ideas about what was important in the data, and several studies minimized adult ideas and maximized child agendas. Some studies encouraged child-led analysis and interpretation, with adult researchers on hand to provide advice (Chen et al., 2010; Ergler, 2011; Johnson, 2008; Lundy et al., 2011; Smith et al., 2002). Others incorporated collaborative analysis between adults and children (Mayes, 2013; Morrow, 2005; Pinter and Zandian, 2014, 2015). Finally, in two child-initiated and directed studies, young researchers took control of data analysis and interpretation (Kellett, 2010; Kellett et al., 2004).

Chen, Weiss, and Johnston Nicholson (2010) engaged girls in all aspects of their research, including the data analysis portion. The adults gave their younger researchers a simple process to conduct data analysis of interview answers. They '(1) read across responses, (2) identified key themes and patterns, and (3) tallied results, calculated percentages, and selected quotes to illustrated normative responses for a particular theme' (Chen et al., 2010, p. 233). They received worksheets to assist them in this process as they worked in groups. The girls divided the answers into smaller groups and talked about their results in front of the larger group, and they received constructive criticism from adults and other youth on the team (Chen et al., 2010). By dividing a relatively new and complicated process into smaller, more manageable tasks, the researchers were able to teach the girls how to conduct data analysis (Chen et al., 2010).

Data Sharing

Methods for data sharing included maintaining a policy of openness with young researchers and sharing and discussing collected data throughout the process. A few adult researchers shared data with their young colleagues, using group debriefing sessions, focus groups, and check-ins (Chen et al., 2010; Ergler, 2011; Mayes and Groundwater-Smith, 2010).

Interpretations

This method relates to the temporality of children's interpretations. Some adult researchers provided examples of capturing children's interpretations in context during data collection (Clark, 2007; Hunleth, 2011; Kellett, 2010; Kirova and Emme, 2008). Other adult researchers only discussed children's interpretations with them at the completion of the project (Chen et al., 2010; Ergler, 2011; Leonard, 2019; Mayes, 2013). Some studies used both methods (Johnson, 2008; Kellett et al., 2004; Mayes and Groundwater-Smith, 2010; Pinter and Zandian, 2015).

Dissemination

Methods for adult/child collaboration in dissemination were concerned with the direction in which a study's findings are shared. Dissemination includes two methods, representing two directions: *inward*, in which findings are disseminated back to the young co-researchers, and *outward*, in which the dissemination of findings outside the project achieves a broader impact.

Sharing Findings

These studies provide examples in which the findings that young co-researchers helped to generate (through their work on the research project) were shared with them (Evans, 2016; Kellett, 2010; Kellett et al., 2004; Pinter and Zandian, 2015).

Broader Impact

In some studies, adult and child research partners together discussed broader implications of their findings (Chen et al., 2010; Evans, 2016; Johnson, 2008; Kellett et al., 2004; Kirova and Emme, 2008; Lundy et al., 2011; Mayes and Groundwater-Smith, 2010; Pinter and Zandian, 2015; Porter and Abane, 2008; Porter et al., 2010).

With proper guidance, young researchers can engage in data collection, data analysis, presentation of data, and writing a final paper. This capacity was shown in a study conducted by Kellet, Forrest, Dent, and Ward, which had three 10-year-olds as co-authors, in addition to the adult researcher (2004). The publication included the research papers that the child researchers wrote about their studies. These children wrote how they conducted their study, their data analysis, and conclusions, showing that children can do more in research than just data collection (Kellett et al., 2004). Though not about tourism, this study has informed our own work. We now turn to our dark tourism case study, which illustrates many of the strategies outlined in Figure 19.1.

Case Study

The National 9/11 Pentagon Memorial receives 750,000 tourists each year, with most being school trip students (A. Ammerman, personal communication, 2019). Despite this large number of young tourists, no research has documented how they experience this dark site. To address this problem, we invited 23 youth co-researchers to design a study. At the end of their ninth year of school, the students, ages 13–14 years, travel for four days to Washington, D.C. for sightseeing. This trip, therefore, provided the opportunity for students to conduct research with their fellow travelers.

In Chapter 18, we described how to establish research–school collaborations to support such research projects. Here, we outline our steps for this specific study, cross-referencing the Price (2018) framework.

Curriculum

Our first challenge was to design a curriculum suitable for 13–14-year-olds who had never conducted research. Twice a week for 40 minutes, eighth graders learned research methods or more general strategies. (These lessons followed one semester of a U.S. History class on the events leading up to the 9/11 attack.) Recognizing the need for actively engaging adolescents, each lesson featured small group activities as well as a video component delivering content from university faculty or undergraduate students. Brief lesson descriptions follow.

Lesson 1: How Researchers Work
This lesson introduced students to how social sciences researchers work, beginning with the identification of a research topic. The students learned how to describe research, how researchers identify research questions, and how their participation in research could relate to their future post-high school goals.
Lesson 2: Tourism Research: Welcome to the Field
This lesson introduced students to tourism research, more specifically. Students learned about a tourism researcher, Dr. Dallen Timothy, by reading excerpts of his CV and completing a worksheet activity. We also surprised the students with a video from Dr. Timothy encouraging them on their research journey. As a result, students began to recognize their membership in a broader tourism research community.
Lesson 3: How Researchers Ask Survey Questions
This lesson introduced students to criteria to consider when writing a survey. Students studied an example of a visitor survey. They used this information to begin to plan and design their measures.

Lesson 4: Stress Prevention and Management

This lesson taught students that stress is a natural response that can impact them both positively and negatively. The students identified potential sources of stress that could occur on the D.C. trip. The students also learned about how to manage stressful situations that arise within research teams.

Lesson 5: Good Health Practices When Traveling

This lesson taught students about healthy practices to follow while on their trip to D.C. They specifically learned the importance of sleep, nutrition, hydration, walking preparation, and skin protection.

Lesson 6: Informed Consent

This lesson taught students about ethics in research, specifically concerning informed consent. They understood what informed consent entails and how to obtain it from their classmates.

Each lesson was provided in the state-required lesson plan format. Though not formally trained in research methods, the teacher skillfully facilitated the lessons and assisted students with the activities. Through these engaging weekly lessons and videoconferences with us every few weeks, the students received the training and support they needed to design their measures and carry out their studies.

Data Collection

After the conclusion of their introductory lessons, the students chose three forms of data to collect: background information, pre-trip and post-trip surveys, and pre- and post-trip interviews on the tour bus. Gathering background data required students to interview school administrators about the demographic data of their school. Pre-trip surveys gathered information about their classmates' prior knowledge about the memorial, prior travel experience, and expectations of visiting the site. The students wrote the questions for their teacher to post on the board during classes. Post-visit surveys invited their classmates to share their reactions to the memorial. The student researchers distributed these surveys on the bus one or two days after the Pentagon Memorial visit.

Adapting the motorcoach interview methods we developed while traveling with prior cohorts (see Chapters 17 and 18), students also designed pre- and post-visit interviews. They first pilot tested their interviews with students who had gone on a previous trip. They then unpacked, set up, and practiced using the small audio recorders. Students also brainstormed scenarios they might encounter, such as someone forgetting a recorder or someone spilling water on a recorder. For each problem they identified, they also identified a solution. The student researchers conducted and recorded individual interviews with peers who volunteered during the trip.

Interestingly, the students required no adult assistance in conducting this part of their study. In fact, in our subsequent study with a different cohort, we turned over this entire process to the students.

Data Analysis

To assist the students whose school year ended shortly after the trip, we transcribed some of the interview data for them. We then uploaded these interview data into a secure server to which they had access. Next, we planned lessons to guide the student researchers through the data analysis process, the writing of a research paper, and the presentation of their findings.

We presented a brief video to the students, which discussed the purpose of a research paper. We also gave them an outline of a research paper, annotated to explain each section. The goal of this was to challenge the students without making the task seem impossible. Below is an example.

Introduction/Background Section

- Briefly introduce the topic you studied. Why do you think people should be interested in what you're studying? *The goal here is to try and get your reader excited about what you accomplished this past year in your research. This paragraph should also tell the reader why the Pentagon Memorial needed this study. (Write about 1 paragraph.)*
- Explain how your school and class got involved in this research. Tell us a little bit about your community and the trip. *This is where you'll tell the reader information about your school and community and your school's partnership with the University of Pittsburgh.*

After explaining the format of a research report, we also led the student researchers through the process of data analysis. This involved creating another lesson for students to follow. Here, we divided the students into different groups and asked each group to look at a different set of data. We asked the student researchers to look through the responses that their classmates gave for each question and then come up with categories. Importantly, the student researchers identified the codes and themes for their qualitative analysis of comments and interviews. The following guidelines illustrate our instruction:

You will be working with the data from one of the questions you asked before the trip.

- 'Is there something you are interested in seeing at the Pentagon Memorial?'
- Step 1: Read through all of the responses and decide how you want to group the different responses. Some group ideas to consider might be specific places in the memorial people wanted to see, or if they talked about specific experiences they wanted to have, etc. But remember that this is your research and data, so you have the authority regarding how you want to group these responses.

- Step 2: Now that you've categorized the responses into specific groups, take a look through the different responses and see if you have any really interesting ones. Discuss these with your group. Why do you find these interesting? Why might the Pentagon be interested in what this person is saying? Make a note of these responses including specific quotes.

Curiously, the students noticed in their analysis obvious elements of the visit that adults completely overlooked. For example, the young researchers highlighted comments about a vivid sunset: 'People should plan what time of day they are visiting. For instance, our group thought the sunset made it more meaningful.' Yet, none of the adult researchers recalled seeing it.

Moreover, their interpretations reflected an understanding of their peers' language that eluded the adults. Consider this discussion of the frequently used word, 'cool':

> The students did not have a definitive thing that they wanted to see, but they expected that the monument would be interesting and something that they would like to look at. We think that 'cool' means something that the kids would want to see, but it is not one specific element. It piqued their interest.

Recalling that adult researchers often forget to ask whether children like or dislike conducting research (Price, 2018), we invited the students to reflect not only on their study methods but also on their personal experiences. They chose the topics and the format for these sections. Here are examples of their insights:

> We realized that interviewing on the bus was the best idea because we didn't want to interview students while they were at the memorial. It would have taken away their time to enjoy and observe the memorial. It was convenient on the bus because there was usually nothing happening on the bus. The trips in D.C. were long because of traffic. Bus rides gave us time to interview all of the volunteering participants. One advantage of doing the interviews on the bus is that the interviewers did not get confused with who has or has not been interviewed yet: we went in order of seats on the bus. We did not want to interview in the hotel because we wanted the [research] volunteers to have time to relax after a long day in D.C.

> We do have a suggestion about the interview scripts. What we did in our interviews was to ask an initial and formal question that lasted an average of one minute. After the study, we realized we should've added follow up questions to get longer and better answers, so we could have more detailed opinions to put into our research.

> The research that we are partaking in differs from others because it consists of 8th-grade students. This is important because research is mainly

conducted by adults. Yet, we are doing this as teenagers. We have made questions to collect our data and put it all together ourselves.

I am a pretty average person with not many extraordinary accomplishments. I have only been at this school for about 5 years and am happy to be a part of this research group. I feel that this research is useful because it is good for people to go somewhere to have a good learning experience that they would be willing to go back to. Simply put, we are trying to help make it enjoyable so people will go back.

I love working on this project and it is very important to me and many others. I think that this project can show and tell people how important the Pentagon Memorial and 9/11 is to everyone. It can show how much of an influence these memorials in Washington D.C. have on young people and how much we care about the families having peace with their family members that passed away on that tragic day.

As these comments reveal, younger researchers can talk among themselves to garner information that is more challenging for an adult researcher to elicit. We turn now to our concluding remarks.

Conclusion

When studying children, we traditionally learn about them through the view of adult researchers (Price, 2018). However, social sciences research has begun to shift toward engaging *with* children in research rather than merely studying them (Canosa et al., 2019; Kellett, 2010; Kellett et al., 2004; Pinter and Zandian, 2014). One of the forces pressing this transition is the 1989 United Nations Convention on Rights of the Child, which delineated entitlements for those under age 18. For example, the Convention concluded that young people construct their own beliefs and can engage in 'freedom of expression,' which encompasses obtaining and communicating their knowledge and ideas (United Nations General Assembly, 1989). When enacted and acknowledged in research, these beliefs reveal a contrast between what children think as compared to adult views (Coppock, 2011; Johnson, 2008; Pinter and Zandian, 2015).

As our introductory review of the literature indicates, researchers *have* begun to collaborate with children as co-researchers, defined as children being a colleague with the adult researcher in a study (Price, 2018). However, adult researchers usually do not include young co-researchers in *all* aspects of their studies, especially data analysis and sharing the outcomes of the study. This approach means that even though the adult and child researchers initially worked together in certain aspects of the study, dissemination relies on adult language with little or no input from the young researchers (Pinter and Zandian, 2015; Price, 2018).

Children do not expect adults to think that their ideas and opinions are important (Punch, 2002). Even though they are meant to be co-researchers in some studies, children may expect that adults will edit what they have said (Pinter and Zandian, 2015). Youth are accustomed to adults having control in every aspect of their lives; therefore, being colleagues may be a foreign idea to them (Punch, 2002). In fact, studies that do have child co-researchers do not always include their unedited remarks (Price, 2018).

In contrast, our case illustrates that researchers have much to learn about young people's perspectives on all phases of research including methods, data analysis, and interpretation. Children and young people have shown their ability to work alongside adults as co-researchers (Kerr et al., 2021). Moreover, their insights have earned them a central role among other dark tourism researchers.

References

Burns, L. (2018) Overnight School Trips: An Overlooked Phenomenon. EdD Thesis, University of Pittsburgh, Pittsburgh, PA.

Canosa, A., Graham, A., and Wilson, E. (2019) Progressing a child-centred research agenda in tourism studies. *Tourism Analysis*, *24*(1), 95–100.

Chen, P., Weiss, F. L., Johnston Nicholson, H., and Girls Incorporated®. (2010) Girls study Girls Inc.: Engaging girls in evaluation through participatory action research. *American Journal of Community Psychology*, *46*, 228–237.

Cheshire, J., and Edwards, V. (1991) Schoolchildren as sociolinguistic researchers. *Linguistics and Education*, *3*, 225–249.

Clark, A. (2007) Views from inside the shed: Young children's perspectives of the outdoor environment. *Education 3-13*, *35*(4), 349–363.

Coppock, V. (2011) Children as peer researchers: A journey of discovery. *Children and Society*, *25*(6), 435–446. doi: 10.1111/j.1099-0860.2010.00296.x.

Ergler, C. (2011) Beyond passive participation: Children as collaborators in understanding the neighbourhood experience. *Graduate Journal of Asia-Pacific Studies*, *7*(2), 78–98.

Evans, R. (2016) Critical reflections on participatory dissemination: Coproducing research messages with young people. In R. Evans, L. Holt, and T. Skelton (Eds.), *Methodological approaches* (pp. 1–30). Singapore: Springer Singapore.

Freeman, M., and Mathison, S. (2009) *Researching children's experiences.* New York: The Guilford Press.

Hunleth, J. (2011) Beyond on or with: Questioning power dynamics and knowledge production in 'child-oriented' research methodology. *Childhood*, *18*(1), 81–93.

Johnson, K. (2008) Teaching children to use visual research methods. In P. Thomson (Ed.), *Doing visual research with children and young people* (pp. 77–94). London: Routledge.

Kellett, M. (2010) Small shoes, big steps! Empowering children as active researchers. *American Journal of Community Psychology*, *46*(1), 195–203.

Kellett, M., Forrest, R., Dent, N., and Ward, S. (2004) "Just teach us the skills, please, we'll do the rest:" Empowering ten-year-olds as active researchers. *Children and Society*, *18*(5), 329–343.

Kerr, M. M., and Price, R. H. (2016) Overlooked encounters: Young tourists' experiences at dark sites. *Journal of Heritage Tourism*, *11*(2), 177–185 doi: 10.1080/1743873X.2015.1075543.

Kerr, M. M., and Price, R. H. (2018) "I know the plane crashed": Children's perspectives in dark tourism. In P. R. Stone, R. Hartmann, T. Seaton, R. Sharpley, and L. White (Eds.), *The Palgrave MacMillian handbook of dark tourism studies* (pp. 553–583). Palgrave Macmillan, London.

Kerr, M. M., Price, R. H., Demore Savine, C., Ifft, K., and McMullen, M. A. (2017) Interpreting terrorism: Learning from children's visitor comments. *Journal of Interpretation Research*, *22*(1). https://lnt.org/wp-content/uploads/2018/10/JIR-v22n1.pdf#page=89

Kerr, M. M., Stone, P. R., and Price, R. H. (2021) Young tourists' experiences at dark tourism sites: Toward a conceptual framework. *Tourist Studies*, *21*(2), 198–218.

Kirova, A., and Emme, M. (2008) Fotonovela as a research tool in image-based participatory research with immigrant children. *International Journal of Qualitative Methods*, *7*(2), 35–57.

Leonard, M. (2019) The teenage gaze: Teens and tourism in belfast. *Childhood*, *26*(4), 448–461.

Lundy, L., McEvoy, L., and Byrne, B. (2011) Working with young children as co-researchers: An approach informed by the United Nations Convention on Rights of the Child. *Early Education and Development*, *22*(5), 714–736.

Mayes, E. (2013, December) *Students researching teachers' practice: Lines of flight and temporary assemblage conversions in and through a students-as-co-researchers event*. Paper presented at conference of Australian Association for Research in Education, Adelaide, Australia. Retrieved from http://www.aare.edu.au/data/publications/2013/Mayes13.pdf

Mayes, E., and Groundwater-Smith, S. (2010, December) *Year 9 as co-researchers: "Our gee'd-up school."* Paper presented at conference of Australian Association for Research in Education, Adelaide, SA, Australia. Retrieved from http://s3.amazonaws.com/academia.edu.documents/31395813/1723bMayesGroundwaterSmith.pdf?AWSAccessKeyId=AKIAJ56TQJRTWSMTNPEAandExpires=14768-98127andSignature=10Zmpdlv6soUZuW1mntqqPgITFE%3Dandresponse-content-disposition=inline%3B%20filename%3DWith_Groundwater-Smith_S._Year_9_as_Co-.pdf

Morrow, V. (2005) Ethical issues in collaborative research with children. In A. Farrell (Ed.), *Ethical research with children* (pp. 150–165). Maidenhead, Berkshire: Open University Press.

Pinter, A., and Zandian, S. (2014) "I don't ever want to leave this room:" Benefits of researching 'with' children. *ELT Journal*, *68*(1), 64–74. doi: 10.1093/elt/cct057.

Pinter, A., and Zandian, S. (2015) "I thought it would be tiny little one phrase that we said, in a huge big pile of papers:" Children's reflections on their involvement in participatory research. *Qualitative Research*, *15*(2), 235–250.

Porter, G., and Abane, A. (2008) Increasing children's participation in African transport planning: Reflections on methodological issues in a child-centered research project. *Children's Geographies*, *6*(2), 151–167.

Porter, G., Hampshire, K., Bourdillon, M., Robson, E., Munthali, A., Abane, A., and Mashiri, M. (2010) Children as research collaborators: Issues and reflections from a mobility study in Sub-Saharan Africa. *American Journal of Community Psychology*, *46*(1), 215–227.

Price, R. H. (2018) *Expectations and Revelations: Children Discuss Conducting Research During a Multi-Day School Excursion*. PhD Thesis: University of Pittsburgh.

Price, R. H., and Kerr, M. M. (2017, January 12) *'I thought it was cool how we were part of research': Youth as co-researchers* [Conference presentation]. *The Qualitative Report 8th Annual Conference*, Fort Lauderdale, FL, United States. https://nsu-works.nova.edu/tqrc/eighth/day1/19/

Punch, S. (2002) Research with children: The same or different from research with adults? *Childhood*, *9*(3), 321–341.

Schäfer, N., and Yarwood, R. (2008) Involving young people as researchers: Uncovering multiple power relations among youths. *Children's Geographies*, *6*(2), 121–135.

Smith, R., Monaghan, M., and Broad, B. (2002) Involving young people as co-researchers: Facing up to the methodological issues. *Qualitative Social Work*, *1*(2), 191–207.

United Nations General Assembly (1989) Convention on the rights of the child. *United Nations, Treaty Series*, *1577*(3), 1–23.

Epilogue: 'Monsters and Mediating Mortality Moments'

Dark Tourism and Childhood Encounters

Philip R. Stone

> And I would have gotten away with it, too, if it weren't for you meddling kids!
>
> (Various Villains, Scooby-Doo, 1969–present)

The children's television franchise 'Scooby-Doo' involved an animated hound solving mysteries by supposedly supernatural creatures through a series of adventures and antics. Yet, every spectral mystery solved by the lovable Scooby-Doo and his juvenile companions taught us the same conclusion – in that, the real monsters always turned out to be human! Whilst the cartoon canine went on to expose monsters in the fictional midst, actual monstrosity from real life is often represented in dark tourism and consumed as mortality moments. Indeed, dark tourism is where calamities and misfortunes of our noteworthy dead are mediated and engineered through a political and cultural process of collective memory. Within global visitor economies, spaces of fatality are commemorated and often commercialised as places of mortality through selective practices of difficult heritage. Sites *of* suffering or sites *associated with* ordeal (re)create contemporary traumascapes, whereby modern tourists are told tragic tales of distress and disturbance. Consequently, narratives associated with 'heritage that hurts' are produced and expose our existential fragility as museums, monuments and visitor attractions trade our significant Other dead.

Of course, 'dark tourism' is an academic appellation given to a taxonomy of touristic sites that commemorate or commercialise acts of death, disaster, or the seemingly macabre (Stone, 2006). However, within these real or recreated traumascapes, adult tourists 'sightsee in the mansions of the dead' where traumatic histories often reflect secular forms of spiritual experience (Keil, 2005; also see Stone, 2021). It is here where children also roam, often with their adult guardians, and where touristic experiences may raise profound questions. These include questions of collective memory and political forgetting; issues of guilt and redemption, meaning, and ownership; and testimony and evidence. It is also where (adult) tourist experiences are polysemic and, consequently, externalizing and disarming traumatic memory within tragic places can be fractured or ambivalent (Stone and Grebenar, 2021; Seaton, 2009; Keil, 2005).

DOI: 10.4324/9781003032199-25

Yet, despite children occupying dark tourism places and consuming (adult) narratives of tragedy and the deceased, the absence of research into childhood dark tourism experiences is matched by a similar scarcity of scholarship of children and tourism more largely. Therefore, as we have noted in this book, childhood and youth encounters within tourism generally, and dark tourism in particular, are poorly understood at present. Thus, this book is the first-ever volume to bring the context of dark tourism together with concepts of childhood and youth experiences. Undoubtedly, the contributing authors to this text are in the vanguard of the subject field and, subsequently, have opened up critical discourse for future critique. Of course, being the first tome within a subject area will mean that omissions have been made. If that is the case, then these omissions should form future gaps in the literature for other scholars to reconnoitre. Nonetheless, this book has firmly shone a light on children, dark tourism, and their experiences. By way of summary and synthesis, each chapter is now succinctly outlined with main concepts, contexts, and challenges identified. Thereafter, I offer some final thoughts on the future of children and dark tourism research.

The book: toward a synthesis

In Chapter 1, Philip R. Stone offers an introductory critique to the state of play of children and tourism scholarship. He sets a critical milieu to relevant scholarship, as well providing a rationale to studying children within (dark) tourism research agendas. By recognising a dearth of attention to children's experiences and tourism, Stone calls for post-disciplinary research approaches whereby future tourism scholars form study alliances with experts in childhood education and youth psychology. With this in mind, the call for action is firmly made, particularly within dark tourism studies with specific challenges of children's conceptions of death and mortality.

It is here that, in Chapter 2, Mary Margaret Kerr, Philip R. Stone, and Rebecca H. Price reiterate a conceptual framework for researching children's experiences at dark tourism sites. In particular, they outline how younger children may not possess an adult-like knowledge of death and, subsequently, may not experience a particular site as 'dark.' Issues of children's limited agency in choosing the site, as well as unique playful behaviour, address the inadequacy of current dark tourism conceptualisations. In turn, they offer a rationale for interrogating children's encounters of dark tourism, including a child's understanding of death, the visit preparation, site interpretation features, and dynamics of a specific visit.

Some of these issues are further examined in Chapter 3 by Sue Dockett. Particularly, Dockett queries established notions of childhood innocence and, subsequently, how children can become competent contributors to conversations about their own lives and experiences. In so doing, she advocates children as co-constructors of meaning, including within the ethically fraught arena of dark tourism. Indeed, Dockett identifies the importance of pursuing research as a means to recognise children's rights, as well as the obligations of adults to actualise them. Meanwhile, Timothy M. Wagner in Chapter 4 takes on the mantle of children as co-constructors of meaning and illuminates pre-adolescence and adolescence experiences in dark tourism. He offers a useful framework for understanding typical pre-adolescent and adolescent development and explains how dark tourism experiences can relate to this development. By doing so, Wagner offers evidence-based recommendations for guiding school-aged tourists through dark heritage sites. He goes on to argue that transformative opportunities exist when developmental research, dark tourism practices, and accounts of lived experiences are coordinated.

In Chapter 5, Andrea Croom and Gopika Rajanikanth also put development, childhood and adolescence at the core of their chapter. In particular, they examine the developmental stages of death concepts during childhood and adolescence. They suggest that understanding the development of death concepts during childhood will help researchers firstly, recognise how children navigate and make sense of dark tourism sites. Secondly, they argue that better understanding can lead to enhanced interpretation and design material, thus helping caregivers, educators, and docents explore dark sites in a more meaningful way. Finally, Croom and Rajanikanth advocate that better understanding of death concept development will lead to informing research designs and improved methodologies of children and dark tourism scholarship.

Meanwhile, in Chapter 6, Cristina Restrepo-Harner, Kristen Marsico, and Mary Margaret Kerr invite readers to consider how children with disabilities might navigate dark tourism sites. Notwithstanding issues already noted in previous chapters, Restrepo-Harner et al. examine various barriers children with special needs face, including exclusion, differing conceptions of death, and mediations of their own mortality as they reflect upon their own maladies. Restrepo-Harner et al. call for future research to use an inclusive model and address the overlooked voices of children with disabilities as they gaze upon the dark tourism dead. The subject of the tourist gaze is investigated further in Chapter 7 by Roy Ballantyne, Jan Packer, Karen Hughes, and Tobias Broughton. They argue for dark tourism interpretative strategies to engender positive responses amongst children and young audiences. Augmenting the notion of 'hot interpretation' and evoking emotion in dark tourism site interpretation, Ballantyne et al. warn against dilution of information for children. Rather, they argue that the 'new moral spaces' of dark tourism (after Stone, 2009) have the capacity to teach empathy, to address stereotypes, and to promote reconciliation. Ultimately, Ballantyne et al. suggest that children should have positive experiences of dark tourism, rather than being distressed, and be empowered with how to address the fights, follies, and misfortunes made by adults.

The idea of empowerment is also examined by R. Scott Marsh in Chapter 8 within the realms of dark tourism placemaking and children's sense of place. Specifically, Marsh advocates that emotional connections and cognitive 'groundedness' of feeling a sense of place at dark tourism sites might serve as powerful teaching tools. In essence, the attachment of meaning and meaningfulness to a difficult heritage space becomes a dark tourism place. By establishing emotional resonance, children can explore traumascapes and tragic tales of dark tourism with their own co-constructed sense of being and self-identity. Yet, while Marsh explores the physicality of sense of place and emotional attachments, Gregory J. Wittig in Chapter 9 examines children's sense of place from digital perspectives. By recognising the interconnectedness to other people, information, as well as entertainment, the 'digital child' is only a click away from 'thana-technology'. Thus, Wittig asks pertinent questions of who is the digital child, how do we prepare them for dark tourism, and how should dark tourism sites respond to this new type of learner? With the modern digital child being trans-fluent between digital and real worlds, Wittig concludes that dark tourism sites should be relevant and engaging, active, and instantly useful to contemporary children. Whilst acknowledging dissonance heritage and, consequently, the framing of contested (hi)stories, Wittig argues that dark tourism can create a digital sense of place, where technology supplements the learning experience for children, rather than substituting it. As a result, technology as a tool can integrate the digital child into real-world (dark) experiences, without obscuring or replacing them.

Issues of engaging children in dark tourism are also explored in Chapter 10 by Margee Kerr. Particularly, Kerr investigates how children use recreation as a means of engaging with potentially overwhelming, often fearful realities – including that of death and dying. In essence, Kerr argues that through a lens of recreation and entertainment, children can learn about difficult inevitabilities, practice how to manage them and, in so doing, cultivate a sense of resilience for adulthood. Adopting and augmenting the conceptual framework of dark tourism as engineered and orchestrated remembrance of mortality and fatality (after Seaton, 2018); Kerr suggests children consume messages of *fearful delight*, where the tone of the touristic environment is not sombre but spirited and animated. By using playful engagement and active entertainment, children can approach dark tourism material on their own terms. It is here that, in Chapter 11, Daniel W. M. Wright offers an insight into the world of entertainment and education – termed *edutainment*. In particular, he critically evaluates how new immersive technologies, such as Virtual Reality and Augmented Reality in dark tourism, may provide fresh intersections of engagement for children. Wright goes on to conclude that dark tourism curators and technology content managers should offer effective *edutainment* experiences as a means to better connect children with past difficult heritage and global anxieties of the future.

The idea of engagement is also the theme of Chapter 12 in which Michael Lovorn takes the reader on a pedagogical journey into deconstructing difficult heritage for high school students. Specifically, Lovorn outlines a case of teaching the assassination of US President John F. Kennedy in 1963. Adopting a historiographical analysis technique, Lovorn narrates the deep connections between historical commemoration and the act of narrative-building for teaching difficult heritage. In so doing, Lovorn offers a personal account of how young people can hone skills to recognise perspective and agency, evaluate sources and evidence, explain causal relationships, and formulate strong debates. Ultimately, Lovorn suggests that tragic events present students with challenging imagery and even greater challenges to grand narratives as history is turned into heritage.

Similarly, Laura M. Burns and Daniel E. Keller in Chapter 13 also examine pedagogy and dark tourism studies and, specifically, explore school trips as a unique form of student learning. They outline key parameters of school field visits and offer the Contextual Model of Learning as a specific framework. In particular, they argue the model provides a defined structure in which to explore the overlapping of personal, socio-cultural, and physical experiences. In turn, Burns and Keller stipulate dark tourism learning within the confines of school field trips occurs as students process, connect with, and make meaning of real-world encounters. Furthermore, real-world encounters are highlighted in Chapter 14 as Jennifer Frost and Warwick Frost explore commemorative events and their impact upon children. Specifically, they examine the centenary of World War One within an Australian context and note how children as the next generation have to bear the responsibility of carrying memory forward. Frost and Frost go on to conclude that children should not be neglected in commemorative heritage interpretation and be provided with a range of voices and narratives. Moreover, they argue (difficult) interpretation should not be sanitised if the aim is to create emotional engagement and stimulate critical reflection in children.

Emotional connections and critical reflections in children's experiences are also themes discussed by Antonia Canosa and Rebecca H. Price in Chapter 15. However, rather than adopting the perspective of the child-tourist, Canosa and Price examine complex politics of identity and belonging at 'dark sites' from the perspective of child-residents. Consequently, they offer a unique conceptual framework in which to explore children and young people living in dark tourism destinations and how they subsequently negotiate identity and belonging. In so doing, they propose a 'spectrum of connection' model to ascertain how children respond to changes and challenges at a local-level post-disaster when (dark) tourists begin to visit their communities. Adopting a child-centred research approach, Canosa and Price argue that children's evolving process of identity formation and connections to place and community are crucial for their social and emotional wellbeing. Yet, personal accounts of loss or distress from child-residents become entangled with experiences of being 'on display' for tourists. Importantly, Canosa and Price advocate that children's 'voices' and involvement in any disaster recovery process are paramount if connections to place and people are to be maintained.

Child-centred research approaches and bespoke child-friendly research methodologies have been a consistent theme throughout this book. Therefore, the final four chapters of this volume focus exclusively on research philosophies and designs appropriate for empirical scholarship with children in (dark) tourism. In Chapter 16, Rebecca H. Price explores ethical research conduct with children by addressing inherent complexities in both scholarship and dark tourism contexts. She focusses on key ethical practices that are unique to children and research(ing), including adult–child power dynamics, consent, reward and compensation, autonomy, and confidentiality. Price goes on to offer a range of useful and practical recommendations for accomplishing child-centred research designs. Ultimately, she argues that by empowering children within research processes, adult-researchers can work in an alliance with child-respondents and, consequently, achieve meaningful understanding of children's experiences within dark tourism.

In Chapter 17, Mary Margaret Kerr, Rebecca H. Price, and Gopika Rajanikanth outline a series of research methods drawn from museology, childhood studies, education, and psychology. In so doing, they also offer discourse on appropriate methods to accelerate ethical research with children and young tourists. With children's developmental stages firmly in mind when planning such research designs, Kerr et al. argue that even the most con-strained research situation will benefit from children's input. Meanwhile, in Chapter 18, Mary Margaret Kerr joins forces with Cecilia Green and R. Scott Marsh to explore research collaborations with schools. As research involving children typically requires collaboration with child-serving organisations – that is, schools – effective partnerships are paramount. As tourism research-ers may not have ongoing partnerships, Kerr et al. highlight a number of fundamental challenges. These include safeguarding issues, engaging parents or legal guardians, promoting children's wellbeing and overall safety, and becoming an unobtrusive participant-observer during children's visits. Other issues are also explored including travel logistics and itinerary planning, funding, research equipment management, and non-disruptive data collec-tion times and spaces. Finally, in Chapter 19, Rebecca H. Price, Mary Margaret Kerr, and Gopika Rajanikanth evaluate co-research with children and young people. In particular, they offer a conceptual model in which to frame co-research as well as a case study to empirically illustrate research partnerships with youths. It is here that the transition of engaging *with* chil-dren in research rather than merely studying them is firmly made. When chil-dren and young people step forward to conduct their own research, childhood insights will finally secure a central role in dark tourism research.

Final Thoughts

I have been writing about people visiting our noteworthy dead within visitor economies for almost two decades. During that period, the adult tourist experience and emergent consequences have largely been the focus of schol-arship. Indubitably, there is still much to do in terms of ascertaining the con-sequences of dark tourism experiences, yet it is clear that children and their dark touristic voices have been ignored.

While adults may be outdated children, it is adults that make our fights, follies, and misfortunes. Meanwhile, it is our children that make history. Therefore, they deserve to be heard. The time has come time for us within the adult academy to adopt child-centred research approaches within dark tourism scholarship. If we do not act, voices of children within traumascapes will be lost, their experiences forgotten, and their understandings obscured. As 'ghosts' of our dark tourism deceased cannot leave fingerprints and are left to selective memory, the mark of children on our significant Other dead needs to be hearkened. Only then will we listen to children to understand, rather than listening to them to reply.

References

Keil, C. (2005) Sightseeing in the mansions of the dead. *Social & Cultural Geography*, 6(4), 479–494.

Seaton, T. (2009) Purposeful otherness: Approaches to the management of thanatourism. In R. Sharpley & P.R. Stone (Eds.), *The Darker Side of Travel: The Theory and Practice of Dark Tourism*. Bristol: Channel View Publications, pp. 75–108.

Seaton, T. (2018) Encountering engineered and orchestrated remembrance: A situational model of dark tourism and its history. In P.R. Stone et al. (Eds.), *The Palgrave Handbook of Dark Tourism Studies*. London: Palgrave Macmillan, pp. 9–31.

Stone, P.R. & Grebenar, A. (2021) 'Making tragic places': Dark tourism, kitsch and the commodification of atrocity. *Journal of Tourism and Cultural Change*. DOI: 10.1080/14766825.2021.1960852.

Stone, P.R. (2006) A dark tourism spectrum: Towards a typology of death and macabre related tourist sites, attractions and exhibitions. *TOURISM: An Interdisciplinary International Journal*, 54(2), 145–160.

Stone, P.R. (2009) Dark tourism: Morality and new moral spaces. In R. Sharpley & P.R. Stone (Eds.), *The Darker Side of Travel: The Theory and Practice of Dark Tourism*. Bristol: Channel View Publications, pp. 56–72.

Stone, P.R. (2021) *111 Dark Places in England That You Shouldn't Miss*. Cologne, Germany: Emons Verlag.

Index

Pages in *italics* refer figures and pages in **bold** refer tables.